Henri Stephanou

Systems, Machines, and Problem-Solving

Epistemic Studies

Philosophy of Science, Cognition and Mind

Edited by
Michael Esfeld, Stephan Hartmann, Albert Newen

Volume 55

Henri Stephanou

Systems, Machines, and Problem-Solving

DE GRUYTER

ISBN 978-3-11-149257-5
e-ISBN (PDF) 978-3-11-149450-0
e-ISBN (EPUB) 978-3-11-149470-8
ISSN 2512-5168

Library of Congress Control Number: 2025945630

Bibliographic information published by the Deutsche Nationalbibliothek
The Deutsche Nationalbibliothek lists this publication in the Deutsche Nationalbibliografie; detailed bibliographic data are available on the internet at http://dnb.dnb.de.

Printing and binding: CPI books GmbH, Leck

www.degruyter.com
Questions about General Product Safety Regulation:
productsafety@degruyterbrill.com

MIX
Papier | Fördert gute Waldnutzung
FSC® C083411
FSC www.fsc.org

Contents

Appendices

List of abbreviations

SVF	Hans von Arnim, ed. [1905] (1964), *Stoicorum Veterum Fragmenta*, Stuttgart: B. G. Teubner
AT	René Descartes [1897–1911] (1996), *Œuvres*, ed. by Charles Adam and Paul Tannery, Paris: Vrin
C	Gottfried Wilhelm Leibniz (1903), *Opuscules et fragments inédits de Leibniz*, ed. by Louis Couturat, Paris: Felix Alcan
D	Gottfried Wilhelm Leibniz (1923–), *Sämtliche Schriften und Briefe*, Deutsche Akademie der Wissenschaften, Darmstadt; Leipzig; Berlin: Akademie-Verlag
G	Gottfried Wilhelm Leibniz [1875–1890] (1962), *Philosophische Schriften*, ed. by C. J. Gerhardt, 7 vol., Hildesheim: Olms
E	Denis Diderot and Jean Le Rond d'Alembert, eds. [1765] (2017–2022), *Encyclopédie ou Dictionnaire raisonné des sciences, des arts et des métiers*, ENCCRE, Académie des Sciences, URL: https://enccre.academie-sciences.fr/encyclopedie/ (visited on 04/03/2025)
AA	Immanuel Kant (1902–), *Gesammelte Schriften*, Deutsche Akademie der Wissenschaften, Berlin; Göttingen: G. Reimer; Walter de Gruyter
EW	John Dewey (1967–1972), *The Early Works 1882–1898*, ed. by Jo Ann Boydston, 5 vol., Carbondale: Southern Illinois university press
MW	John Dewey (1975–1983), *The Middle Works 1899–1924*, ed. by Jo Ann Boydston, 15 vol., Carbondale: Southern Illinois university press
LW	John Dewey (1981–1990), *The Later Works 1925–1953*, ed. by Jo Ann Boydston, 17 vol., Carbondale: Southern Illinois university press
LTI	John Dewey [1938] (1986), *Logic: The Theory of Inquiry*, ed. by Jo Ann Boydston, vol. 12, The Later Works, 1925-1953, Carbondale: Southern Illinois university press
SA	Herbert A. Simon [1969] (1996b), *The Sciences of the Artificial*, 3rd ed., Cambridge, Mass: MIT Press
OED	*Oxford English Dictionary* (2010–), Oxford: Oxford University Press, URL: https://www.oed.com/?tl=true (visited on 04/03/2025)

Aristotle's works

Met.	*Metaphysics*
Nic. Eth.	*Nicomachean Ethics*
APr.	*First Analytics*
APo.	*Second Analytics*
Top.	*Topics*
Rh.	*Rhetoric*
Pol.	*Politics*

https://doi.org/10.1515/9783111494500-203

Acknowledgments

This book was inspired by two places. The first is in Paris, rue Du Four, at the IHPST (Institut d'Histoire et de Philosophie des Sciences et des Techniques, Université Paris 1 Panthéon-Sorbonne / CNRS), where I was a student some 30 years ago and which welcomed me back to engage in this research. I would like to thank Pierre Wagner, who patiently guided me during the six years of this research. Exchanges with Alberto Naibo and Marco Panza on logical reconstructions of geometry stimulated me. I sparred with David Waszek on Herbert Simon and with Matias Osta Vélez on cognitive science. Adrien Champougny and Perceval Pillon discussed logical questions with kindness and patience. As part of the IHPST, the ANR GOA (ANR-20-CE27–0004) allowed me to work closely with several of these people, as well as with Walter Dean and Filippos Papayannopoulos.

The second place central to the writing of this book was Bertinoro, in Emilia-Romagna (Italy), whose beautiful *rocca* (castle), property of the University of Bologna, hosted the PROGRAMme team (ANR-17-CE38–0003–01) each year to reflect on the nature of computer programs and the diversity of possible perspectives on them. Liesbeth De Mol welcomed me into this community with great openness, right from the start of my work. Many of the ideas presented here matured with this group, especially Nicola Angius, Troy Astarte, Maarten Bullynck, Selmer Bringsjord, Felice Cardone, Martin Carlé, Marie-José Durand-Richard, Jean-Baptiste Joinet, Cliff Jones, Jean Lassègue, Baptiste Mélès, Simone Martini, Pierre Mounier-Kuhn, Elisabetta Mori, Maël Pégny, Tomas Petricek, Giuseppe Primiero, Julian Rohruber, David Schmudde, Máté Szabó, Ray Turner, and Nick Wiggershaus.

I would also like to thank Orsolya Farkas de Longevialle, Gianni Gastaldi, Matthias Girel, David Hildebrand, Daniel Jackson, Luc Pelissier, Dominique Pradelle, Sophie Roux, David Zualaga, Sjoerd Zwart, all of whom helped me at key moments in this research with their kind advice, encouragement, or close reading.

My translation of the original French text applied some suggestions from Deepl Pro. The final translation and any error are mine. Translations of quoted non-English authors are generally mine, unless otherwise stated. All Greek texts are cited according to their edition in the *Thesaurus Linguae Grecae* (Pantelia 1985 –).

https://doi.org/10.1515/9783111494500-204

1 Introduction

1.1 Motivation and guiding question

This book is part of a project motivated by my enduring puzzlement at the pervasive presence of computers in our human world. These machines are in our pockets, hide behind all our screens, and are now intermediaries in the satisfaction of most of our needs and leisure activities: shopping, transport, communications, social relations, games, reading, entertainment. Computers are also central to the operation of human institutions and organisations, whether businesses or public administrations. The global financial system is entirely computer-based – so much so that new currencies, independent of the banking system, are now emerging. Computers have taken control of factories and supply chains, and are embedded in all kinds of robotic artefacts or everyday objects that they make reactive to their environment, like a parking meter in the street or a valve in a water system. They pilot aircraft and gradually take control of cars. For the same reason, they are transforming weapons and the nature of warfare. Finally, they are transforming science itself, where progress in many areas is now entirely dependent on measuring instruments, simulations and calculations that are impracticable without considerable computing power. The progress of large-scale knowledge projects, such as the analysis of the heavens, of particle interactions or of genomics, has become entirely dependent on their computerisation. Computing is even taking over our very conception of knowledge. For example, the notion of proof in mathematics has been the subject of bitter debate ever since mathematicians proposed computer-driven proofs. The scale and depth of these changes has led computer scientist Gérard Berry (2017) to speak of the "hyperpower of computing".

The singularity of the computer is obvious when it is compared to other great inventions of the industrial revolution. If it is easy to determine the usefulness of each – communication for the telephone, transport for the car, energy for electricity – what can be said of the computer's? The English term refers to calculation but, for example, the French terms *ordinateur* and *informatique* refer to those of order and information. These notions have no obvious connection with each other, and seem to reflect a multiplicity of possible perspectives on the function of these machines. The reference to the notion of information also brings computing closer to telecommunications, but the computer seems to do much more than transmit messages. It seems to operate at an entirely different level from all these inventions.

If the car, electricity and the telephone have taken over and transformed human activities as diverse as industry, commerce and leisure, it is because transport, energy and communication are such essential functions that their transformation

https://doi.org/10.1515/9783111494500-001

could not fail to open up new prospects for these activities. Can the same be said of computers? What function do they fulfil? If one says that they automate computation, does that explain the scale of the changes just mentioned? It would imply that calculation invisibly underlies all human activities in which these machines have come to play a major role. This is not obvious in, say, social networking or health care. Another common explanation is that computers are “information processing” machines, but it is difficult to pin down exactly what information means and to articulate it with an epistemology of practices. It is not the purpose of this book to criticise these two views of the nature of computing machines – the *calculation* view and the *information* view. I will do so in the second book in this project, *Reasoning with Machines*, which is forthcoming. My project is to explore another hypothesis to explain the effectiveness and versatility of computers, and this book is about its epistemological grounding.

This hypothesis does not make use of the difficult concepts of *computation* and *information*. It does not even require the question of the primary function of computers to be resolved at once. It starts indeed with the remark that computers differ from other technologies in that they need to be *programmed* to meet the various needs that call upon them, and that they are therefore essentially *incomplete* artefacts without the software or applications that determine their particular operation, in this or that context. Computers are first and foremost *programmable* machines, and it is this property that makes them versatile, through their adaptability to solving many kinds of problems.

What does it mean that solutions to a problem can be *programmed*? An answer which presents itself naturally is that a problem solved is not only about an end-result, but also about *the procedure* for producing it. It is this procedure which may be coded and programmed. It is by mimicking and expanding the procedures by which *we, humans*, solve our problems, that computers become useful. But are all procedures programmable? And in which sense do computers “mimick” human procedures?

Consider the example of choosing a route to reach a familiar destination. Intuitively, we usually think of two or three options that come to mind and then compare them against two or three criteria to make our decision. For example, if I am cycling in Paris, I might consider taking the *grands boulevards*, a longer route but with dedicated cycle lanes. Or I might take a straighter route which involves going up and down the hill of Montmartre. Or I may pass through narrower streets where car traffic may be dense, but the route flat. Although I may look at a map to generate a few more options to consider, this will not radically change the deliberation process.

The procedure used by a location-based service such as Google Maps is somewhat similar in some respects, and strikingly different in others. What is similar is the general shape of the process: generating options and ranking them according to criteria (duration, safety, effort, etc.). What is different, however, is that the service will use precise information, possibly collected in real time (as with traffic), on all possible routes to make an accurate calculation. We are often baffled by the complex routes that these services sometimes lead us to take, because they have determined that we would save a few minutes on a two-hour route, neglecting the comfort that could be gained by taking simpler ones. Such a procedure is *systematic* and blind to any other consideration unless specified.

Moreover, looking under the hood of such an algorithm, one finds that it works quite differently from intuitive reflection. For example, in familiar places we generate the potential routes from start to destination in one go, even if we do not know exactly which roads we might take in this or that place. Even in unfamiliar places, when we look at a map, we first try to connect the two places with a straight line, and then consider the possible detours through major roads and make additional checks. Algorithms have very different logics. For example, the famous Dijkstra algorithm (named after its designer, computer scientist Edsger Dijkstra) starts by equipping each "node" of the map (a node is any place where at least two different roads meet) with information about which neighbouring node gives access to the best route to reach *any other node* on the map. This is done by having neighbouring nodes continuously exchange their available information about other nodes. For example, node A can say to neighbouring node B: *"I can reach node F in 6 km by going through my neighbour C"* (to which B has no access). And B can say to A: *"I can reach it in 4 km via my neighbour D"* (to which A has no access). Then, since the distance between A and B is only 1 km, A will update its routing table to note that it could reach node F in 5 km via its neighbour B, discarding the C-through route. Dijkstra demonstrated that, after a few iterations, all nodes would "know" the best route to any other node on the map, or rather, *where to go next* to take this best route.

This way of thinking, although invented by a human mind and in theory applicable by a human mind, is completely at odds with our intuitive ways of making decisions. Not only would we stumble at the number and complexity of the calculations to be made, but we would not even think of such an approach to find our way. It seems that the design of such a procedure requires not only superior intelligence and creativity, which Dijkstra undoubtedly possessed, but also a very specific attitude to problem-solving, which I call *systematic*, for reasons that will gradually become clear. It suffices now to say that I use this term in the primary sense that it has today, *i.e.* "arranged, or conducted according to a system, plan or organized method" (OED, "systematic", § 3). I prefer it to *methodic* because of its reference to a *system* which I believe crucial, and through this term also to the ideas of completeness and coherence. A treatise is systematic when it concerns some subject of study "containing or presenting a complete exposition [...or pursued as...] a coherent body of ideas or principles" (§ 2).

It is not evident that all problems can be solved through this systematic attitude. For example, the general problem of vision, *i.e.* of recognizing a specific object in an image, has long eluded algorithmic solutions. The breakthrough came only lately through machine-learning techniques, which refine a generic statistical procedure through massive training data. The systematicity here lies in the training procedure; it still tackles only indirectly the vision problem itself.

Conversely, many problems are solved algorithmically which are completely beyond the reach of human capabilities. Examples are so numerous that they barely need mentioning. In sciences, the sequencing of the genome relies for a significant part on a computing procedure. In mathematics, some landmark proofs, such as the Four Colour Theorem, were made possible thanks to computer processing.[1] In production, additive manufacturing – commonly known as 3D printing – is a process that creates objects by building them up layer by layer, enabling shapes that are beyond the reach of traditional manufacturing methods. It uses computer control to deposit and solidify material, often in the form of powder, using tools like extruders, laser beams, or inkjet-style print heads, following a digital file that encodes the desired shape.

So it seems that, in some areas, intuitive and systematic ways of solving problems meet, but not in many others. How should we understand these areas of overlap and divergence? In other words, what are the conditions, limits and consequences of the *conversion* of procedures from being human-centred, intuitive and adaptable to being machine-centred, systematic and automated? How can we explain the incredible success of effective procedures in taking by storm almost all areas of human practice and reshaping them into systems of rules, and what are its anthropological implications? The purpose of this book is the elucidation of these questions. It is an inquiry into the nature of problem-solving, with a particular emphasis on its systematic modality and its intimate relationship with the notion of *machine*, which makes all its effectiveness in the material world.

It must be said however that this book will not investigate the specific nature of computers, *i.e.* machines capable of performing *any* effective procedure, and the consequences of this possibility. That is the purpose of the second book of this project, *Reasoning with Machines*, where I question the very concept of a *universal machine.*

Understanding the nature of systematic problem-solving may help illuminate at least two series of fundamental epistemological problems.

1 A good introduction on this topic is MacKenzie 2004: ch. 4.

The first series of problems relate to the concepts of *rule* and *rule-following*. Although they were extensively discussed in the 1980s', and important distinctions made at the time, such as that between *constitutive* and *regulative* rules (Searle [1965] 2000), the debate cannot be considered as settled. In a recent book, historian of science Daston (2022) brings to the forefront a dynamic perspective on the polysemy of the *rule* concept. According to her, it is only since the 19th century that rules have come to be understood primarily as algorithmic elements that deliver infallible prescriptions to the agent. My comparison of two modes of problem-solving may highlight the close relationship that this understanding of rules as algorithmic has with another one identified by Daston, that of rules for *guiding judgement* and providing *models* for action and production.

Looking at rules through the prism of problem-solving also helps to make sense of a second series of problems related to the unity of knowledge. Fantl (2017), enumerates a number of related oppositions that are thematised in philosophical discourse: between *know-how* and *know-that* (and sometimes *know-what*), between *techné* and *epistemè*, between practical and theoretical knowledge, between procedural and declarative knowledge, and so on. These oppositions are often the *locus* of heated debates about the nature of knowledge, either by claiming the primacy of one of the oppositions (e.g. *all knowledge comes from practical know-how*) or by dismissing the opposition as merely relative or cultural. Problem-solving cuts across all disciplines of intellectual activity: from the outset (and I shall return to this point shortly) it seems to have a similar meaning whether in mathematics, science, engineering, health care, business, and so on – *i.e.* in theoretical and practical disciplines alike. Moreover, a procedure for solving a problem in scientific matters may appear as an action "sneaking in" theoretical considerations; it is therefore often relegated, in classical epistemology, to the *context of discovery*, perceived as the antechamber of science. Conversely, an engineering procedure can often be translated into a logical or mathematical proof that makes it look like a direct consequence of theory, *i.e.* a theorem. These puzzles relate to the very concept of procedure, which seems to be indifferent to the distinction between rules of inference and rules of action. I will therefore try to show that the analysis of problem-solving is fruitful for injecting new arguments into these debates about the unity of knowledge.

Taken together, I hope that these two series of discussions contribute to showing how useful a pragmatist conception of knowledge – taken broadly – can be to understand the deep relationship between mathematics and technology, a topic which is seldom addressed.[2]

---

2 See however Hansson's collection (2018).

1.2 Procedures and problems: preliminary framing

One important difference between problems solved intuitively and problems solved systematically is that the second ones always require some conditions to be met at the outset. A daily life problem, such as: *"Where did I put my keys?"* cannot be solved systematically unless I know, for example, that they are somewhere in my apartment; then a systematic method could be to look for them room after room, in every nook and cranny. *"What will I cook for dinner tonight?"* is a very open question. One systematic approach could be to look through a cookbook and select the first recipe for which I have all the ingredients. The prerequisites here are having such a cookbook and a list of ingredients in my kitchen.

These conditions help to define a *problem space* that guarantees that the proposed method will end successfully, in all cases. The method is then general relative to this problem space. A systematic search will find my keys, regardless of which room they are hidden in, *if they are indeed in the apartment*. Similarly, my search for a dinner will succeed, provided that the list of available ingredients is sufficient for at least one recipe in the cookbook. This last example shows that one way to achieve systematicity is to identify problematic or difficult cases at the outset (for example, if my kitchen is poorly stocked). To take another example: a simple algorithm for getting out of mazes is to keep your right-hand on the wall and let it guide you: it should take you to the exit, *unless* the topology of the maze involves loops: in that case you could be circling around a block forever. Nevertheless, this algorithm can be said to be systematic if such mazes are excluded from the outset.

This exhaustiveness condition of systematicity implies a second one: to ensure the *exactness* of the problem definition. A vague generality would allow for unexpected exceptions. For example, building a new factory is not a problem that can be solved systematically, even if its function, capacity and technology are well defined. Attempting to follow a systematic procedure in its design and construction would be constantly marred by the discovery of unexpected new conditions, such as country regulations, social requirements or land constraints, which would require the procedure to be constantly modified. Even the attempt to list all such contingencies in advance (e.g. those that might occur in any country) usually fails miserably. Note that the exactness condition applies not only to the specification of an acceptable solution, but also to the means of action. Solutions that rely on non-replicable expertise or on-the-spot opportunities would not be systematic because the goal of generality would be compromised. Systematic problem-solving requires that both specifications and solutions are described in a predetermined, unambiguous language. Specifications should be written in terms that allow uncontroversial verification of the correctness of the solution, and procedural solutions should be written in terms that allow their straightforward execution by a qualified agent.

Exact, uncontroversial, straightforward: all these terms show the difficulty of qualifying exactly the type of rules that Daston calls *algorithmic*. In his colourful style, the philosopher Gilbert Ryle (2009b) had suggested calling them *Procrustean rules*, as opposed to *canonical* rules, which only indicate the directions in which a good solution might be sought, but leave much to judgement. Procrustean rules – named after the famous brigand of Attica who, after inviting travellers to sleep at his house, sawed off the limbs of those who were too large for the bed he offered them, and elongated the ones of those who were too small – are clear-cut rules, generally of the yes-no kind, that allow us to judge immediately whether a suggestion conforms to the rule or not. These are the rules of grammar, logic and games in general.

These two conditions, exhaustiveness and exactness, that I lump into the single term of *systematicity*, shape the traditional definition of an effective procedure, *i.e.* one whose execution can be automated (performed by a machine). According to computer scientist Donald Knuth (1997: 1: 4–6) it is "a finite set of rules that gives a sequence of operations for solving a specific type of problem", which must terminate in a finite number of steps. Additionally, each of the instructed operations must be "specified rigorously and unambiguously" so that the agent can "understand it exactly" and executable by them "exactly and in a finite length of time".

These conditions are quite stringent, and it is not at all obvious that many human practices – business, medicine, engineering, military, etc. – can be described in this way. A further surprise is that the concept of *effective procedure* did not emerge in all its clarity before the 1930s, shortly before the advent of computing machines. Although there have always been recipes for cooking, methods of construction, military tactics, computational algorithms, etc., it is only recently that the essential relationship between all these objects of knowledge, as well as the possibility to describe them according to a common logic have been recognised. Why is it so? Knuth (1974: 323), for whom computer science is the study of procedures, can rightly exclaim:

> Computer Science should have existed long before the advent of computers.

The term *procedure* itself appears in French legal vocabulary as early as the 14th century to mean "the set of rules and formalities which must be observed in the investigation and resolution of a judicial affair."[3] However, one must note that such procedures stay external and formal relative to the problem at hand. They do not specify sequences of exact legal moves that will decide the case, but only the general form that legal debates and decisions must follow. Following procedural law is a

3 Trésor de la Langue Française interfactif, article "procédure", http://atilf.atilf.fr/tlf.htm (last visited March 28, 2025).

necessary condition for successfully solving a legal case, but is in no way a sufficient one.

Outside law, *procedure* has had until the 1940s a very generic sense. For example, the Webster dictionary of 1936 presents the word as follows:

> PROCEDURE, n. [F. procédure.] 1. Manner or method of proceeding in a process or course of action; also, a particular way of proceeding. 2. The continuance of a process or operation; progress. 3. Customary method of conducting business in a deliberative body; parliamentary order; as, rules of procedure.
> PROCEED, v. i. [OF. proceder, fr. L. procedere, -cessum, to go before, fr. pro- + cedere to move.] 1. To move, pass, or go, forward or onward; to advance. 2. To issue or come forth as from a source or origin; to come (from). 3. To go on in an orderly or regulated manner; to prosecute a design. 4. Law. To begin and carry on a legal proceeding. Syn. See ADVANCE. Ant. Retrogress, recede, withdraw; stop. (*Webster's collegiate dictionary* (1936), 790)

In order to clearly perceive the gap between this definition and the one in use today, one can refer to the definition given by the same dictionary today:

> PROCEDURE. 1. A particular way of accomplishing something or acting; a step in a procedure. 2. A series of steps followed in a regular definite order: a legal procedure, a surgical procedure; a set of computer instructions that has a name by which it can be called into action. 3. a traditional or established way of doing things; (protocol) a code prescribing strict adherence to correct etiquette and precedence (as in diplomatic exchange and in the military services).[4]

Thus, in 1936, the meaning of *procedure* is still largely dependent on that of *to proceed*, and therefore on the general notion of progression; it can easily be substituted for *process*. However the emphasis is already on the manner, custom and method of proceeding, no doubt under the influence of the legal connotation of the term. Its current central meanings, as a goal-oriented (1) and step-by-step (2) process, are not yet apparent. This transformation can be seen in the intersecting evolution of the use of the terms *procedure* and *proceed.* The verbal form, whose meaning has changed little, sees its relative frequency divided by around 2.5 between 1900 and 2000 according to the Google Ngrams tool. The use of *procedure*, on the other hand, whose meaning gradually becomes autonomous from it, explodes between 1900 and 1945, rising from 30 to 115 occurrences per million words, stabilising at this level afterwards.

The explanation of the late elucidation of the concept of *procedure* may lie partly in the fact that the overarching concept of *problem-solving* itself, although very old, has only gradually settled in its current meaning. The verbs *to solve* and

4 Dictionary *Merriam-Webster* online, https://www.merriam-webster.com/dictionary/procedure (last visited March 28, 2025).

to resolve come from the Latin. Meaning originally *to dissolve, to untie, to break*, they were quite naturally associated with the idea of an inquiry into questions, difficulties or problems. *Resolutio* is indeed the Latin translation of ἀνάλυσις (*analysis*), a term which Greek mathematicians associated with σύνθεσις (*synthesis*, or *compositio* in latin) as two complementary movements in inquiry.[5] According to Pappus, analysis starts from the term sought and regresses towards its conditions until things already given are reached; and synthesis binds them together in good order to get properly to the term sought (*Collection*, book VII, 1986: 377). Pappus points out that this dual method applies to construction problems as well as to demonstrations. Indeed, for Greek geometers, *problems* had a restricted meaning that was in sharp contrast and complementarity with *theorems*. Whereas a theorem requires a proof regarding a property of a geometric object, a problem requires one to construct, from given initial objects, an object related to them, such as the construction of the triangle circumscribed inside a given circle. Ἀναλύειν πρόβλημα in ancient Greek would therefore not mean *to solve a problem*, in today's sense, but *to analyse the specification of a required construction in order to find where some critical elements (e.g. points) should be located.*

Although a detailed account is lacking, it was probably only gradually that *solving* acquired its more general meaning of *finding the solution*, while also retaining its original *breaking apart* sense. This movement was certainly helped by the praise of *analysis*, in the 17th century, as the "true path by which the thing [is] methodically discovered",[6] against synthesis, seen as a mere order of exposition of method. It is also at this time that the notion of *problem* gains wider acceptance within mathematics. The rise of Viète's and Descartes' algebra shows the equivalence of construction problems with finding unknown quantities in a system of equations. However, the idea that proving a theorem can also be seen as a kind of problem is definitely more recent. George Pólya's popular and didactic book *How to solve it* (1945) still remarks that mathematics deals with two kinds of problems: *problems-to-find* and *problems-to-prove*.

Outside mathematics, *problem* also had a dialectical meaning in Greek philosophy, denoting the consideration of two contradictory propositions (e.g. *"Is two-legged animal the definition of a human being, or not?"*), potentially in any field of philosophical inquiry – metaphysics, ethics, or physics. Here the problem is primarily one of challenging *opinions* or presuppositions, and the analysis aims at deriving such an opinion from accepted truths or principles. In his *Analytics*, Aristotle pro-

5 On analysis and synthesis in mathematics, see the collection Panza and Otte 1997, and especially Panza 1997.

6 Descartes, *Responses to the second Objections*, AT 7: 155–156, ed. and transl. Moriarty 2008: 99.

poses the syllogism as a universal inferential form for making such connections. In this case, too, analysis can only be partially identified with problem-solving, depending on whether it also takes responsibility for finding the middle term that allows the premise to be connected to the conclusion. Aristotle's position on this question will be examined in section 4.3. But for the mainstream tradition in the Middle Ages and Renaissance, this responsibility was rather devoted to *invention*, which relied on *loci* provided by the *Topic* or the *Dialectic* rather than the *Analytics*.

Practical problem-solving as we would understand the term today – finding solutions to technical problems, deciding on a course of action in a difficult situation, choosing between equally tempting options, etc. – does not fit well into either of these conceptual frameworks. Although Aristotle notes the kinship between geometric analysis and practical deliberation,[7] he does not put this remark to use, and rather promotes a generic syllogistic model. Deliberation, according to Aristotle, begins by considering the means necessary to achieve an end. It then sets them as intermediate goals and proceeds iteratively until it reaches means, all of which are immediately available, and takes action. If it reaches an impossibility, then the goal is abandoned. As we shall see, such a scheme still informs Herbert Simon's early views of problem-solving, and is at the heart of the first general program in artificial intelligence, the General Problem Solver.[8]

The means-ends view of problem-solving has some interesting features, but also many limitations. On the positive side, it is quite general and cuts across all disciplines, as one would expect from problem-solving. This framework can even be accommodated to describe geometric analysis *à la* Pappus. It also shows that problem-solving has a compositional structure: solved problems can serve as means for unsolved ones. Finally, it suggests the idea of a procedure: a practical plan that links all identified means through rigorous syllogisms does indeed look like a prototype of an effective procedure as defined above.

However, its rudimentary form cannot express the solutions to non-trivial problems. It does not explain how the means themselves can be identified, leaving this core task to the good judgement of the inquirer. Additionally, a syllogistic chain does not allow for open-ended iterations (e.g. *"Allow people to enter until you have let ten knights in"*) or exceptions (e.g. *"Stop immediately when you recognise the Duke; then call the butler"*). Finally, it is directional rather than configurational, *i.e.* it assumes that there is a linear or arborescent path back to the critical elements, rather than conceiving of these elements as satisfying a system of constraints (e.g. a set of equa-

7 *Nic. Eth.* III.3. This text is quoted and commented later in 4.1.1.
8 See below, 2.1.2 and appendix A.

tions) to be handled by a particular method.[9] *Reductions in art*, which are Renaissance attempts to describe liberal and technical disciplines in an ordered, synthetic way, are marred by these conceptual limitations (see below, section 7.3).

It is therefore through other routes that problem-solving takes its prominence in the 20th century, to finally become associated with the concept of effective procedure. Although a precise historical account is lacking here too, it seems that its generalised use in psychology and philosophy to describe intelligent inquiry and decision-making in any area of human practice, theoretical or practical, has its roots in the pedagogical thinking of the 19th century. This promoted the active learning of the child through exercises, as opposed to the rote assimilation of knowledge. It is indeed only when *learning* came to be understood as a basic skill of children and humans in general that it became clear to psychologists and educationalists that *problem-solving* was its measurable manifestation that one could aim to steer and improve. Still, in pedagogical books of the 19th century, such as Bain's or Herbart's, the expression *problem-solving* refers strictly to mathematical activities. John Dewey's 1910 book *How we think* (MW 6) is certainly pivotal in the expansion of its use.

Dewey is in some way aware of his *coup de force* when he describes a walking man noticing the cool air and dark clouds in the sky as experiencing a problem:

> To say that the abrupt occurrence of the change of temperature constitutes a problem may sound forced and artificial; but if we are willing to extend the meaning of the word *problem* to whatever – no matter how slight and commonplace in character – perplexes and challenges the mind so that it makes belief at all uncertain, there is a genuine problem or question involved in this experience of sudden change. (MW 6: 188)

It is only when such an experience is recast as a problem to be solved that *thinking* properly occurs:

> *Demand for the solution of a perplexity is the steadying and guiding factor in the entire process of reflection.* Where there is no question of a problem to be solved or a difficulty to be surmounted, the course of suggestions flows on at random; [...] a question to be answered, an ambiguity to be resolved, sets up an end and holds the current of ideas to a definite channel. Every suggested conclusion is tested by its reference to this regulating end, by its pertinence to the problem in hand. This need of straightening out a perplexity also controls the kind of inquiry undertaken. *The problem fixes the end of thought and the end controls the process of thinking.* (MW 6: 189)

9 Mäenpää (1997) makes this useful distinction.

In this text one can see the potential connection of problem-solving with procedures, through the idea that it *channels, guides* or *regulates* the process of thinking. But this connection is not made explicit.

From the 1930s onwards, *problem-solving* imposes itself as the main subject-matter for psychologists of cognition and educationalists alike. In particular, a book by gestaltist Karl Duncker that was translated in 1945 to English with the title *On problem-solving*, became a strong inspiration for early cognitive scientists such as Simon and Newell. But whereas Duncker and *Gestalt* theorists aimed at capturing the moment of *insight* when all pieces of the puzzle fell into place, the new cognitive scientists, inspired by the computer metaphor, thought that what mattered was the whole process of inquiry, of which they tried to delineate the rules.

This historical background is significant. At least since the Gestalt movement, *problem-solving tests* (or *task-completion tests*, as they are also often called) have had a runaway success in framing the experimental methods of cognitive psychologists in all kinds of investigations, whether tests of intelligence, memory, perception, decision making, and so on, whether in practical or scientific matters. However, this success must not hide the plain fact that their current meaning has its origins in a classroom atmosphere where scientifically minded teachers would try to frame controlled experiments to derive the laws governing learning. It is significant that, despite this practical and experimental success, and its importance in some areas of cognitive science, problem-solving remains a second-class citizen in most philosophies of knowledge. A dialogue with Dewey, who is the first philosopher to give it a central role in his *Logic*, is therefore central to my enterprise.

Early cognitive scientists boldly posed the hypothesis that some sort of programs were actually driving animal and human behaviour alike, an idea that was expressed in all its clarity – and with a certain naivety – by Miller, Galanter and Pribram ([1960] 2013), in a seminal work with the telling title *Plans and the Structure of Behaviour*. It argued that the functioning of the mind is based on the execution of *Plans*, which are "for an organism, essentially the same thing as a program for a computer" (2013: p. 16). It even summoned the whole architecture of the modern computer to explain human intention and decision, as it can be perceived in this short extract:

> Something important happens when a decision is made to execute [a Plan]. It is drawn from dead storage and placed in control of a segment of our information processing capacity. It is placed in the focus of attention and, as we begin to execute it, we also engage in a series of minor but necessary tasks, such as gathering information, reminding ourselves of the progress in carrying out the plan at each moment, etc. (Miller, Galanter and Pribram [1960] 2013: 64)

I have summarised elsewhere (2022) the historical vicissitudes and intellectual difficulties of this thesis, commonly known as the *computational hypothesis of the mind*.

Although it is no longer dominant, or at least taken for granted, a key assumption has kept a very wide, often implicit, appeal: that there is an essential continuity between computer procedures and procedures in the generic sense, and that the former are to be distinguished from the latter only by their greater precision and rigour. In this perspective, the transformation of any procedure into an effective, computable one, is only a matter of further refinement. The success of computers can be explained by the explication and formalisation of a huge quantity of informal material present in virtually any human practice. Know-how, embodied into reflexes and habits, consists of implicit procedures that just need to be patiently described.

This apparently innocuous hypothesis raises conceptual problems, since it has indeed two possible and conflicting interpretations. The first one, which has often been called the *intellectualist* interpretation, posits that any know-how can in principle be explicated in a procedure (or set of procedures) which captures it accurately. To act, following this interpretation, means to represent oneself a goal as well as the means for achieving it – even if only fleetingly. Aristotle therefore presents decision-making as a kind of syllogizing, that could even be available to some animals. Whereas the theoretical syllogism starts from the cause and concludes with the effect, the practical syllogism is simply an inverted presentation of the same content of knowledge: Agent A desires effect E. The cause C makes it possible to obtain E. So A seeks to achieve C. In this interpretation, practical reason is nothing else than reason *tout court*, and it is no mystery at all why in essence, any plan of action can be formalized into an effective procedure, given time enough to describe exactly the situation and the possible moves of the agent.

This reading of the nature of rational decision-making and action is still common today, especially in analytical philosophy. It is however strongly challenged by cognitive scientists and social scientists alike, who balk at the idea that rational action could be reduced to chains of inferences. There are endless cases of expert performance, where explicit reasoning is crucially complemented, if not completely superseded, by pattern-recognition and by embodied, contextual skills.[10] Encoding these skills into explicit rules of inference has proven to be an endless and even senseless task, as the repeated failure of logicist artificial intelligence (AI) programs has shown. Today's AI is massively oriented towards statistical induction, which allows the machine to learn by example, rather than by explicit rules.

There is therefore a second interpretation of our ability to systematise know-how and informal problem-solving procedures into effective ones, taking logic itself as a natural outgrowth of our practices. For example, according to the Wittgenstein

---

10 For an overview of the debate, see Clark 1997.

scholar Juliet Floyd, procedures are the elementary forms by which we chain our thoughts and our comprehension of the world, and logic is only their regulation:

> Chaining [of thoughts] evinces possibilities and also necessities of movement and non-movement, shortening and parsing, rearrangment of links, the possibility of tracing from end to end, backwards and forward. It is a synthesis, holding a train of thought together, and also analyzable into parts – which may, however, be substituted into by other parts, other linkings and chainings, other syntheses, and so on. [...] We are binders, working against a universal backdrop, a whirl of life in which procedures, routines, chainings, have already been imbricated. Logic is concerned with establishing and interweaving procedures and possible procedures, as well as plumbing the limits of certain kinds of procedures. (J. Floyd 2016: 40)

Rules and procedures would thus not be abstract objects, detached from life, but the very logic of our forms of life, the logical forms of life, as she again points out:

> The philosophical achievement here is a dynamic and procedural conception of the logical, of logical form as something *in* and *of* life itself. It should be pointed out that it is only in his mature thought that Wittgenstein succeeds in anchoring logic *in* life, as something *of* life, by making it ubiquitous through criteria and procedures of our everyday life. (J. Floyd 2018: 68)

It is precisely because logic is part of forms of life that procedures can be refined and regulated into effective procedures. There is therefore no mystery in the fundamental continuity of procedural habits and mechanical procedures. In the same vein, philosopher of computing Bruno Bachimont writes:

> The natural tendency of the method is to pursue its formalisation in such a way as to be able to function as a program and therefore to mobilise the human agent not as an interpreting and improvising agent but as a mere executor. Sometimes the method eliminates the mobilisation of interpretative mediation by objectifying itself completely into a technical and machinic device, as we see in management systems and the industrialisation of our cognitive techniques. (Bachimont 2010: 64)

This position is inspired by the Derrida–Stiegler interpretation of technology as a fundamental rewriting of one's behaviour in order to adapt to the environment and free oneself from one's own determinisms. In Bernard Stiegler's philosophical anthropology, the notion of *program* plays a central role: genetically programmed, the uniqueness of human beings is that they can reprogram themselves – in their body by learning new chains of action, but also socially through institutions, above all through writing, which enables them to structure and reshape their own mental capacities.

> New programs appear, suspending the effectiveness of existing programs by reintegrating them and subjecting them to a new order. These suspensions generate epochs. Incessant *epochè*, the history of man would be nothing but such suspensions. (Stiegler 2018: 391)

These broad, evocative frescoes of the essential continuity of technology, beyond all its apparent ruptures – the revolution in writing, the steam engine, the computer, etc. – are in part inspired by the anthropologist Leroi-Gourhan, who developed the thesis of the *projection of organs*. According to Leroi-Gourhan, human development took place through the gradual externalisation of humans bodily possibilities – tearing with the teeth, scraping with the nails, grabbing with the hand, etc. – into tools that amplified them, increasingly freeing intellectual capacities from their enslavement to the control of the elementary gestures of life. No doubt inspired by cybernetic ideas, Leroi-Gourhan presents information technology as the ultimate stage in this process, with the computer taking charge of memorising and replicating humans' technical gestures, which he calls *operating chains*, much as Floyd in the quote above talked about *chains of thoughts*. In a striking shortcut that seems to anticipate the famous introductory sequence to *2001: A Space Odyssey*, he writes:

> Mechanical automation is the penultimate possible stage in the process initiated by the Australanthrope armed with his chopper. The liberation of the territories of the cerebral motor cortex, definitively acquired with the vertical station, is completed at the moment when man externalises his motor brain. (Leroi-Gourhan 1965: 52)

These pragmatic and biological views of technique face a problem however. Considering systematic procedures as mere refinements of pragmatic know-how and prescriptions does not explain the movement of *objectification* mentioned by Bachimont in our quote. The further precision of a subjective procedure of action will never make it a mechanical one, since it will still depend on the skills and prior knowledge of the agent. Any transposition of a procedure from one agent to another requires its *redesign* to fit the constraints and possibilities of its new agent, and it is this redesign that forbids the hasty assimilation of human habit-making with machine programming that is suggested by Floyd, Bachimont and Stiegler. Explaining how a machine culture may continuously arise in a tool culture is a recurrent problem of biological philosophies of technology (Clarizio 2021: 149).

A third possibility is open to consideration, which this book will explore: that there is a radical discontinuity between the ordinary procedures of human practices and the effective procedures designed for computers. This design, although it could take practical know-how as a model and inspiration, would start from a fundamentally different epistemic attitude than the one that originally abstracted rules and prescriptions from practical life, the attitude that I call here systematic. This hypothesis is actually quite intuitive and has often been put forward in the past by scholars: that there is a radical break between traditional techniques of craftsmanship and modern technology driven by science and the scientific method. What does this break signify? If it is not "natural" to humans, where does it come from?

How does it really change our human practices? And where will it stop? The recent advances of artificial intelligence show that the limits of systematic problem-solving can be extended in surprising ways.

My aim in this book is not to answer these questions, but to lay the foundations for a proper understanding of their meaning. My main thesis is that the demand for exact problem-solving, although probably originating in earlier endeavours, took its definite form in the geometrical thinking of the 17th century. What is surprising is that such a stringent requirement could be successfully applied to large areas of human practice. The reason for this success, I will argue, does not lie in some metaphysical truth about the mathematical shape of our universe or the description of reality in terms of quantities and numbers. While this hypothesis may be valid in other respects (I don't intend to discuss it here), it misses the central transformation at stake. The central change concerns the framing of inquiry when it sets out to solve problems systematically, *once and for all*, as mathematicians do. Solving "once and for all" problems in the uncertain world of physics and human affairs requires dividing the phenomena involved in the situation into two groups: those for which one can make certain claims about their behaviour, called the *machine*, and those for which only minimal claims are available, called the *context* or *environment*. The machine is where constructive procedures of geometric rigour can be applied. The environment is a simple but exact description of the problem, the constraints that its solution must satisfy, and the few resources that the machine can use to achieve it. It must have a minimal systematicity that makes it possible to reason about it when designing the machine, to build exact chains of inference, and ideally to verify the exhaustiveness of the proposed solution for all possible configurations of the problem.

One can see that in this formulation a machine does not primarily mean a material artefact, nor does a system primarily mean an organised set of representations or symbols. The distinction between "material" and "representational" is a subtle one that must be drawn carefully, lest one fail to see that a machine is primarily a logical construction, and that systems are often effective as embodied infrastructures or institutions. I argue that the concepts of *system* and *machine* are best thought of as categories of experience, in a Kantian sense, *i.e.* as principles of intelligibility of phenomena.

This perspective helps to explain why systematic inquiry often involves a complete reconfiguration of the problem, of the situation itself: to make the operation of machines safe (by which I mean: working as intended in all cases) requires preparing environments for such operations, removing residual ambiguities and keeping unexpected events at bay. This preparation is often material – as when a factory shop floor is redesigned to allow further automation – but it always begins with a mental reconfiguration: making our ideas clear, exact, searchable, and verifiable, so that chains of inferences can be safely designed.

Although it is only since the invention of the computer that we can clearly see how far the scope of systematic inquiry can go, the first machines, clocks, mills, and later the water frame, the steam engine, or the Jacquard loom, show that such an attitude was at work long before. Computers are just one moment in the history of systematic inquiry, albeit a crucial one. They mark the full understanding of the concept of mechanism itself and the design of a universal one. By *mechanism* I mean the central part of any machine, *i.e.* the set of components that controls the transmission of motion or energy to the other parts. Today's computers are best described as universal transmission mechanisms – a proposition that I only sketch in this book and develop fully in *Reasoning with Machines.*

1.3 Approach of the book

I begin my exploration in chapter 2 with Herbert Simon's bold assertion that computer programming is the *exact* form of problem-solving. In Simon's grand vision, programming is destined to become the central discipline in a new science of design that will enable a new dialogue and cross-fertilisation between the professions, the sciences and the arts. I show the many difficulties of this appealing thesis. In particular, it seems to stumble against the very idea of a *practice* as it has been articulated, for example, by MacIntyre or Bourdieu. A rich human practice, involving traditions, institutions, standards of excellence and competition between its members, can never be reduced to the sole dimension of its technical rules.

In chapter 3, therefore, I limit my inquiry to highly technical work environments: can they be said to be governed by systematic procedures? Workplace studies point to situations – e.g. maintenance of photocopiers, navigation of warships – where this is obviously not the case. In fact, systematic procedures are quite alien to human beings, even where traditional craftsmanship is not involved. Not only do procedures seldom provide useful guidance in solving problems, but they can also be in direct conflict with intuitive ways of doing things. This can be seen by examining the various motivations that drive automation in industrial environments; they go well beyond productivity gains. The main epistemic break is not between practice and technical knowledge, but between craft and industry. Industry is a notoriously difficult notion to pin down; but its narrative, since the Industrial Revolution, has always linked it to science and the scientific method – with technology now defined as science applied to industry. What is required thus is to understand the complex links between science, technology and systematicity.

Chapter 4 directs this question to Aristotle, who first framed the meanings of these terms. Aristotle closely associated science with systematic speculation, but was less clear about technical knowledge or *technè*. Sometimes he presented it as the first step towards science, but at other times he sharply distinguished it from the theoretical mind, associating it rather with practical prudence and the deliberative mind. I suggest that technical knowledge, being about finding the right means to an end, differs from prudence insofar as it can generalise some practical deliberations; and it differs from science insofar as general means are only hints to universal causes, stopping short of establishing them from first principles. For Aristotle, then, technical knowledge is incapable of systematic thinking, as it is concerned with contingent ends and means in an uncertain sublunar world. However, I show that Aristotle was very much interested in systematizing inquiry itself, and devised a method, famous in the Middle Ages under the name of *pons asinorum*, for systematically finding middle terms for a syllogism.

The three next chapters aim to adjust the Aristotelian framework to explain the possibility of systematic problem-solving in all areas of human practices. Chapter 5 is devoted to the exploration of the essential complementarity and convertibility between theoretical and practical knowledge. It is not only that science and technology are in incessant dialogue and exchange. Technical procedures can be converted into descriptive statements, and conversely, there are *constructions* at the heart of the most theoretical discipline, geometry. This convertibility was already theorized by Aquinas himself, but it is only with Kant that its fundamental reason is made clear, that all practical precepts and theoretical judgements have only one fundamental form, which is that of the *hypothetical rule*. Rules are both acts of liaison and relations between phenomena. They are constructive. Martin-Löf's theory of types allows us to make sense of this fundamental convertibility between theorems and problems.

Chapter 6 establishes the identity of nature between practical deliberation and theoretical problem-solving. Aristotle had himself noticed their proximity and called them both *zetesis*, or *inquiry*. Inquiry is the core term of Dewey's philosophy of knowledge, and this chapter is devoted to its close examination; it provides a convincing framework for establishing the nature of procedures, as *reconstructed inquiries*.

Chapter 7 considers the nature of systematic thinking. A historical approach tracks its roots in the Renaissance, and especially in *reductions in art* which aimed at ordered, rules-based descriptions of the arts, liberal and practical alike. Drawing on Kant and Dewey, I analyse these kinds of projects, including modern theory-building, as *global inquiries* which allow local inquiries (or problem-solving) to benefit from verified and coherent systems of inference for a given field of problems.

I then spell out the required characteristics of systems powerful enough to express *any* effective procedure, according to the Church-Turing thesis.

After this groundwork, it is finally possible, in chapter 8, to understand the conditions for systematic problem-solving. What allowed it to leave the realm of pure mathematics and to become applicable to ever larger areas of human practice is the concept of *machine*, which gets its modern sense in the 17th century as a *geometrical object*. A machine is any dynamic phenomenon whose interactions with its environment can supposedly be represented by exact rules of inference, and which can therefore also be perfectly controlled, if one of these interactions can serve as a handle for steering its behaviour. It is in the process of extracting and assembling such ideal micro-spaces of inference and gradually finding out how to make them ever more flexible to express larger sets of procedures, that the principle of programmable machines was finally discovered.

* * *

The patient reader is encouraged to follow the progress of this exploration in the order just outlined, which is roughly how my own journey to understand the effectiveness and versatility of systematic procedures took shape. I believe that it is the most convincing path, as it patiently deconstructs many preconceived notions regarding practical and theoretical knowledge, a prerequisite for arriving at a satisfactory explanation of the effectiveness of machines in our human world, while avoiding metaphysical questions.

A quicker route may go through chapters 3, 6, 7 and 8, skipping the discussion of Herbert Simon's epistemology of problem-solving in chapter 2 and the theory / practice discussions of chapters 4 and 5.

An even quicker one would simply starts at the end – *i.e.* at chapters 7 and 8 – to confront directly the main theses of the book. If the reader is convinced, they can stop here. If, however, they are puzzled by some of the theses, or even by the meaning given to some of the terms, such as *inquiry, procedure, system, ratiocination*, then they should go back to earlier chapters to engage in the discussion – which is the real aim of this book.

2 Design and deliberation (Simon)

My main research question is about *systematic* problem-solving, and about the conditions for this systematicity. Can one just say that all procedures stem from an identical approach to problem-solving, that can only be made more or less precise or rigorous? And if so, what do one means by *precision* and *rigour*?

In the beginnings of cognitive science, Herbert Simon gave a resounding "Yes!" to this question, asserting, in his famous work on the *Sciences of the Artificial*, that programming was the *prototype* of all problem-solving, which had to be seen as an act of *design*. The advent of computer science, which finally gave this form of rationality its own language, would enable all practical disciplines to finally constitute themselves as sciences.

This chapter is devoted to the discussion of this thesis. The assertion that programming is not a form of practical knowledge among others, but its essential form, goes directly to the heart of this investigation, where the similarities, but also the gaps and even the contradictions between systematic problem-solving and other forms of knowledge can best be seen.

The chapter begins with an exposé of Simon's general doctrine on practical knowledge as design (section 2.1). It explicates the exemplary role that programming should play if professions are to be constituted as science (section 2.2). Section 2.3 turns to the critique of this thesis. The decision to apply a rule-based procedure requires the correct identification of the problem in the first place, which cannot itself be based on rules, as else an infinite-regress would occur. This critique leads to a review of the entangled conceptual relationship between practical sense and technical rules in human practices.

2.1 Design and practical deliberation

Herbert Simon, in his famous work *The Sciences of the Artificial*, calls *professions* useful human activities, constituted around bodies of autonomous and specialised discursive knowledge necessary for their proper exercise. Professions may include engineers, who deal with the production of artefacts in the strict sense, but also doctors, lawyers, administrators or managers of economic activities, and so on. The point of this circumscription of practices is that it excludes leisure, the creative arts, practices of the body (especially sports) and social practices (rituals, conversation, etc.) – all practices where it is reasonable to think that their procedural reduction is complex or even impossible. At the same time, it is broader than the simple domain

https://doi.org/10.1515/9783111494500-002

of production (represented by engineering and architecture) since it includes, for example, medicine and law.

Simon's ambition, indicated by the title of his work, is to defend the autonomy and particularity of this knowledge relative to the academic sciences, and to show that it has become possible to envisage their constitution as scientific disciplines in their own right, according nevertheless to a new standard of scientificity appropriate to their objects, insofar as these are not *observed* but *designed*. In other words, through the professions, and thanks to the paradigmatic role that programming is called upon to play in them, the specific character of technical rationality can now be brought to light.

This section begins by identifying the notion of *design* that Simon considers central to all professional thinking (2.1.1), before characterising it as deliberation (2.1.2) and finally highlighting its reflexive character (2.1.3).

2.1.1 Design and the artefact

The origin of *The Sciences of the Artificial* (1996: hereafter abbreviated to SA) is composite: its first edition, in 1969, consists of the publication of three lectures given at M.I.T. in 1968, to which he adds an article formerly published in 1962, *the Architecture of complexity*. In 1981, a second edition appeared, in which he interwove the texts of the original lectures with those of three other lectures, given in 1980 at Berkeley. Finally, in 1996, on the occasion of the 3th edition, he interposed before *the Architecture of complexity* an eighth chapter designed to take account of new conceptions of complexity since the distant 1962 article. However, he makes only marginal changes to the texts, mainly adding references in footnotes.

Simon updated his book until the end of his life (he died in 2001). This fact attests to his constant conviction regarding their validity, but also reflects the growing success of the work. Indeed, with Christopher Alexander's *Notes on the synthesis of form* (1964), it is considered as a major source for design as an academic discipline.[1]

It is in the three original lectures that the general argument of *Sciences of the Artificial* is most clearly perceived (chapter numbers are those of the latest edition):

- Chapter 1: Understanding the natural world and the artificial world.
- Chapter 3: The psychology of thought: inserting artifice into nature.
- Chapter 5: The science of design: creating the artificial.

1 See Buchanan 2009: 421–426 and Cross 2006: 96.

The first chapter sets out the conceptual framework of what Simon means by "artificial", the third explores its epistemological foundations, and the fifth culminates in a prescriptive program concerning the academic discipline he calls for. From the outset, he perceives this objective as paradoxical, and the first two chapters should be understood as aiming to justify it: for indeed, how could design, a creative process, be the object of scientific study?

Chapter 5 of *Sciences of the Artificial* begins with a passionate call for the renewal of academic teaching of the professions (SA: xii-xiii). Indeed, Simon is not satisfied with their reduction to the status of applied sciences since the end of the Second World War. Although there are still engineering, architecture and medicine courses, students are no longer really taught how to build, design or heal. Instead, they are taught analytical skills that are supposedly useful for these professions, such as applied mathematics for mechanics, materials science for architects, or biology for doctors. According to him, this displacement had an institutional cause: the standardisation of university curricula and diplomas after the Second World War, which required disciplines to be subject to objective assessment, doctoral theses, research articles, etc.

> Such a universal phenomenon [had] a very obvious [reason]. [Professional schools...] hankered after academic respectability. In terms of the prevailing norms, academic respectability calls for subject matter that is intellectually tough, analytic, formalizable, and teachable. In the past much, if not most, of what we knew about design and about the artificial sciences was intellectually soft, intuitive, informal, and cook-booky. (SA: 112)

Now if Simon thinks that technical expertise cannot be derived mechanically from the sciences, he also thinks that it is better than experience:

> Nowhere do we need to return or retreat to the methods of the cookbook that originally put design into disrepute and drove it from the engineering curriculum. For there exist today a considerable number of examples of actual design processes, of many different kinds, that have been defined fully and cast in the metal, so to speak, in the form of running computer programs. (SA: 135)

So what does Simon mean by this term *design*, and why does he think it legitimate to place it at the foundation of disciplines as varied as engineering, law or business? After all, it seems to relate to the production of artefacts, and therefore to technology rather than to the practical realm in general. And why should programming be called upon to play such a central role in its rational refoundation?

Design is about artefacts, but an artefact is not primarily a material object; it is defined functionally, *i.e.* by the use to which it is put in a given environment. This means that it is not primarily defined by its constituent elements, its materials for example. Rather, these are constraints that design must take advantage of, or overcome by substituting other elements. The purpose of design is to find solutions for the desired ends, taking into account the constraints imposed. These solutions take the form of interfaces between an internal system, endowed with specific laws of behaviour, and an external environment, which imposes constraints and at the same time assesses the system's suitability for the objectives set. Simon takes the example of the invention of marine clocks, which were insensitive to the disruption of mechanisms by the rolling of the sea. A "good" clock is one adapted to its specific conditions of use.

> An artifact can be thought of as a meeting point – an "interface" in today's terms – between an "inner" environment, the substance and organization of the artifact itself, and an "outer" environment, the surroundings in which it operates. If the inner environment is appropriate to the outer environment, or vice versa, the artifact will serve its intended purpose. (SA: 6)

This means that the proper object of any technique is to design *interfaces*:

> the best of all possible worlds – at least for a designer – we might even hope [...] to be able to characterize the main properties of the system and its behavior without elaborating the detail of either the outer or inner environments. We might look toward a science of the artificial that would depend on the relative simplicity of the interface as its primary source of abstraction and generality [...] Description of an artifice in terms of its organization and functioning – its interface between inner and outer environments – is a major objective of invention and design activity. (SA: 9)

This conception of technology as happening on the interfaces between two environments, one internal and the other external, was Herbert Simon's major intuition when he wrote the *Sciences of the Artificial.* The technician only ever works on natural systems (materials, organic bodies, artefacts used in turn as sub-systems). What is properly artificial in the artefact is the arrangement of its parts to perform a function in its environment.

> We must be careful about equating 'biological' with 'natural'. A forest may be a phenomenon of nature; a farm certainly is not. [...]. A plowed field is no more part of nature than an asphalted street. [...] Those things that we call artifacts are not apart from nature. They have no dispensation to ignore or violate physical law. At the same time they are adapted to human goals and purposes. (SA: 3)

Artificial is not "unnatural". In particular, the opposition between the natural and the artificial must not be equated with that between the living and the machine, since both follow the same natural laws. This does not mean however that design can be reduced to the application of scientific results – physical laws, materials, taxonomies of phenomena. It can mobilise them, but cannot be deduced from them, because its very object is their arrangement to achieve the desired ends. Indeed, an interface is an arrangement of environments subject to distinct natural laws in order to make them cooperate to bring about a desired situation. Even more fundamentally, Simon recovers Aristotle's definition of deliberative rationality as occupied with contingent things, as opposed to theoretical rationality occupied with necessary things:[2]

> Engineering, medicine, business, architecture and painting are concerned not with the necessary but with the contingent – not with how things are, but with how they might be – in short, with design. (SA: xii)

The paradox of Simon's position is all here: he thinks that design is just another name for deliberative rational activity in general, that it has its own principles and laws, and that it can therefore be the object of a science.

2.1.2 Design and deliberation

It is worth taking a closer look at this assimilation of design and deliberation. At first sight, it is not at all self-evident. Design, in the usual sense, concerns material objects. It may be visualised by plans and blueprints of the artefact to be produced. Deliberation, on the other hand, is about action, about a course of events that one wants to happen in order to achieve their goals. It may be visualised by sequence diagrams that represent the actions.

For Simon, this distinction is superficial, for what he means by design is indeed deliberative rationality in general:

> Engineers are not the only professional designers. Everyone designs who devises courses of action aimed at changing existing situations into preferred ones. The intellectual activity that produces material artefacts is no different fundamentally from the one that prescribes remedies for a sick patient or the one that devises a new sales plan for a company or a social welfare policy for a state. Design, so construed, is the core of all professional training. (SA: 111)

Simon's appropriation of the term *design* can be traced to a diptych of two texts, *the Logic of Rational Decision* (1965) and *the Logic of Heuristic Decision-Making* (1967) thus shortly preceding the lectures at M.I.T., which refer to it.[3] The term already appears in Simon's earlier writings, in its traditional engineering-related sense. He already notes the parallel between action plan development and engineering design (Simon [1947] 1997: 312–33) but it is in the first article of 1965 (Simon [1967] 1977:

2 *Nic. Eth.* VI.1. Aristotle's position is studied in chapter 4.

3 Text reprinted in Simon [1967] 1977: chp. 3.1 and 3.2.

148–149) that he begins to use it as a synonym, and it is in the 1967 article that one finds its first thematisation as a central concept of practical rationality.

The subject of the two articles is the "theory of action", and Simon prides himself on basing his ideas on the observation of concrete cases of decision-making by the businessman, the engineer, the doctor, the soldier, and so on. He distinguishes two phases in decision making: a phase concerned with framing the problem to be solved – for example, choosing between several competing priorities, and a phase concerned with strict deliberation, *i.e.* determining the means for solving the problem. The latter is introduced in the following terms:

> Design decisions
> The second stage in decision making is to devise or discover possible courses of action. This is the activity that in fields like engineering and architecture is called 'design'; in military affairs 'planning'; in chemistry, 'synthesis'; in other contexts, 'invention'; 'composition'; or that most approving of labels 'creation'. In routine, repetitive decision making, design may play a small role, for the alternatives of action may be already at hand, to be taken off the shelf as occasion requires. In slightly less structured and repetitive situations, design may require only relatively unproblematic assembly of action alternatives from prefabricated components. In most human affairs, however, the design stage in decision making occupies a far larger part of the mind's information processing capacity than either the stages of attention directing or of choice. Design is concerned with devising possible means for ends. (Simon [1967] 1977: 160)

The schema of *ends and means* is thus indeed Simon's starting point concerning practical rationality. This schema goes back to Aristotle, who equates it in *Nic. Eth.* III.3 with the very process of deliberation: when one seeks to achieve a given end, one examines the means by which it can be achieved, and each necessary means in turn becomes a new end whose feasibility must be explored by the same method (see below, 4.1.3). Simon has long thought of this schema as the general form of practical rationality, and it is on this basis that he builds, with Shaw and Newell, one of the first emblematic programs of artificial intelligence, the General Problem Solver (GPS) (Newell, Shaw and Simon 1959). The program has an recursive structure and uses an evaluation function of system states in relation to the final state sought to test each possible action and choose the one that gets closest to it (Simon [1967] 1977: 167).[4]

Simon was aware of a certain incongruity in the new meaning he wished to give to the term *design*, since one does not usually say that an engineer is "planning a line of action" when they make a blueprint of a building. Simon therefore had to explain this unity between productive action and action in general:

4 More detail on this important program is given in appendix A.

> It may be objected that our proposed meaning for 'design' excludes almost everything called that in engineering and architecture. The typical solution of an engineering or architectural design problem is not a program in the action space, but a description (or picture) of an object in the state space -a blueprint or an elevation, say. But the reason why this description is regarded as a solution to the design problem is that the steps completing the translation from that description to an action program (a manufacturing or construction program) are nonproblematic. The sole action the engineer or architect has to perform (if I may indulge in a bit of heroic oversimplification) to realize the design is to turn it over to a manufacturing department or construction contractor. (Simon [1967] 1977: 164)

The distinction between material object and action plan is replaced here by a distinction between action space and state space, a formulation inspired by mathematical physics where one often speaks of transition matrices on state spaces.

The case of engineering is therefore specific in that it is already endowed with production procedures that explain how to make an object from a given blueprint. The engineer can therefore focus on the design of the blueprint alone. But in the general case, the object of practical reflection is to relate possible actions to desired situations.

This also explains why, conversely, deliberations that appear to concern actions are also design. On the face of it, a doctor prescribes actions to be taken: taking such and such medication at such and such a frequency, doing such and such exercise, and so on. But the aim of this treatment is to change the current situation (the disease) into a desired situation (health) by modifying some of its characteristics (such and such hormonal balance, such and such muscle strength, etc.). Unlike the engineer, who directly describes the desired end state and leaves implicit the actions to be taken to achieve it, the doctor only describes these and leaves implicit the intermediate and final situations he is actually aiming for. The army general and the chess player are in the same situation as the doctor: they only spell out the actions to be taken or the moves to be made, but what they have clearly in mind are the situations they would like to see happen on the battlefield or on the chessboard.

In the general case of deliberation, however, nothing is obvious – neither the situations to be brought about nor the actions to be taken, and possibilities have to be conceived of from both sides:

> In these complex, real-life situations, we may think of the design process as concerned with weaving a tight web of connections between actual physical actions (the actions of lathe operators and carpenters) on one side, and realized states of the world (electric motors, office buildings) on the other. Many parts of this web can be woven using traditional patterns, without new problem-solving activity that goes beyond choice among known alternatives. At other points in the web, however, weaving together the fabric of means, coming from the left with the fabric of ends, coming from the right, requires exploration for new alternatives. It is at these

> interfaces (there may be more than one) that design activity takes place. (Simon [1967] 1977: 164–165)

One can see in this text Simon very precisely and even very visually substituting, through this textile metaphor, the schema of interface for the schema of ends and means. In addition to projecting actions into a predefined space, Simon makes a second shift in relation to the Aristotelian schema of deliberation, by substituting the linear sequence of analysis ("if there is only one to arrive at this end, we seek how it makes it possible to arrive at it and also by what means this means itself can be attained..."[5]) with the figure of a network that must be connected at several points. The Aristotelian diagram could maybe take the form of a tree, since the text of *Nic. Eth.* III.3 may imply that several means are necessary to achieve an end, and that the same analysis must therefore be applied to each of them. But in the Simonian network diagram there is complete symmetry in the arrangement of ends and means. It is indeed an illusion to think that one who deliberates is looking for one single thing; this illusion is undoubtedly linked to the focus of attention on the salient property of the state of the world that one is trying to achieve. But there are always constraints, often implicit, that one imposes on any target situation, for example that of not depleting the available resources beyond a certain threshold. These resources are not "means" but also properties of the state space. Strictly speaking, "resources" are the actions that one can take. These can sometimes be described by plans or written instructions, whose execution routines are then known – which is what *a tight web of connections* means.

This expansion of the linear or tree-like schema of ends and means into a mesh schema of actions and states makes it possible to reformulate the notion of problem; it is no longer a question of finding the path from means to end, but of solving a systemic equation, where several constraints must be satisfied in the space of possibilities. The path, if one can still call it that, is an optimum in a space of possible paths, which best resolves a system of constraints; by this Simon brings the traditional scheme of ends and means down to a more general scheme, that of the search for solutions in a complex space.

A problem space is therefore nothing more than a set of possible interfaces between an agent capable of actions and a situation characterised by states. Designing an interface comes down to finding the rules that systematically adjust the agent's actions to the state of the situation. This general diagram shows the profound identity of design and deliberation, as well as of production and action.

---

5 This text is quoted and commented on below, 4.1.3.

Designing an artefact is simply a certain way of envisaging the reconfiguration of a material situation with a view to an end, it is nothing more than a plan of action.

2.1.3 The reflexivity of design

If Simon keeps *design* rather than *deliberation* to designate the activity of practical reason, it is because he is going to use this term to designate the second-degree deliberation that consists of identifying the right search strategies in the problem space and simplifying this space itself. Design can be reflective, and therefore can establish itself as an autonomous science.

The characterisation of design as a search strategy comes from operational research (OR), *i.e.* the study of mathematical methods applied to decision problems, such as resource allocation, which developed during the Second World War, where it was a critical planning tool. Simon incorporated OR into his curriculum as early as 1949 (SA: 139) and would often point out its proximity to artificial intelligence (Simon 1987). Although OR is only a very specific form of search, requiring a representation of the problem by a system of equations over continuous quantities (SA: 28), Simon first notes that practical problems can be objects of computation:

> What was genuinely new for economics in operations research was the concern for procedural rationality – finding efficient procedures for computing actual solutions to concrete decision problems. The demands of computability led to two kinds of deviation from classical optimization: simplification of the model to make computation of an "optimum" feasible, or, alternatively, searching for satisfactory, rather than optimal choices. (Simon 1976b: 76)

Deliberating on a simple problem, indeed – for example, how to get from Etoile to Nation by Paris public transport – is not what really interests the science of design that Simon wants to bring to light. Complex problems have spaces of possibilities of such magnitude that it very quickly dooms random search to failure. What is required – and what is really the object of design according to Simon – is to develop search strategies that enable acceptable solutions to be found in the problem space as efficiently as possible. This space is itself made up, remember, of authorised sequences of states and actions. Design science is therefore about second-degree search.

This reflexive aspect is clearly presented in the aforementioned 1965 article on the theory of action. This article distinguishes three levels of imperatives guiding any deliberation:

> In problem-solving activity, we have to consider the relations of three sets of imperatives: (A) imperatives prescribing performance goals and constraints; (B) imperatives prescribing

> performance for attaining the goals (A); and (C) imperatives prescribing solving programmes for discovering and evaluating the programmes (B). (Simon [1967] 1977: 149)

By this distinction, Simon wants to point out that in general there is no deduction from (A) to (B), *i.e.* from objectives to means. Most of the time, programs (B) are proposed on the basis of past experience, programs that are already known and which one assumes, without demonstration, can be transposed to the present case. There are cases where one has a method (C) for deducing exactly (B) from (A), as in certain operational research problems; but in the general case these methods (C) do not guarantee the optimality of the solution, or even that they always find one; they are *heuristic*; and it is in the design and improvement of these methods that the place of practical reflection – the place of invention – properly lies.

In a problem space with very large number of combinations, certain heuristics (C) improve search efficiency by eliminating possibilities. For example, if the problem is to solve a system of equations involving 10 unknowns representing the numbers 0…9, the size of the space is 10! if all possible combinations are tested. Managing to fix the value of just one through an observation on constraints reduces the size by a factor 10, then 9, etc. (SA: 54–57)

The heuristics therefore relate not only to the search strategy, but also to the simplification of the problem space. This is where the reflexive dimension of design comes into its own. It is not constrained by the problem space, because on the contrary its object is precisely to represent it correctly. Simon goes so far as to say that, in a way, "solving a problem means representing it in such a way as to make its solution transparent" (SA: 132).

A classic strategy is to manage to partition the problem space into independent or quasi-independent subspaces,[6] each of which is of a reasonable size for a classical search. For example, if someone is looking for a restaurant among the 12,000 that exist in Paris, it is not efficient to examine them one by one. By first answering two questions each with around twenty possible answers, such as *"in which neighbourhood do I want to dine?"*, *"what type of cuisine do I fancy?"*, the problem is reduced to around thirty possibilities (on average), by being broken down into three simple problems to solve.

It is therefore common in solving complex problems to proceed by hierarchical abstractions, which simplify the action and state spaces so as to focus search on a few broad options, leaving aside all the detail. For example, for an engineer building a railway through a mountainous country, it is a matter of choosing between three or four broad options, and only then drawing up more detailed plans on the basis

6 On this notion of *quasi-independence*, see SA: 197.

of precise calculations: multiple *planning spaces* can thus be distinguished in the course of a design process. The example of road design is one of Simon's favourites, no doubt because of its naturally metaphorical character for any course of action.[7] There is therefore no difference in kind between methods and procedures; methods are sketches of procedures, indications of markers that will be specified as procedures when applied to concrete situations. Their indeterminacy is only the condition of their adaptability.

A general plan thus becomes progressively more detailed; each higher decision frames the problems posed at lower levels and the decisions that will have to be taken there, and in turn each complex path appears as the composition of a more elementary hierarchy. This view is somewhat schematic, since Simon is well aware that it is rare to be able to decompose a problem neatly into independent subproblems; but he thinks that it is possible, in most cases, to arrive at *quasi-modular* representations of the problem space, where the subspaces have weak interactions with each other, which makes it possible to study the subspaces separately, *at first order*, and to study their composition as a separate *integration* problem, which is precisely that of the higher level. Here the full relevance of the notion of design appears clearly, because the real issue is to achieve such an effective representation of the problem space and to identify the cutting lines that will radically simplify its treatment. Knowing how to solve a problem is essentially to have the right representation of it. This means that each field of problem requires its own specific language:

> We know that problems can be described verbally, in natural language. They often can be described mathematically, using standard formalisms of algebra, geometry, set theory, analysis, or topology. If the problems relate to physical objects, they (or their solutions) can be represented by floor plans, engineering drawings, renderings, or three-dimensional models. Problems that have to do with actions can be attacked with flow charts and programs. (SA: 133)

Simon is not, however, satisfied with the present state of what should be a *theory of representations*:

> An early step toward understanding any set of phenomena is to learn what kinds of things there are in the set – to develop a taxonomy. This step has not yet been taken with respect to representations. We have only a sketchy and incomplete knowledge of the different ways in which problems can be represented and much less knowledge of the significance of the differences. (SA: 133)

7 See for example Simon [1947] 1997: 109, 1972: 172, SA: 125.

This can be seen in that sometimes the standard strategy, *i.e.* the search for a modular decomposition of a space, is unsuccessful. It is then necessary to *extend* the representation of the problem space. A classic example to which Simon often returns is that of "the mutilated chessboard", from which two squares have been removed at two diagonally opposite ends (see figure 2.1). The question is then whether one can pave the remaining 62 squares of the chessboard with 31 dominoes. An empirical search for such a tessellation leads nowhere. We have to change our representations to find the solution, which then seems obvious. If we pay attention to the alternating colours of the squares on the chessboard, we see that each domino must necessarily cover two squares of the opposite colour. The mutilated chessboard contains 32 squares of one colour and 30 squares of the other. This proves that there is no solution to the problem. This method involves *seeing* the colour of the squares, no longer as an accidental property, but as an essential property of the chessboard. By

Figure 2.1: The problem of the mutilated chessboard

introducing a new concept, square parity, the whole problem space is radically simplified. This operation takes place in what Kaplan and Simon (1990: 380) describe as the "meta-space of possible problem spaces for the mutilated chessboard problem".

Design therefore has an essentially reflexive dimension. It becomes the subject of scientific study not through the resolution of concrete problems, but through the study of problem-solving strategies, either to optimise search in a given problem space, or to reconfigure this space itself in order to better represent it. This reflexive aspect also shows its creative dimension, which is naturally evoked by the term *design*.

> Classical decision theory has been concerned with choice among given alternatives; design is concerned with the discovery and elaboration of alternatives. [...] The evaluations and comparisons that take place during this design process are not, in general, comparisons among complete designs. Evaluations take place, first of all, to guide the search the elaboration of the design itself. They provide the basis for decisions that the design should be elaborated in one direction rather than another. (Simon 1972: 72)

The activity of design is the search for and invention of the right heuristic and the right problem space, which occurs through a creative process in which each step creates new problems and opens up new possibilities. In "ill-structured" problems, *i.e.* problems for which the initial specification is only vague or incomplete, the design process consists of constantly modifying, partitioning and restructuring the overall problem space until the problem is well posed. Simon here studies the architect's creative process, which progresses by alternating between solving local problems and reconfiguring the global problem (Simon [1967] 1977: 312–317).

2.2 Programming as exemplary design

This section considers the essential role that programming must play in the advent of a science of design. It begins by establishing the paradoxically non-prescriptive character that a logic of design must have, according to Simon, which allows it to rely on the same logical tools as sciences and mathematics in general: *designs* are positive objects available in principle for analysis (2.2.1). Nevertheless, it is computing science that makes this possibility real, thanks to programs which are nothing else than the completed expressions of methods (2.2.2). Discussion 2.2.3 shows how Simon envisioned in concrete terms this science of design based on computing frameworks.

2.2.1 A descriptive logic for intentional objects

The former discussion makes it clear that design will be able to constitute itself as a science if it is capable of reflexively taking itself as an object, that is to say, of observing methods and problem spaces as objects. But such a possibility is not self-evident.

Simon was aware of this difficulty, for this knowledge is "subtle" in the literal sense, *i.e.* quick to subtract itself from observation:

> The kernel of the problem lies in the phrase "artificial science". [...] A science of artificial phenomena is always in imminent danger of dissolving and vanishing. The peculiar properties of the artifact lie on the thin interface between the natural laws within it and the natural laws without. (SA: 112)

As a result, the invasion of the methods of the classical sciences into the teaching of the professions follows a certain natural tendency:

> Engineering schools gradually became schools of physics and mathematics; medical schools became schools of biological science; business schools became schools of finite mathematics. The use of adjectives like "applied" concealed, but did not change, the fact. [...] It did not mean that design continued to be taught, as distinguished from analysis. (SA: 111)

The term *analysis* refers here to an original interpretation of the classical opposition between synthesis and analysis, which Simon associates from the beginning of the *Sciences of the Artificial* with that between the prescriptive character of techniques and the descriptive character of the natural sciences:

> "Synthetic" is often used in the broader sense of "designed" or "composed." We speak of engineering as concerned with "synthesis," while science is concerned with "analysis." [...] The engineer, and more generally the designer, is concerned with how things *ought* to be how they ought to be in order to *attain goals*, and to *function*. (SA: 4–5)

Simon here reduces the classical epistemic opposition between analysis and synthesis to that between *is* and *ought*, which dates back to Hume.[8] This distinction establishes the independence *de jure* of the sciences of the artificial, for *ought* designates to inspection a point of view on the objects of the world that no science attains, that of their functional properties. However, doesn't this normative charge pose a problem? Can propositions that should be formulated in a modal logic be described as scientific?

> Natural science has found a way to exclude the normative and to concern itself solely with how things are. Can or should we maintain this exclusion when we move from natural to artificial phenomena, from analysis to synthesis? (SA: 5)

At this point, a footnote refers to a subsequent discussion of Chapter 5 – which itself refers, on this question, to the two articles already cited and to Chapter 3 of *Administrative Behavior* (1997). These texts support two points that may at first sight

8 Hume ([1740] 1990), *A Treatise of Human Nature*, Book III, Part I, Section 1.

seem antithetical: 1) Establishing facts about artefacts requires no more than classical, non-modal, first-order logic (SA: 114–115). Simon gives the example of a multi-constraint optimisation problem:

> The logic is exactly the same as if we were to adjoin the goal constraints and the maximization requirement, as new "natural laws," to the existing natural laws embodied in the environmental conditions. We simply ask what values the command variables would have in a world meeting all these conditions and conclude that these are the values the command variables *should* have. (SA: 118)

2) The *ought* is not, however, reducible to a factual datum. Here Simon follows Ayer ([1936] 1952: 105) and the logical empiricists in the thesis that one cannot define values or duty from natural magnitudes such as pleasure and utility. He does not go so far as to follow Ayer in asserting that ethical propositions would be mere emotional expressions; Simon is content to maintain the ethical domain in radical logical independence from matters of fact:

> Factual propositions cannot be derived from ethical ones by any process of reasoning, nor can ethical propositions be compared directly with the facts-since they assert "oughts" rather than facts. (Simon [1947] 1997: 56)

And, in his commentary on the 1997 edition:

> The fundamental point is that you can't get an "ought," by any manner of careful reasoning, solely from a set of pure "is's." To reach an "ought" in the conclusions, at least some "ought" must be lurking in the initial premises. No amassing of knowledge about how the world really is can, entirely by itself, tell us how the world ought to be. For the latter, we must be willing to say what kind of a world we would like to have; we must posit some values that go beyond the facts. (Simon [1947] 1997: 68–69)

To study artificial phenomena scientifically therefore amounts to purifying them of their deontic component, to pushing it back in order to bring to the fore the relations and positive facts it presupposes. Simon ([1947] 1997: 57) takes up a passage from the US Army infantry manual: "Surprise is an essential element of a successful attack". Hidden in this proposition is an "deontic" proposition: "Succeed in your attack!" and a technical proposition: "An attack is successful only when it is carried out under conditions of surprise". The second proposition is in the purely positive domain of design, from which can develop the means necessary for surprise, such as uncertainty about time and hour, avoidance of stereotyped procedures, speed of execution, etc. (SA: 58)

The *ought* thus motivates the existence of problems, procedures and artefacts, but must be excluded from their positive study. The possibility of such a study has only appeared since the advent of computing.

2.2.2 Computation and deliberation

The science of design is made possible by the advent of computing, because computing endows it with the reflexivity that is essential to it. For the first time design is capable of *reflecting* on its objects: the productions of computational activity – programs – are both pure methods and artefacts. The quotation already given at the beginning of the previous section (2.1.1) continues thus:

> There exist today a considerable number of examples of actual design processes, of many different kinds, that have been defined fully and cast in the metal, so to speak, in the form of running computer programs: optimizing algorithms, search procedures, and special-purpose programs for designing motors, balancing assembly lines, selecting investment portfolios, locating warehouses, designing highways, diagnosing and treating diseases, and so forth. Because these computer programs describe complex design processes in complete, painstaking detail, they are open to full inspection and analysis, or to trial by simulation. They constitute a body of empirical phenomena to which the student of design can address himself and which he can seek to understand. There is no question, since these programs exist, of the design process hiding behind the cloak of "judgement" or "experience". (SA: 134)

Design science is therefore an *empirical* science. Simon insists all the more on this point as it may seem counter-intuitive, given that the object of this science are pure conceptions of the human mind, and that it might be thought to be a mathematical discipline (Simon [1967] 1977: 214).

He thus places computing among the major advances in human history. He is not so much interested in the technical prowess that the computer represents, as in the conceptual progress that it heralds. So Simon does not compare computers, say, to electricity or aviation, the other major technical discoveries of the 20th century, but rather to writing or algebra:

> In histories of human civilization, the invention of writing and the invention of printing are always treated as key events. Perhaps in future histories the invention of electrical communication and the invention of the computer will receive comparable emphasis. What all of these developments have in common, and what makes them so important, is that they represent basic changes in man's equipment for making rational choices – in his computational capabilities. (Simon 1978a: 14)

Indeed, one deliberates all the better when one reduces the decision to a computation. This is not a petition of principle. Simon does not start from the notion of computation to reduce deliberation to design, but rather it is his analysis of deliberation as design that leads him to it. The reference to writing and printing here is not accidental, since computation must be understood in the modern sense given by Hilbert's formalist program, as the rewriting of formulas according to rules. The notion of *sytem of symbols* establishes the link between computation and deliberation.

> [Symbols] are physical patterns (e.g., chalk marks on a blackboard) that can occur as components of symbol structures (sometimes called "expressions") [...] A symbol system also possesses a number of simple processes that operate upon symbol structures – processes that create, modify, copy and destroy symbols. A physical symbol system is a machine that, as it moves through time, produces an evolving collection of symbol structures. [...] The hypothesis is that a physical symbol system of the sort I have just described has the necessary and sufficient means for intelligent action. (SA: 22–23)

The "processes that create, modify, copy and destroy symbols" may evoke the definition of a Turing machine. It is actually closer to Post's production systems, and which Simon, under the influence of Newell, began to see from the late 1960s as the most general and purest form of computation, and of mental life in general.

Simon and Newell (1972: 44–46) informally describe a *production* as a pair *"Condition → Action"*. The condition relates to the current state of the system in play, and the action is a modification of this state. A production is therefore nothing more than a coupling between a state and an action, *i.e.* the elementary object of deliberation (see above 2.1.2). A production system (or symbol system, in the terminology of the *Sciences of Artificial*), on the other hand, is an ordered set of such rules, which makes it possible to determine sequences of state transformations. The formalism of production systems has the same power of computability as that of Turing machines. That Post's formalism makes it possible to define the notion of computation on the basis of this coupling must certainly have struck Simon and encouraged him to adopt it. Rational decision-making is nothing more than the application of a rule of action specified by an environment. Acting rationally, in this framework, is nothing else than executing such a system of rules, and deliberating is nothing else than elaborating this system, *i.e. programming* it.

2.2.3 Programming and design science

Programming therefore makes it possible to fix design discoveries as objects that can subsequently be reused by other design enterprises. An analogy Simon takes concerns learning in mathematics, which is a kind of *self-programming*:

> When a human being has learned to do algebra, if an equation is written down, he or she can solve that equation, as the saying goes, hardly thinking about it. When the human being has an algorithm which says, "Oh, now I subtract 4x from the left side and the right side, and now I do this and now I do that," by that time one can say that the intelligence has been squeezed out of the process. All the intelligence was put to inventing and then to learning that algorithm.

> That algorithm becomes a matter of rote to anyone who is skilled in the art. That is sort of the way that a compiler is. (Simon 1986: 727)[9]

The reference to compilers in the quoted passage is not accidental. The text from which it is taken, a lecture given in 1986 to an audience of computer scientists, is one of the most comprehensive developments Simon has given on his vision of programming and its relationship to knowledge. A compiler, he says, is an algorithm that translates instructions written in a high-level language into elementary instructions that the machine can execute directly. This translation is deterministic, meaning that one knows exactly how the machine should behave when the high-level instruction is invoked; it is a problem that one knows how to solve, just as a student knows how to solve an algebra equation. There is no intelligence involved in this activity, which is why it can be automated. This allows the programmer to concentrate on a difficult problem, for which no algorithm yet exists or for which there is no satisfactory one. Programming means finding such satisfactory solutions; the resulting program can therefore be reused to solve similar problems. When cases vary, it is usually possible to factor portions of the program that partially solve the problem in its general dimensions. These general solutions can be integrated into the compiler through ready-to-use libraries of functions.

It can therefore be expected that the proportion of automation will increase over time. The programmer's role will remain focused on new problems, or new situations, or new combinations of problems and situations.

> We have been moving that boundary of automation. We have been putting an increasing load of the task on computers. (Simon 1986: 730)

However this is only the first step. The challenge for Simon is to move this boundary even further, aiming at the very formalisation of the problem and the search for a solution:

> What compilers do not do is to solve the problem of converting a situation that is presented in an unstructured form, as an informal problem statement. The compiler does not do the task of converting that problem statement into something on which one of these systematic, maybe complex, algorithms can go to work. (Simon 1986: 727)

Simon's interest lies precisely here, in inserting the computer into the real problematic situation that the programmer faces, which consists of describing the problem and its solution in actionable elements, *i.e.* to write the specification.

9 See also Simon 1978b: 17.

The difficulty is that before one even sets about the task of finding a satisfactory solution in the space of possibilities, one has to design that space, to move from an unstructured situation to a "formalised situation" (Simon 1986: 727) – something that only humans know how to do. For Simon, the solution should be no different than for well-posed problems; heuristics must be applied to meta-problem spaces: "in such and such an (informal) situation, try such and such a (formal) representation of the problem". Simon doesn't think that every informal situation one encounters requires a "magical" operation of the mind to formalise it, but that a large part of this work can be described by procedures – for example, interviews with the "client", identification of ambiguous situations, exceptions, etc. These procedures, like the heuristics, are designed to help the "client" to understand and clarify their needs. Like solution-seeking heuristics, they never offer a guaranteed solution, but they do form methods that can be at least partially automated.

Simon sums up this vision by identifying three knowledge bases that should enable computers to assist people in their design undertakings:

1. A base of procedures that are already automated and can be mobilised for problem-solving. This "procedural heritage" (in today's parlance: libraries of functions or services), as it might be called, enables the available operators in the problem space to be represented.
2. A base of data representations that allows the problems themselves to be represented, *i.e.* the possible states of the situation, its constraints, the objectives sought.
3. A base of "strategies" or "higher-level procedures" (Simon 1986: 729) – the procedures (C) discussed above (see above 2.1.2).

Simon adds to these three requisites the need for a formal specification language, which would be a kind of *lingua franca* providing a framework for expressing these bases, and in which this work of formulating problems and solutions could therefore be carried out (Simon 1986: 730).

A programming language, which offers at least a coherent set of the first two knowledge bases, is therefore a theory of representations or of problem spaces, since the two terms are equivalent (Kaplan and Simon 1990: 376).

It is therefore in programming languages that one should best see the progress of representation theory which, according to Simon, is most lacking in design science today. The passage on this subject, quoted above, continues as follows:

> But even though our classification is incomplete, we are beginning to build a theory of the properties of these representations. The growing theories of computer architectures and programming languages – for example, the work on functional languages and object-oriented languages – illustrate some of the directions that a theory of representations can take. (SA: 134)

This sentence refers to the observation, common in the 1990s, that the various families of programming languages – functional languages, object languages, logic programming, etc. – radically alter the programmer's relationship to their problem, *i.e.* the power they have to express it, their heuristics, even the algorithms they will favour.[10] It may be noted that the idea of a theory of representations is indeed already to be found in the original 1969 edition, even if Simon seems less assured about it.

What Simon thus describes in this vision of the three knowledge bases of design is a cumulative dynamic of practical knowledge made possible by computers: they should make it possible to formalise procedural, representational and heuristic knowledge in databases which would in turn enable people to automate or semi-automate problem-solving. Programmers could then concentrate on more difficult problems – and in doing so capture new procedures, uncover new representations and discover new heuristics which would in turn enrich these databases.

It is therefore not a question of "mechanising" intelligence and design – at least not on the visible horizon, but of building interactive dynamics between man and machine on which Simon places particular emphasis. He thus takes the example of the chemist E. J. Corey, who won the Nobel Prize thanks in particular for the introduction of computational methods calculating possible synthesis paths for organic molecules.

> If you look at the work Corey has done in recent years, you will see that it is not based primarily on automatic *design*, in which case they would have had to give the Nobel Prize to the computer, but on interactive *design* in which both the computer and the chemist play substantial roles (Simon 1994: 6).
> I really want to talk about software as the mind of the system, but not the exclusive mind, because the people who are developing the system and the people who are using the system, of course, also contribute a very important component of the mind (Simon 1986: 727).
> We talk about a man-machine interactive system and about gradually moving that boundary so that more of the arduous and complex tasks are taken over by the machine part of that system. [about the telephone system:] The system has always been a man-machine system. Ways were found to move that boundary imperceptibly from a system that was almost all men or women to a system that today is almost all computers. (Simon 1986: 731)

The science of design must therefore enable a new symbiosis between man and machine (Simon 1976a), the shift of an interface within a global system of intelligence that already exceeds the limits of the human mind.

10 See Backus 1978; R. W. Floyd 1979 and Priestley 2017: 453–455 for the history of notions of *style* and *paradigm* programming.

That these effects were not apparent, and that people could be dubious about artificial intelligence, annoyed Simon: design automation is progressing, but it immediately becomes invisible. Thus he said in a lecture given to design students:

> If we went around this room I am sure we would get a recital of many examples of computer programs that participate in the design process along the whole spectrum from programs, like most CAD programs, that are drawing aids for the engineer, to programs that carry out substantial design processes automatically. I found at Cornell a few months ago, a program that was doing a fine job of designing the basic processes for chemical processing plants. Those of you in the room could provide many examples of automatic and semi-automatic design of computer circuits and computer chips and whatnot. (Simon 1994)

Programming is therefore the general form of design practice, as it allows to discover and list the various forms of representation, the elementary procedures, as well as the strategies of each field of design in a directly operable way. Thus it can be understood why it enables design to become a science. The exact meaning of this proposition, which is the entire project of the *Sciences of the Artificial*, must now be explored. Is it not paradoxical to aim at a theory of practical knowledge?

2.3 Difficulties in Simon's theses

Sixty years on, can one say that the program laid out in the *Sciences of the Artificial* has been realised? Are we witnessing the emergence of a science of design? Are professions being reconfigured by programming? Are the knowledge bases that Simon foresaw being built up and playing their universal role?

On all these questions, the answers are mixed. The study of design, in its methodological component, has become an academic discipline in its own right, but the reference to Simon, while important, is imbued with a critical outlook. More dialectical, interrogative approaches, which give greater importance to the interaction of the designer with the situation and its history, developed from the *Design Methods* movement, contemporary with Simon's thinking.[11]

To the credit of Simon's position, one can observe the major transformation that computing brings about in all professions. Computing is leading the professions to come up with new solutions to old problems, to radically reformulate them, or to set themselves new problems. Programming languages, as Simon observed, have become increasingly specialised and capable of describing precise classes of problems according to their own logic, such as Modelica in engineering or Catala in law.[12]

11 On these subjects, see in particular Buchanan 2009 and Cross 2006.

12 For Modelica, see Wetter et al. 2011 and for Catala see Mérigoux 2021.

Computer Aided Design (CAO) software, meanwhile, incorporates more and more design tasks originally reserved for humans, as he predicted.[13]

However, there are many discrepancies between the current state of affairs and Simon's programmatic vision. First of all, it does not seem that there is any convergence of the methods and representations specific to each discipline in the direction of a universal science of design. Certainly, the pedagogical movement of computational thinking (Wing 2006) attempts to inculcate in students a general culture concerning the key concepts and methods of programming. However, the positive content of this knowledge and its organisation are still rather unclear (Bocconi et al. 2016: 6). Taxonomies, to use Simon's word remain empirical and fairly generic – it is not yet possible to connect the procedures and representations they describe to their specialised applications within each discipline. When Denning, Martell and Cerf (2015) attempt to list the "great principles of computer science" they end up with an empirical and open-ended list of some forty propositions, whose systematic unity is not clear. Programming languages, meanwhile, could be seen as *de facto* theories, but their classification itself remains problematic.

Secondly, Simon's vision of "intelligent compilers", which would integrate the function of problem analysis and formulation, has certainly made spectacular progress very recently thanks to the large language models of AI. However, this progress is not based on the procedural rationality promoted by Simon, but rather on the statistical simulation of decision functions obtained by machine-learning.

Now, Simon was very sceptical about the importance of statistical machine-learning, and in particular that of neural networks.[14] The only form of machine-learning that he promoted, and which he associated with the discovery of new knowledge, is rule induction (Langley and Simon 1995). This which forces a statistical algorithm to present the correlations found between series of data in the form of propositional relations, such as *"If the MP voted for the budget, against the pay freeze and for the Education Act, then he is (must be) a Democrat"* (Džeroski et al. 1997: 98) – a constraint that most modern machine-learning algorithms dispense with. In other words, for Simon practical rationality must necessarily be formulated in the form of unitary rules, *i.e.* productions in Post's sense, as was seen above. Since the predictions of neural networks cannot be formulated in this way, they are called *sub-symbolic* in the 1980s debates about them. The current orientation of science and the professions towards the intensive use of statistical machine-learning thus turns its back on Simon's "symbolic" – it would be better to say *discursive* – vision.

---

13 On the history of computers in the field of architecture, see Gaudillière-Jami 2022.

14 See for example his judgement on the first neural network model, the Perceptron (Simon 1983: 32).

This is a key point. The fact that information technology is moving towards statistical methods to capture the correct answer to certain practical problems, neglecting to explicate any general procedure, suggests that rational human practice is discursive up to a point only, and cannot be fully explicated in rules. This intuition formed the core critique of Simon's theses by Hubert Dreyfus in his famous book *What Computers can't do* (1992), and by the Wittgenstein-inspired rule-following critique in the 1980s. This critique often relied on argument of infinite regress which actually dates back to Kant (see below 5.1.3). Pierre Bourdieu expresses it in this way:

> It is significant that any attempt to base a practice on obedience to an explicitly formulated rule, whether in the field of art, morality, politics, medicine or even science (think of the rules of method), comes up against the question of the rules defining the manner and the opportune moment – *kairos*, as the sophists used to say – of applying the rules or, as we so aptly put it, of putting into practice a repertoire of recipes or techniques, in short of the art of execution. (Bourdieu [1972] 2000: 301)

Discussion 2.3.1 develops another argument against the reducibility of any human decision-making to effective procedures. It claims that, in complex cases, it diverges significantly from the actual human experience of decision-making. Is it however correct to locate this gap between practical sense, on the one side, and "recipes and techniques" on the other, as Bourdieu suggests? In order to clarify this opposition between *practical sense* and *technical rule*, a terminological analysis of the terms *practical* and *technical* is carried out (2.3.2), extended to an analysis of MacIntyre's famous definition of *human practices* (2.3.3), and finally to Bourdieu's *practical sense* (2.3.4). The conclusion is that *practices* and *techniques* are strongly embedded into one another. It is not on the basis of this distinction, therefore, that the continuity or irreducibility of human practices to computations should be evaluated.

2.3.1 The illegibility of books of procedure

The idea that programming is the completed form of problem-solving is actually paradoxical. To understand why, one needs to imagine what would be an exhaustive description of the positive knowledge of a practice in the form of a book of procedures – like the immense script enabling John Searle (1980), in his Chinese room, to hold any dialogue in that language.

It is useful to start with the example of law, where such a conception of knowledge is possible, since these procedural books actually exist. The organisation of the law into specialities (commercial law, labour law, property law, etc.) with their related specialised codes, shows the practical challenge of making the immense content of cumulated knowledge accessible: the organisation of laws follows quite naturally the typology of the problems encountered. For example, if there is a dispute between employer and employee, a specialist in labour law will take up the case. The law also organises priorities between the different codes if more than one is applicable to the dispute: for example, in case the dispute involved violence, the penal code takes precedence over the labour code.

This state of affair results from the ability of lawyers to *organise* their positive knowledge into books of procedure. If such an organisation were not possible *i.e.*, if the law were an immense and disordered list of rules, its practice, as it exists today, and even its development, would be impossible. A book of procedure is not enough to record and transmit human knowledge. It is also necessary to give its *user's guide*, *i.e.* to explain how to access the relevant rules according to the problem posed and to manage the hierarchies of rules when several may apply. But there are practices for which, even if all its useful rules could be stated, such an order and hierarchy would be problematic.

Strategy games provide a good example of this problem. Every situation in the game of chess or go is a problem to be solved, but it is not possible to classify these problems by typology, apart from the very elementary one of "beginning, middle and end game" problems. Players classify problems by recognising typical patterns on the board. A chess expert, according to Simon (1984: 3), has more than 50,000 patterns of positions that they can recognise immediately on a chessboard, rather like a reader reads whole words directly on the page, without mentally spelling out the letters that make them up. Each of these patterns is indeed a rule, even if it is memorised diagrammatically. The player associates it, for example, with a specific move, or with a tactical objective, or quite simply with a positive or negative assessment of the situation.

Let us imagine that a chess grandmaster manages to record all these rules in a book, and even finds a clever principle of classification that makes it easy to find each rule relevant to a given situation (for example: *"rules where the King is on the initial line"*, *"rules where the Queen is in the centre of the board"*, etc.). It will remain necessary to give the reader additional rules for ranking the applicable rules against each other, in complex ways, such as *"this pattern is bad, unless such other pattern is also the case, then a mate objective is possible"*.

Such a configuration is not imaginary; it roughly describes, in a very simplified way, the operation of the Deep Blue program, which defeated Kasparov at chess in 1997 (Campbell, Hoane and Hsu 2002), undoubtedly one of the pinnacles of symbolic artificial intelligence. In particular, the program was equipped with an evaluation function consisting of more than 8,000 parameters describing patterns ranging from the simplest (such-and-such piece on such-and-such square) to the most complex (rook on open line, king protection). The importance given to each parameter was itself influenced by various factors, such as the number of pieces remaining on the chessboard, or another pattern. The rook on open line scheme, for example, is described by 192 parameters, and its importance is partially interdependent with that of king protection (Campbell, Hoane and Hsu 2002: 73–76).

It could be argued that Deep Blue is indeed a "book of procedures" for playing chess, since it undeniably helps to codify, in an absolutely positive way, the solutions to the problems posed on the board.

However, such an argument poses a twofold problem. On the one hand such a book would be properly *monstrous* for a human person to apply, and so one arrives at the paradoxical notion of a *non-human knowledge* whose subject (the computer?) is problematic. On the other hand, it still remains *too human*, in that it encodes patterns catalogued by the experience of traditional players, before recombining them through estimated weightings. It is significant in this respect that, between each game against Kasparov, the coefficients of certain evaluation sub-functions were adjusted manually by the programmers, according to their analysis of the previous game. Eliminating this tinkering is tantamount to taking the logic behind the Deep Blue evaluation function to its conclusion *i.e.*, rather than fixing in advance the library of relevant patterns and their rules of priority and combination, giving free rein to the statistical association of situations encountered and their consequences to identify the patterns to be considered and their importance. It may be, for example, that given a certain pattern applicable to the situation, two other distinct patterns need to be combined together to derive a relevant rule; or that one pattern needs to be adjusted by removing a part that has become irrelevant, before being recombined with another pattern to which another part has been added, and so on. This is how a neural network works, through successive phases of learning used to associate different positional patterns with each other, according to relative weightings, taking into account only their statistical relevance, and all this at several levels of abstraction.

In other words, it seems logical (with the benefit of historical hindsight) that machine-learning was, for this type of problem, the outcome of symbolic artificial intelligence. When the human logic of enunciation, classification and recombination of relevant rules is taken to its limit, it gives way to their statistical self-calibration. This is how the world Go champion was defeated in 2016 by a program built entirely using this logic, AlphaGo.

A neural network is indeed a book of procedures, and it is just as symbolic as any other mathematical algorithm, as Simon likes to remind us.[15] Yet it would be

---

15 See the debate on this subject in the articles Vera and Simon 1993a,b, 1994.

even more monstrous to apply to human beings than the Deep Blue algorithm. The computations it commands are not only of a length inaccessible to us, they are also opaque to comprehension.

It would seem, at least for practices dealing with "entangled" situations, as is the case with strategy games, that the exhaustive explication of the rules relevant to their exercise would result in a book of procedures inaccessible to human intelligence. So it is certainly not the practical *human* knowledge in question that would be represented there. This helps to understand why Simon was instinctively hostile to statistical machine-learning: it was because this approach challenged the very idea of a general science of design and of procedural rationality that is to say, basically, the idea that the *positive* knowledge prevalent in professions and practices can in general be reduced to a sum of technical rules.

There is an important result here. All the good practices of a chess grandmaster could be written down, because they could always explain why they played this or that move. The problem pointed out here concerns rather the *use* of these rules. Written down as they are, in no particular order, they would be unusable. Reorganised so that their priorities and all their interdependencies are taken into account, they would result in an algorithm calibrated by numerical gradients that would no longer tell anything, and whose execution would itself be problematic without the help of machines. This algorithm would no longer have much to do with what is called human knowledge. Humans, of course, have different ways of making informed decisions in complex disciplines.

Are most human professions like chess or like law? Actually, the practice of law itself is arguably much more complex than chess. Legal codes, despite their strong structure and explicit hierarchy, never give a clear and unique way to solve a case, except for the simplest ones. Multiple perspectives or starting points may often be invoked that change dramatically the case's outlook. These starting points and perspectives are not themselves written down as rules. Once found and validated by success, they enrich case law.

To answer this question in a definite manner, *human practices* must be more clearly understood, and this begins by looking again at the multiple meanings of *practice* and its close shadow *technical knowledge*.

2.3.2 Practice and technique: a terminological analysis

The radical *practice*, which appears in English around 1420, has as its main derivations the verb *to practise*, the noun *practice* and the adjective *practical*. This discussion looks first at the rather restricted meanings taken by its verbal form, and then at the many connotations added to them by its nominal and adjectival forms, before turning to *technique, technics* and *technical*.[16]

To practice is a transitive verb. So what does one practise? On the one hand, according to the dictionary, an art or profession, but also a sport or a language. There is regular practice in recognised areas of competence, where expertise is possible, so that a disciplinary field can emerge. These exercises lead to tangible and appreciated results. In this sense, practice is linked to knowledge from the outset.

There are many connotations to this first use of the verb. First of all, it is not synonymous with "to act", because it immediately denotes an action that is regular over time and therefore has a recognisable form, despite the possible singularity of circumstances. One cannot practise a singular event, a heroic action on the battlefield, for example. But one can practise war.

Secondly, it is more general than the professional practice of a trade. One can practise the art of ironwork without being an ironmonger, or practise astronomy without being an astronomer. In fact, this expression is used specifically to designate the enlightened amateur, capable of frequenting professionals without being a professional. Here surfaces an old meaning of practise, that of *knowing how to deal with somebody* to persuade or influence them.

In this sense, the term takes on a connotation of *experience*, as opposed to a merely theoretical (in the pejorative sense) connection with its object. Practising a language means having more than just a scholastic knowledge of it.

But *practise* can also be coloured with the opposite connotation, when it means *putting into practice* the teachings of a discipline. *Practising chemistry*, for example, evokes the idea of laboratory work, which complements rather than opposes the theoretical dimension of this discipline. This sense of *applying* is seen most clearly when *practise* is used in the social and moral domain: *to practise a religion, charity*, etc. It then becomes synonymous with *observing* commandments and rules.

The verb *to practise* has another large category of object complements, which does not connote skills or disciplines, but ways of doing things. The most common examples are pejorative: *to practise usury, blackmail, etc.*. But it can also refer more broadly to customs, as in *"this community practices circumcision"*.

These two main categories of complements of the verb organise the use of the noun *practice*. When referring to a discipline, *practice* is an uncountable noun. *"The practice of football"* cannot be used as a count noun without changing its meaning. The expression *"the practices of football"* refers to the habits and methods particular

---

16 This terminological survey is based on the *Oxford English Dictionary* (2010–) and the *Trésor de la Langue Française* (ATILF - CNRS and Université de Lorraine 1994).

to the socio-economic field of this sport. In this second sense, one also speaks of *good practices* to designate the state of the art of a discipline.

In the first sense, the noun *practice* accentuates the connotations of the verb, notably through the common expressions *to put in practice* (application) and *lack of practice* (experience). Even when it is a question of application, as in *"the practice of a science"*, the use of the noun insists more than the verb did on the *positive difference* that this practice adds to one's knowledge of the discipline in question. There is here the idea that practice always goes beyond the simple application of theories.

Above all, the concept of "practice" (in general, without an object complement), associated with "theory", becomes the theme of a recurring and trans-disciplinary opposition, a kind of *topos* both scholarly and popular, which opposes two forms of being in the world, one contemplative and the other active, or two forms of knowledge, one derived from pure reflection or learned from books, the other from the experience of action and life. This *topos* is therefore value-laden in many ways, which can give an advantage to one side or the other. One can, for example, as Nietzsche does, glorify life and reject the dry, disembodied knowledge of theory as a ruse of life itself. Conversely, one can condemn the *violence* that all practice does to the things it deals with, in contrast to the attitude of disinterestedness and respect towards the very things from which all theoretical endeavour starts.

It is in the adjective *practical* that the threads of these multiple connotations are most clearly unravelled. If one sets aside the moral meaning given to it by the philosophical tradition alone, particularly after Kant, *practical* qualifies all things having to do with action, with two major connotations: the first, the orientation towards the useful, the concrete and the material, particularly when it qualifies people or their actions (*"someone who has practical sense"* = someone who has a sense of their concrete interests), and the second, efficiency, particularly when it qualifies tools or methods (*"a practical knife, a practical shortcut, etc."*).

2.3.3 Human practices according to MacIntyre

This terminological inquiry helps to contrast three famous interpretations of the notion of *practice*, that of Hannah Arendt, Alasdair MacIntyre and Pierre Bourdieu.

The first salient point is that a practice relates to regular action. It has therefore a certain form and generality. In this sense, it is opposed to *action*, particularly in the sense that Hannah Arendt gives to the latter term, when she insists on the singularity of the events that human action provokes, essentially unpredictable because revealing an even deeper singularity, that specific to each human person. Action, with its associated speech, is the foundation of the public space where people meet *as* individuals, where they learn to speak to each other in their essential otherness (Arendt [1958] 2012: 200). By naming *action* this third domain of the *vita activa* (along with *labour* and *work*) Hannah Arendt implicitly acknowledges the inappropriateness of *practice* in modern language to signify this reign of singularity, despite its Greek and Aristotelian origin to which she wants to return: so she is forced to use the Greek root as it stands, *praxis*, when she wants to refer to it.[17]

The second point is that *practice* conceals a semantic duality, admittedly with blurred contours, but nevertheless clearly perceptible in the opposition already cited between *the practice of football* and *the practices of football*. Using this opposition as a guide, MacIntyre's and Bourdieu's approaches to practice can be sharply contrasted.

This contrast can be seen in MacIntyre's decision to contravene usage by making *practice* a count noun, calling *practices* (plural) anything which can be practised, *i.e.* disciplines, arts, crafts, skills. His definition is famous:

> By a 'practice' I am going to mean any coherent and complex form of socially established cooperative human activity through which goods internal to that form of activity are realized in the course of trying to achieve those standards of excellence which are appropriate to, and partially definitive of, that form of activity, with the result that human powers to achieve excellence, and human conceptions of the ends and goods involved, are systematically extended. Tic-tac-toe is not an example of a practice in this sense, nor is throwing a football with skill, but the game of football is, and so is chess. Bricklaying is not a practice; architecture is. Planting turnips is not a practice; farming is. So are the enquiries of physics. chemistry and biology, and so is the work of the historian, and so are painting and music. In the ancient and medieval worlds the creation and maintenance of communities – families, cities, nations – is also considered a practice in the sense I have defined it. So the scope of practices is wide: arts, sciences, games, politics [...], the formation and maintenance of family life, all fall under this concept. (MacIntyre [1981] 1985: 187)

This definition forces the use of language. While we may accept that *practice* means directly a professional activity, as in the expressions *"he set up practice in London"* (he set up (professionally) in London) or *"he has a large medical practice"*, MacIntyre extends this meaning far beyond the strict professional occupation, to include hobbies, science, politics. We have no difficulty in accepting the expression "the practice of football", but we are more circumspect about the assertion that "football is a practice" as MacIntyre proposes. We would rather say *an occupation, a hobby, an activity, a trade*, etc., but none of these terms covers the full scope intended by the Scottish philosopher.

17 See for example Arendt [1958] 2012: 70,79.

MacIntyre's lexical move is not trivial. It is not just a convenient synecdoche that allows to move from the exercise of an activity to the activity itself. MacIntyre's aim here is obviously ethical and anthropological: by forging a *new* concept of *practice*, he seeks to reactivate the old idea of the *liberal* exercise of a discipline in the old sense, *i.e.* freed from the aim of external goods (money or fame) in favour of norms internal to practice, thus disinterested, endowed with a social inscription and an historical depth. The concept he proposes is, in a way, arbitrary. It is less broad than *activity*, which can be trivial, but broader than the professional occupation, or even the *discipline*, a term reserved for intellectual activities. The example of the game of tic-tac-toe is significant: if it is not a practice, it is because it is too simple: there is a procedure for systematically winning or drawing it. This game no longer has any possible development, and therefore cannot be an arena for the exercise and growth of human excellence. Winning depends on mere technicity. For a similar reason, basic skills in the strict sense of the word – such as language skills, which are part of the semantic scope of the word *practice* – seem also to be excluded from this new concept. While one can speak a language more or less well, there is no question of its simple use becoming a field of excellence – unless it becomes an art, like rhetoric.

The possibility of excellence that characterises a practice is relative first and foremost to its own internal standards, which can evolve. Painting, for example, never ceased to redefine itself, as also did sciences and mathematics. In this regard, and by way of illustrating MacIntyre's conception, we might cite the recursive definition of mathematics proposed by the mathematician Thurston:

> Mathematics is the smallest subject satisfying the following: 1) Mathematics includes the natural numbers and plane and solid geometry. 2) Mathematics is that which mathematicians study. 3) Mathematicians are those humans who advance human understanding of mathematics. In other words, as mathematics advances, we incorporate it into our thinking. As our thinking becomes more sophisticated, we generate new mathematical concepts and new mathematical structures: the subject matter of mathematics changes to reflect how we think. (Thurston [1994] 1998: 340)

These internal norms are translated into external rewards (honours, wealth, etc.) in proportion to the public recognition available to the discipline, but it is characteristic of a practice that it is not essentially guided by their pursuit, in which case it would be no more than mercenary, to take Plato's word (*Republic*, Book I, 346b). This *autonomy* of practices, in the literal sense, leads one to confer on them a kind of trans-generational historical individuality, to make them subjects – this can be seen from the syntax of MacIntyre's developments (1985: 189–193), where the term appears mainly as a grammatical subject, and often of verbs connoting activity and life.

MacIntyre ([1981] 1985: 194) contrasts practices with the technical skills and the institutions that support them. Institutions (clubs, laboratories, unions, universities) organise the interface between practice and the public arena. They are primarily concerned with the negotiation of the external goods of the practice; they are necessary in the long term to ensure its continuity, but they can also be a factor in its corruption, when games of power and money take precedence over its autonomous development.

Technical skills are one of the means by which a practice is learned and passed on. But a practice is always more than a set of technical skills, because these are conditioned by its internal norms which, as has been seen, may evolve. If painting had become so sclerotic as to be reduced to the skill of perspective, abstract painting would never have been invented. So the transmission of a practice is not just about these skills, but also about a tradition and models of excellence that their heirs set themselves the task of surpassing (MacIntyre [1981] 1985: 193–194).

From this perspective, it is easy to see that codifying a practice into procedural rules can never exhaust the practice itself, for at least three reasons. Firstly, its codification in explicit rules is only one possible form of transmission; it may be that learning is more easily achieved through apprenticeship, imitation, and rote exercise. Secondly, transmission is never reduced to know-how, as it also involves the inculcation of norms and models. Finally, the very historicity of these norms and models (as opposed to dogmatic immutability) invites the apprentice to seek to surpass them – and it is in this search that excellence and mastery are defined, *i.e.* its authentic knowledge *at this precise moment in time*, because it recapitulates all of its past development and opens up towards its future possibilities.

2.3.4 Practical sense (Bourdieu)

It is interesting to contrast MacIntyre's ethical perspective with the more anthropological and critical one of Pierre Bourdieu. For the sociologist, *practical sense*, or *habitus*,[18] is not specific to any particular discipline, nor reducible to the sum of their silos, but on the contrary allows their emergence as specialisations of the social field.

What precedes practices in MacIntyre's sense are therefore practices in the usual sense that this term has when it is used in the plural form, *i.e.* ways of doing and conducting oneself. They are manifestations of the *habitus*, which designates the disposition to act in a certain way in the face of various circumstances.

---

18 The equivalence of the expressions is attested by Bourdieu (2015: 292).

This does not mean, however, that an individual's conduct is entirely determined by external factors and fixed social laws. On the contrary, habitus is precisely that capacity to adapt to unforeseen circumstances, to play by the rules or around them, to develop strategies within a given social field. However, this adaptation itself takes place within a certain space of possibilities that is co-determined by the individual habitus and social reality. New practices may thus emerge, but they are part of the history of past practices, with which they can never completely break. The habitus is inculcated in the individual through all their past experience, which is always social (family, school, professional environment), whether through explicit learning or (most importantly) through silent observation, or the unconscious mimicry of body and speech. The opposition between the conscious and the unconscious is of no importance here, because the awareness that accompanies a behaviour, a strategy or a decision is always inscribed in an unconscious background that lends them their meaning and their very possibility.

In this framework, practices in MacIntyre's sense are the result and sediment of past strategies, those of social groups that have successfully campaigned for the delimitation of their own field whose norms, duties and privileges they (the experts) would establish. So it is always the same practical sense that is at work through them and all their prescriptions. For the elementary practical sense is that of knowing how to recognise a norm *as such*, of judiciously assessing its coercive power, of knowing when and how to circumvent it, when and how to comply with it, of knowing how to mobilise it to one's advantage, and so on. Practices (in MacIntyre's sense) are practices (in Bourdieu's sense) that certain groups have managed to impose as norms. They correspond to what the French sociologist rather calls a *field*, *i.e.* the objective space in which these practices can develop in the autonomy of their own norms.

Bourdieu thus radically challenges what he would perhaps call the irenicism underlying the notion of internal goods promoted by MacIntyre. Practitioners of a discipline do not seek to conform to or challenge the norms of that discipline in sincere disinterestedness. Admittedly, these internal goods cannot necessarily be monetised in external goods, but they still constitute symbolic capital that is a power issue internal to the field, which is a structure of struggle and social domination. The evolution of norms does not therefore follow some kind of purely ideal internal necessity, but is the result of a confrontation between actors concerning the definition, and therefore the control, of the entire field – particularly in the semi-structured fields of the arts and humanities.

Beyond this criticism concerning the "true nature" of the so-called internal goods of a practice, Bourdieu's conception is singularly similar to that of MacIntyre.

Firstly, it is just as critical of the idea that practices – whatever they may be, disciplinary or otherwise – can be reduced to their codification in rules. Codifying and making explicit a practice is not a trivial gesture: it means that the practice (or rather the agent codifying it) is in danger: learning it is no longer self-evident, other norms are emerging, it is being contested, etc. It may therefore be part of the defensive consolidation of a field, or in a strategic struggle between two social groups within that field – but the result is that this explanation is always partial, if not one-sided. It says, in negative terms, only what is no longer self-evident, but leaves unmentioned all the essential structures that continue to regulate the habitus and the general space in which the rule quite simply makes sense.

Then, as for MacIntyre, excellence is a real mark of authentic practical knowledge. Commenting on Durkheim's position, that reflection can illuminate art, but is never its essential element, Bourdieu writes:

> It could not be better said that the learned ignorance that is at the root of everyday strategies should not be expressed in the lexicon of the rule, but in that used by all societies to describe excellence, that is to say, the manner and manners of the accomplished man: this "art without art" [...] is never so completely realised as on socially arranged occasions where, as in jousts of honour, playing with the rule is part of the rule of the game. If this particularly successful embodiment of the particular way of being a man recognised by a given group is almost always defined as indefinable, because any formulation would reduce it to the level of a mere process or mechanical trick, it is because virtuosity has no use for the rule, a safeguard or reminder, barely capable of making up for the shortcomings of the habitus; if it is recognisable by its "naturalness", it is because it establishes that magical mastery of one's own body which, as Hegel observes, characterises the dexterity or, in the language of Kabyle honour, the "grace" of the *sarr* [...] (Bourdieu [1972] 2000: 304)

The rule is therefore only "half-learned" (Bourdieu [1972] 2000: 308) (the sense of the Pascalian *demi-habile*) not only because it is a clumsy in-between between the complete ignorance of a practice and its "learned ignorance" which is in reality its true mastery, but also because, at the epistemological level, it is an incomplete transmutation of practical knowledge into theoretical knowledge:

> It is easier to understand why this half-scholarly production that is the rule constitutes the obstacle *par excellence* to the construction of an adequate theory of practice: by falsely occupying the place of the two fundamental notions, the theoretical matrix and the practical matrix, the theoretical model and the practical meaning, it forbids us to ask the question of their relationship. (Bourdieu 1980: 176)

Beyond their differences, then, MacIntyre and Bourdieu agree on this essential property of practical knowledge, that it can never entirely, or even essentially, be codified into rules. Rules seem to be just a way of transmitting technical skills, or of enforcing a particular way of doing things as opposed to another which is forbidden, but these issues do not define the practice itself, which is always evolving in a historical process driven by a community and its leading members.

This being said, technical skills and the rules that describe them appear nonetheless as essential features of human practices. The adjective *technical* (17th century) and the later noun *technique* (19th century) are closely associated to practices, as the *OED* definition of the latter term makes it clear:

> Technique. 1. The formal or practical aspect of any art, occupation, or field; manner of execution or performance with regard to this. Also more generally: way of doing something.
> 2. Practical skill or ability in the formal or practical aspect of a particular field.
> 3. As a count noun: a particular way of carrying out an experiment, procedure, or task, esp. in a scientific discipline or a craft; a technical or scientific method. (*Oxford English Dictionary* 2010–)

These three close meanings try to capture an essential nuance. We can speak of *a pianist's technique* to refer to their *know-how* or *skill* in performing an activity, but also of *techniques of the piano* which are well-described methods that indicate the correct ways of playing that instrument. It is not obvious that the first one could be entirely described by the second. However, the specific use of this term, instead of *know-how* or *expertise* may connote the idea of a *possible* description of this skill in processes. If a pianist's technique can be virtuoso, it is because it sits at the top of a *continuous* gradation of difficulty, and is therefore potentially accessible to other people. Moreover, many teaching methods, particularly in the arts, bear the name of masters who have thus succeeded in formalising and organising their own skills, such as the Hanon method in piano playing.

It is important in a technique, in this sense, that the conditions for its successful practice are made explicit, for example that the apprentice has mastered this or that gesture, or even that their body, such as their hand, has this or that characteristic, before they can successfully apply the prescribed rule. A technique promises results to any agent applying a rule *under certain explicit or explicable conditions*.

The uses of the adjective *technical* add a complementary meaning to that of *technique*, which emphasises the *specialisation* of the activity or knowledge in question: *"a technical work, vocabulary, expression"* are understandable only by specialised people versed in the discipline.

The terms *practice* and *technique* are therefore closely linked, at least in three ways: 1) Across the spectrum of their uses, *practice* and *technique* always relate to actions endowed with a certain regularity and publicity. Singular and entirely private action is not within their purview. 2) In both terms there is the same tension between a meaning circumscribed to certain regular activities, well identified socially, and a broader meaning that refers more generally to skills or ways of doing things. On this scale, *technique* is used much more often than *practice* in the narrow context of a special discipline. *Technicity* even means *the very fact* of specialisation. But it can also be used in quite a generic way, particularly in popular locutions.

3) Both terms are traversed by a second tension, between the personal experience of the action, which cannot be transmitted, and its possible external description, which can be standardised and transmitted. This tension is seen in the opposition between *the practice* of a discipline, always individual, and *the practices* which designate common ways of doing things. It is also present, to a lesser extent, in the use of *technique* as (personal) know-how, which contrasts with its main meaning as *the set of procedures* that form the transmissible background of a discipline. In the term *technique*, this contrast goes so far as to designate an objective meaning, which covers not only the conduct prescribed to agents, but the very arrangement of things in such a way as to determine their behaviour.

Beyond the nuances of the use of the terms *practical* and *technical*, it is therefore essential to question these two tensions, on the one hand between the general *or else* specialised (disciplinary) character of regular action, and on the other hand between its private and subjective *or else* its public and objective character.

Let us project each term towards its main polarity, using the singular for the first one, and the plural for the second. *Practice* then designates the *practical sense*, that *general* faculty of adapting to the commerce of the world, of gaining experience from it, of applying one's knowledge to it, a faculty that then *specialises* locally into *particular practices* (professions, arts, sciences, games, etc.) *transmitted by techniques* and structured by institutions. *Techniques* relate to knowledge, since they are procedures or methods which, in order to be transmissible, must be able to be represented. Techniques can be described and taught. They have to do with those partial and provisional sets of rules to which practical knowledge cannot be reduced. They can determine how people behave.

From here, it is easy to draw a series of oppositions or rather articulations associated with these two terms, which go back to their Greek origins: practice is inscribed in human affairs, whereas techniques aim at the "government of things" (a technique dealing with the human, like advertising, treats people as a standardized objects); practice has a history (personal experience, collective tradition) whereas technique aims at the ever-present effectiveness of methods; practice adapts to circumstances, technique deals only with regular cases.

It is relevant to speak of articulation rather than opposition because practice and technique, in the meanings just crystallised, appear as the two terms of a single movement of externalisation and social structuring of regular action, which is of course reciprocal, since the practical sense is largely, if not entirely, made up of internalization and habituation to the external and public structures of the world. There is the technical description of a practice (externalization) and, in return, the practical exercise of a technique (internalization).

Where do effective procedures fit into this picture? At first glance, it may seem obvious that they belong to the technical side, since they have to do with making explicit rules for action. The versatility of computers in so many disciplines, which is a central problem in this book, could simply be explained by the fact that programs *code* their techniques. This programming work would allow these rules to appear in all their clarity and completion. It could therefore be hypothesized that if practices cannot be reduced to computational procedures, at least their techniques could be. It is this hypothesis that is explored in the next chapter.

3 Technical practices and industrial logic

I have called technical the reflection that aims to make explicit the knowledge specific to a given practice so that it can be examined and transmitted. It is therefore natural to try to restrict Simon's thesis to this aspect of practice. According to what could be called the *computational hypothesis of the technical mind*, technical thinking, once it has reached maturity, is a work of design, the general form of which is programming. This hypothesis would perfectly explain the versatility of computers across many practices and would resonate with the common view – found, for example, in Heidegger's thought – that computing is the culmination of technology.

To test this hypothesis, I propose to restrict my study to a set of professional practices that may be grouped under the general heading of *operations in technical environments*. What is meant by *operations* in the business literature is made clear through examples: production in a factory, logistics, activities of a call centre, of an accounting department of an air traffic control tower, the steering of a large ship, the maintenance of infrastructures or of a fleet of machines, etc. – all are commonly referred to as *operations*, in contradistinction to a company's commercial, management and project activities, including research and development. Operations are recurring activities with well-defined, measurable results (e.g. production for a factory, number of calls handled for a customer service department) at fixed periods (daily, monthly, annually) that can be set against production factors (quantity of work, materials, etc.). The expression *in a technical environment* means that these activities deal with technological artefacts or are mediated by them.

The choice of this area of practice is interesting because it is here that one expects to find the greatest proximity between procedures intended for human operators and systematic procedures. Isn't it often believed that operators, in an industrial environment, should behave "like robots", and don't we all have in mind the striking images of the Tramp in Charlie Chaplin's *Modern Times*, caricaturing and denouncing Frederick Taylor's "scientific management"? Therefore, it is probably in these professional practices that the kinship between these two types of procedures could be best determined.

The modern factory is the technical environment *par excellence*. As direct observers of the Industrial Revolution Ure, Babbage and Marx noted, it is the place where operators are dependent on the machine to do their work: the procedures they are asked to follow are the remnants of what could not yet be automated, for technical or economic reasons. Yet the advent of modern, mechanised industry is also the moment of the conceptual invention of the computer, by Charles Babbage who closely associated the two ideas.

https://doi.org/10.1515/9783111494500-003

Before exploring the historical and conceptual proximity between computational and industrial procedures, it is first essential to observe the role of procedures in non-industrial technical environments, where mechanization does not serve as a central or guiding factor. Two studies in the ethnology of work, one on the operations of the crew of a military ship, and the other on the maintenance of office machines, both conclude that written rules and procedures play a very specific and exceptional role in the organisation of human work. If collective action achieves standardised effectiveness, it is primarily through the implementation of complex *cognitive ecologies*, mobilising material (spatial layout, visual and sound aids, tools, instruments, etc.) and social (apprenticeship, companionship, hierarchy, etc.) resources that implicitly constrain agents' behaviour. Within these ecologies, rules and procedures function as reference points rather than active norms, and focus on specifying, by exception, particular important points that *complement* the constraints already imposed by the cognitive ecology rather than describing them. To borrow an image from the cognitivist and philosopher Andy Clark (2010: 75), they are just the visible tip of an underwater cognitive mountain, embedded in the social and bodily depths of human behaviour (section 3.1).

There is therefore a quasi-frontal opposition between the logic of procedures governing human behaviour and that of industrial and computational procedures. For the former, ambiguity is a resource that enables adaptation to new situations, whereas for the latter, it signifies a lack of control and planning. The two are therefore distinguished by a different trade-off between the *duration* required for adaptation and the *guarantee* of the result. The human procedure may be ambiguous because it is based on an underlying cognitive ecology that enables human operators to adapt easily to new situations, through their "practical sense". However, it does not guarantee the client that this adaptation will meet their expectations, or even be consistent from one operator to the next, which becomes a major obstacle to the transition to large-scale production. Industrialisation, on the other hand, is entirely focused on eliminating any unforeseen factors or uncontrolled initiatives outside the work process. It requires a thorough analysis of the space of possibilities. It is the *systematic* mobilisation of technical thinking that makes it industrial. This requires in turn to place the machine at the centre of the work process, and the person at its periphery (section 3.2).

In this context, computing appears to be the direct continuation and completion of the logic of industrial transformation at play since (at least) the 18th century, which aims to objectify the work process and make it independent of subjective human constraints, whether physical or cognitive (section 3.3). Observing the uses of computers allows to identify, in a more general way than observers of the first industrial revolution could, the advantages provided by mechanisation from the point of view of effectiveness, of which gains in productivity are only one aspect.

Computing and industry thus share a *systematic* approach to technical thinking, which requires observing human practice as an object, *i.e.* theoretically. I conclude this chapter in a problematic manner, on the nature of this distinctly *systematic* and *theoretical* modality of problem-solving, as opposed to the *practical* and *technical* modalities highlighted in this chapter.

3.1 The role of procedures within human practices

This section is intended to show that procedures, as they appear in human techniques, are not sketches of programs. They play an entirely different role from outlining a possible automation of practice. 3.1.1 is centred on a workplace study by Hutchins (1995) about the operations on a warship. It shows that procedure books do not attempt to describe the practical knowledge of sailors step by step, but are part of a material and social *cognitive ecology* that underpins practice. Procedures are costly cognitive artefacts for the human mind, which serve precisely to direct its attention to delicate points (3.1.2). 3.1.3 leverages another workplace study by Orr ([1996] 2016) about a technical maintenance team shows that technicians, when they want to share their knowledge, often prefer the narrative form to procedures; indeed, stories allow to describe the very process of inquiry, whereas procedures only summarize their result.

3.1.1 Practical reason: a matter of organisation?

Running a large ship is a rational activity that requires organisation and method. In a military setting, tactics and manoeuvres are written down explicitly in procedure books to which the crew is obliged to conform. This activity can therefore be described as a system of rules. Hutchins (1995: 199) even goes so far as to describe the operation of the ship as a production system,[1] a concept he borrows from computer science. He asserts that the overall interaction of the crew, their instruments and the situation can be described as a computation, if this term is taken in its Simonian sense, *i.e.* as a transformation of representations according to rules. This is a "metaphor" that Hutchins makes his own:

> Although it is not customary to speak of the computational properties of social institutions, the navigation team in Sea and Anchor Detail can be seen as a computational machine. This

1 On production systems, see above 2.2.2.

> section explores this metaphor, looking at the ways in which aspects of the behavior of the system can be interpreted in a computational framework. (Hutchins 1995: 185)[2]

Hutchins' work, which dates from 1995, is interesting for the history of cognitive science. Already a critic of the computational hypothesis of the mind, as will become clear, he nevertheless maintains another important Simonian thesis, namely that (rational) organisations are also information-processing systems, *i.e.* compute like the human mind and the computer. However, Hutchins's observations undermine the whole computer metaphor from within. They carry some of the seeds of the "4E" cognitive sciences,[3] which were emerging at this time and which would later frequently cite his work.

This tension can be seen in Hutchins' judgment of procedure books:

> Written procedures are not used by the members of the navigation team as structuring resources during the performance of the task, nor do they describe the actual tasks performed. Furthermore, if a system was actually constructed to perform as specified in the procedures as written, it would not work. Still, the normative procedures are a good starting point and provide a stable framework within which the properties of the system can be described. (Hutchins 1995: 178)

How is it possible for a computing machine to function without the use of procedures, and for their use to lead, on the contrary, to its failure? One aim of this discussion is to solve this paradox.

A central concept for Hutchins is that of *cognitive ecology*, which navigation illustrates well. Sailing involves representing the situation of one's ship and its destination. However, there are several possibilities for doing this. Locating them on the fixed space of a two-dimensional map is just one of them. For Micronesians, for example, it is the ship's position relative to the stars that is directly significant, as well as other spatial cues, such as the hues of the sea, which indicate the underwater relief. Once they have set their course, Micronesian navigators estimate the distance to their destination by associating it with the course of a given star (Hutchins 1995: 92). For Western sailors, on the other hand, navigation is an activity mediated by a complex set of instruments and symbolic artefacts. For example, in order to know the ship's position and place it on a map, readings must be taken using a compass and landmarks which are visible from the ship (a lighthouse, a promontory) and can be located on a map. For this technique to be accurate, a complex transformation of representations must take place between several spaces, the real space of observation of the marker on the horizon thanks to a rotating telescope, the space

2 See also the concluding chapter (353) where he returns strongly to this thesis.
3 On this topic, see for example Clark 2013.

of the register on which the angle of view is recorded, then that of the compass on which it is transferred, and finally that of the nautical chart where a line of position passing through the marker can be drawn (Hutchins 1995: 119–123). Thus the position is computed without any computation having been performed by any sailor:

> These tools transform the task the person has to do by representing it in a domain where the answer or the path to the solution is apparent. Second, the existence of such a wide variety of specialized tools and techniques is evidence of a good deal of cultural elaboration directed toward avoiding algebraic reasoning and arithmetic. [...] These tools permit the people using them to do the tasks that need to be done while doing the kinds of things the people are good at: recognizing patterns, modeling simple dynamics of the world, and manipulating objects in the environment. (Hutchins 1995: 155)

This cognitive ecology is not just made up of instruments and symbolic artefacts. Observable natural elements are also mobilised (visual landmarks, stars), albeit to a lesser extent than in Micronesian navigation. Above all, Hutchins insists on the distributed and cooperative nature of this cognitive ecology. Processes are driven by several crew members communicating with each other, allowing everyone to focus on simple tasks. This creates a system of cognitive constraints, the satisfaction of which solves the problem:

> The human component of the system [acts] as a malleable and adaptable coordinating tissue, the job of which is to see to it that the proper coordinating activities are carried out. In their communication and in their joint actions, the members of the navigation team superimpose themselves on the network of material computational media. They provide the connecting tissue that moves representational state across the tools of the trade. (Hutchins 1995: 219)

The essential feature of this system is that it is flexible and robust, since each crew member can stand in for a colleague who is late or has difficulties, and can detect errors of measurement or communication. The circulation of representations between people allows their validation. Sailors do not behave like automatons carrying out fixed tasks, but as proactive agents committed to the success of the whole process, and therefore interested in the performance of their colleagues:

> In addition, they dynamically reconfigure their activities in response to changes in the task demands. This amounts to a restructuring of functional systems that transcends the individual team members. The individual team members do their jobs by constructing local functional systems that bring media in their immediate environment into coordination. They also must coordinate their activities of helping one another achieve coordination. The computation is implemented in the coordination of representational states, and the human participants coordinate their coordinating actions with one another. (Hutchins 1995: 219)

It is therefore the entire crew, together with the tools specific to its navigation activity, that is the "machine". There is no central instance (the commander, for example) to which all the information would go back so that it could draw a decision that the peripheral entities would only execute, as in Simon's physical symbol system (see quote p. 37). The system of rules is distributed and takes the form of a network of communicative constraints to which the crew submits in order to solve problems:

> Far from being a simple one-way trajectory for information, communication is in fact the assembly of several types of bidirectional constraints, bottom-up and top-down. The meanings of assertions and questions are not given in the statements themselves but are negotiated by the participants as part of their understanding of the activities in progress. Participants use assumptions about each other's tasks to resolve ambiguities in communication. Particular meaningful interpretations of statements are simultaneously proposed and presupposed by the courses of action that follow them. Each participant's evidence of successful communication is the flow of joint activity itself. (Hutchins 1995: 236)

Rules therefore take the form of *constraints*. Generally speaking, a constraint does not tell the process what it must do, but what it cannot do. For Hutchins, this is the general form of human procedures: they rarely determine a precise order of task execution, but establish looser sequential dependencies. They say, for example, that a task cannot begin until this one has been completed.

Sometimes it is not even necessary to specify this, when this dependency is inscribed in the physical characteristics of the environment, for example, that socks must be put on before shoes, because the reverse is not possible. Human procedures often rely on such external features to present themselves naturally to the agent. For example, the *check-list* for inspecting an aircraft before take-off lists the items to be checked in the order in which they will appear to the inspector as he walks around the aircraft. Or, in the case of cooperative cognitive tasks, the activities of one person may be constrained by those of another, who provides the necessary information. It is in this fundamental sense that it is relevant to speak of a cognitive ecology: the entire environment of each crew member – the natural environment, the instruments on board, their own position on the ship, their colleagues as well as their solicitations, the military hierarchy, other social and cultural symbols – all of this naturally constrains and *rules* their activity, relieving the need for explicit rules. Procedures thus find themselves focused on the *non-obvious* points of the activity, particularly those where several justified choices are possible, such as how often a senior officer should check the position (Hutchins 1995: 180).

The coherence of Hutchins' two apparently opposing assertions can now be made clear. On the one hand, the entire system, made up of the navigation team and the entire cognitive ecology that envelops it, can quite properly be said to compute. Characterising procedures as a network of constraints is possible in a certain programming paradigm, aptly called *constraint programming*.

However, this system computes while minimising the computations and complex representation transformations *for each member of the crew*, and more generally, as well as their need to look up written procedures to perform their task. Executing a written procedure (what will be later in this book (see below, 6.2.2). called a *ratiocination*) is in fact a costly artefact for human cognitive capacities. This is a point on which Hutchins is particularly convincing, anticipating here current findings in neuroscience (Hutchins 1995: 295–310).[4] Following a written procedure, indeed, requires to coordinate different actions: to read the current instruction, to memorising its position in the procedure (to be able to attend to the next instruction afterwards), to represent both the situation it describes and the actions it prescribes and finally to translate it into motor actions. At each stage, the agent has to coordinate at least three spaces: the space of symbol sequences, the space of mental representations, and the space of action in the immediate environment. This "spatial" coordination (from one space to another) naturally prevents "temporal" coordination (within each space), in particular motor coordination and comprehension (visualisation of the dynamic flow of representations). Expert execution, on the other hand, achieves this by being able to link actions and representations directly, because these flows no longer need to be guided, step by step, by the symbolic grid of the procedure.

Procedures are therefore a small but essential part of the cognitive ecology of navigation: they serve to draw the crew's attention to delicate points in the operations to be carried out, or to constraints to be respected which are not already materialised in the instruments, or in the spatial and cultural relationships between sailors. In particular, they are used to allocate tasks between crew members during procedures requiring a high degree of coordination: each crew member knows what role they must play, but above all what roles their colleagues must also play, which act as constraints for them. A phenomenon of coordination by convention is thus established (Hutchins 1995: 375).

It is for these reasons that Hutchins can argue that if sailors simply followed the written procedures, nothing would work. Taken on their own, procedures describe a largely *under-constrained* network of rules, which leaves open a thousand possibilities for running the ship aground. They are implicitly underpinned by the behaviours imposed by the material environment and above all by the social situation of the sailors (the military hierarchy, the desire for recognition by the group). For example, Hutchins cites the custom, nowhere written down, of offering their

4 On neuroscience, see the next discussion.

support to the crew as soon as a sailor has completed their task because "doing the bare minimum when others know that one has the time and means to do more is a clear message [that one is sending them]" (Hutchins 1995: 225).

It is this whole set of "rules" that makes the ship a system of production. But it is clear that these are not rules to which *the sailors* obey, but *regularities* noted by the ethnologist who describes their behaviour. Hutchins thus concludes firmly his inquiry as follows:

> My original hypothesis about working in a military environment was that behaviour there is explicitly described and that people act more or less like automatons. It should be obvious by now that this is far from the case. (Hutchins 1995: 225)

3.1.2 The notion of cognitive ecology

Hutchins' analysis allows to make an essential step forward. By highlighting the very limited role of written procedures in navigation operations, he shows that explicit rules are rarely a privileged guide for action, which one could blindly obey to be brought safely to destination. It is better to compare them to the markers found on hiking trails, often spaced several hundred metres apart, which allow hikers to regularly check that they have not lost their way. These markers are set only when the path becomes uncertain, and therefore suppose a certain ability of hikers to find it alone.

Procedures are therefore *aids* to human practical reflection rather than its substitute, as it is too often believed. Like markers, which are inserted into the environment in order to prevent the walker from going astray, they invite the practitioner not to trust their instinct or intuition (in case they are still an apprentice, or there is a difficulty or trap specific to the present situation), and to *take control* of their action. Cognitive scientist Clark expresses this quite well:

> The role of structured language as a tool for scaffolding action has been explored in various literatures [...] Examples drawn from everyday experience are abundant and range from memorized instructions to shoe lacing to repeated formulas before crossing a street, such as "look left, right, left again, and if the way is clear, cross with caution!" [...] In such cases, the agent who uses language in this way [...] may engage in a simple kind of behavioural self-scaffolding, using the phonetic or spatial sequence of symbolic codes as guides to the temporal sequence of acts. (Clark 2010: 48)

The procedure for crossing a street does not aim to automate the agent's action, but instead invites them to think before making a decision – in this case paying attention to their visual environment. Similarly, if someone chooses to do a calculation rather than a head calculation, it is because they want to pay particular attention to their operations. Written rules and procedures are artefacts specially designed to guide the agent's attention and force it to *control* its action consciously, to disable automatic and habitual action in certain specific circumstances.

Some contemporary cognitive psychologists interpret these phenomena in terms of the opposition between *system 1*, *i.e.* habitual and automatic intelligence, and *system 2* consisting essentially in our capabilities of attention and reflection that "deactivate" system 1 in circumstances where habitual action fails (Evans and Frankish 2009; Kahneman 2012). Following a symbolic step-by-step procedure is clearly an operation driven by system 2, potentially as part of system 1's training to acquire new automatisms.

Procedures are particularly well-suited for this function, not because they are easy, but precisely because they have a high cognitive cost, mobilising the entire mind. Following a procedure is the opposite of automatic, blind action, as the machine metaphor falsely suggests. It is on this idea that neuro-scientist Dehaene concludes his study of learning mathematics:

> I will hardly surprise the reader by confessing that I suspect the computer metaphor of being insufficient and excessively limiting. On a purely empirical level, the comparison does not hold. The preceding chapters abound with counter-examples suggesting that the human brain does not perform calculations in the manner of a logic machine. Rigorous calculations do not come easily to *Homo sapiens*. Like so many other animals, he comes into the world with a vague and approximate idea of numbers, which has nothing in common with the digital representation of computers. The invention of a language of numbers and exact calculation algorithms is part of humanity's recent cultural history, and in many ways it is an unnatural development. While our culture invented logic and arithmetic, our brains remained unchanged and resistant to the most elementary algorithms. The only proof of this is the difficulty with which our children assimilate arithmetic tables and the rules of arithmetic. [...] The failure of the brain-computer metaphor is almost comical. Where computers excel, in the flawless execution of a series of logical steps, our brains prove to be slow and fallible. (Dehaene [1995] 2010: 257–258)

Practice is often contrasted with reflection, with the latter frequently regarded as a hindrance to the former, either because it slows down routine action, or because it prevents the informal intuition of the right gesture that comes with true mastery. However, between these two poles of practice (routine and intuitive mastery), attention and reflection accompany the action along a continuum of modalities. In particular, it is sometimes important for action to be guided by the explicit control of attention. This is one of the essential roles of procedures as cognitive artefacts.

3.1.3 The inadequacy of books of procedures

A second ethnological workplace study by Orr ([1996] 2016), who observed the activity of a maintenance team for Xerox professional photocopiers in the early 1990s, helps to understand the limitations of procedure books for solving problems. Orr finds that these rules provide limited guidance for human action in a technical environment. They are not even the primary form through which agents express their know-how, whether for advancing one's own inquiry or for engaging in dialogue with others. Orr presents indeed a result very similar to that of Hutchins concerning the impossibility of explaining the behaviour of technicians by simple obedience to rules:

> Technical work is generally believed to consist of the routine repair of similar machines. Routine repair is indeed common. However, individual machines sometimes behave in very peculiar ways. New failure modes appear regularly, and a routine procedure cannot solve unknown problems. The practice of technicians [is best understood as] a response to the fragility of the available understanding of problematic service situations and the fragility of control over their definition and resolution. [...] Work in such circumstances resists rationalisation, because the expertise essential to such a contingent and improvised practice cannot be easily codified. (Orr [1996] 2016: 2)

As a result, their relationship with procedure books is ambivalent. They are only one tool among others in the performance of their work, and they declare that they distrust them:

> The technicians' talk about using the service documentation is full of cautions about the perils of following the diagnostic procedures. Some of them are said to be circular; some are easily misread. Even those technicians who profess the strongest attachment to the diagnostic procedures warn that, without understanding the intent of each procedure, one can easily make the wrong choice and get hopelessly lost. (Orr [1996] 2016: 129)

Orr observes that the essential form of knowledge sharing between technicians is not the procedure but the story: in team meetings, technicians recount their service adventures, the difficulties encountered, their reaction and strategy for resolutions, what worked and didn't work, the customer's attitude, etc. right up to the final resolution. For Orr, the narrative form certainly has a socialisation value (showing one's competence, exchanging emotions, sharing common opinions about a particular customer) but it cannot be reduced to that. It has also its own epistemic value, which is recognised by technicians and even team leaders, who always start their meetings late to allow time for informal discussion (Orr [1996] 2016: 140). It is these stories that are mobilised when a technician presents their colleagues with a problem that they cannot solve.

The preference of the narrative mode over that of the general procedure for conveying information about problem-solving may be intriguing. Indeed, technicians are very keen on general explanations and, of course, capable of abstract reasoning. In particular, while they are certainly tinkerers, they are not content with solutions for which they have no explanation. Their shared values revolve around the ideas of control, order, and systematicity in their work.

Orr places particular emphasis on these latter points:

> Two principal issues for technicians are the fragility of understanding and the fragility of control. [...] Technicians' discourse reveals that they value most highly those attributes that contribute to the preservation of order and understanding. Reputations are built on technical skills, memory, ability to gather information, verbal performance, and the general ability to retain control of the situation. [...] The counter to this threat of chaos is thought to be a systematic approach to the work. [...] A systematic approach to troubleshooting is thought to be the best assurance that everything will get done that should be done, and that everything will be learned about the situation that can be learned. (Orr [1996] 2016: 163)

This systematic approach is not just about the diagnostic method, which must not rule out *a priori* any hypothesis. It's also about precision and cleanliness of gestures, tidying up your belongings and spare parts, following a certain visit protocol, and managing your customers and your area of operation.

It may therefore come as a surprise that such a group mindset is content with the exchange of stories and never seeks to draw systematic lessons from them. It is the resolution of this paradox in Orr's study that is of interest here.

The reason for the complexity of the diagnosis is quite simply the number of parts that make up a professional photocopier, as well as the mix of mechanical and electronic parts. With the whole machine immersed in a continuous stream of ink dust, heated to very high temperatures from time to time, and subjected to high speeds, it is easy to understand that wear and tear and clogging can occur rapidly and affect any part of the machine. In addition, the frequency of maintenance, the extent to which users follow the instructions for use, the incidents already experienced by the machine, the variability brought about by previous repairs and by spare parts – all these factors explain why each situation can be different and why each photocopier has its own history. Orr reports the fact that technicians know their machines *individually*, as well as their "perversity" or "docility", as a shepherd knows their sheep (Orr [1996] 2016: 89).

Each diagnosis of a complex fault is therefore an inquiry which must take account of this background constituted by the context of use and the history of the machine. This inquiry must lead to the *understanding* of the fault, *i.e.* a representation of the faulty mechanism and its causes:

> Technicians must be prepared to solve new and unanticipated problems, which requires them to develop as complete an understanding of the machine as possible. [...] The question of interpretation is critical; technicians working on incomplete diagnostics often say that they think they know the crucial facts but don't recognise them. The most difficult part is recognising the importance of a given fact in relation to the observed or reported behaviour of the machine. (Orr [1996] 2016: 109; 119)

This remark makes clear that the epistemic virtues of the narrative are particularly well suited to the nature of the problems encountered. Firstly, it allows two stories – that of the machine and that of the inquiry – to be continuously intertwined and set against each other. It also makes it possible to retain both relevant and irrelevant details that may have delayed the diagnosis. Finally, it puts the listener in a situation where they can play an active role in the inquiry that unfolds in the story, sharpening their own capacity for judgement. Ultimately, the narrative becomes a diagnostic tool *in situation*. It makes it possible to compare, in all the detail required, the present situation with the narrative. Better still, it constitutes a *model* that one can attempt to project *on* the present situation:

> The stories are artifacts to circulate and preserve. Through them, experience becomes reproducible and reusable. At the same time, each retelling is, in a sense, a re-representation. The stories originate in problematic situations and are told or retold in diagnosis when the activity they represent becomes problematic again. [...] Such stories combine facts about the machine with the context of specific situations. The contextual information demonstrates the claimed validity of the facts of the story and provides a framework for testing those claims against the hearer's model of the machine. The context also constrains the application of those claims by defining the situation in which the facts are known to be true. (Orr [1996] 2016: 127–128)

Stories are therefore the means of understanding and of transmitting understanding. But the understanding itself concerns the internal mechanism of the machine and the precise point at which it failed. What the technicians are aiming for in all their stories is the identification of the faulty mechanism, the cause of its appearance, and the explanation of its symptoms. In other words, the process of inquiry is a work of representation of a reality hidden *behind* the phenomena of which it is the cause or reason, which could be compared to that of a scientific observer of natural phenomena. It is more accurate to liken the technicians' posture to that of the engineers who built the machine:

> This ability to break down the machine into its constituent parts, or compose the whole machine from them, is characteristic of the way technicians talk and maybe think about them. (Orr [1996] 2016: 92)

This posture can also be seen in the general admiration technicians have for well-designed machines, and their desire to see them work well. Indeed, they are often enthusiastic about new machines (Orr [1996] 2016: 25). It is this posture that is partly behind their systematic approach to work. Their inquiry, ideally, should not only solve the problem, but also give the reason for it. The worst problems are indeed the recurrent ones, because they show that the previous repair had only treated the symptom and not the cause (Orr [1996] 2016: 71).

It is now clear why narratives and a systematic approach are not antithetical in the context of a complex diagnosis. Narrative is the very form of diagnosis that *seeks* to be systematic, to *circumscribe* the problem:

> For difficult diagnoses, the representation of the problem takes the form of a narration of the diagnostic process, a verbal consideration of what is known or thought to be known or what could be known about the situation, to see whether it might be interpreted in any coherent way. The narration includes summarizing what is known about the machine, questioning whether what is known is a coherent representation of the situation or could be if interpreted differently, and determining what else needs to be known and how to learn it. (Orr [1996] 2016: 109)

All this helps to explain the complex and ambivalent relationship that technicians have with maintenance manuals. On the one hand, even if they don't like the idea of being reduced to the role of simple executor, they appreciate the systematic approach to procedures, which corresponds to their work ethic. But what interests them most in the manuals are the diagrams of the machine and the lists of spare parts. This enables them to build up a mental image of the mechanism and its potential points of failure. In contrast, the actual repair procedures are viewed with suspicion, because they very often presuppose a far too naïve breakdown situation, which is moreover often than not not made explicit: they are content to characterise the problems by the symptoms (such and such an error message, etc.) and to set out a series of actions to be taken. The life of technicians with their machines is much more complex than that.

So what technicians criticise procedure books for is not that they are too prescriptive, but that they are not systematic enough. Moreover, this is not really a criticism, as they believe that it is impossible to foresee every possible scenario in advance (Orr [1996] 2016: 111). The only possible systematicity is that of inquiry, and this requires a clear understanding of the machine in order to *reason* about it.[5] The false systematicity offered by documentation is followed reluctantly, when the technician finds themselves bereft of any other solution: at least it shields them from professional blame (Orr [1996] 2016: 110).

These observations are very revealing for this inquiry into the role of procedures and rules in human action. Leaving aside for the moment their subjective as-

5 Orr uses the word *reasoning* p. 106 and 108. More often he uses *to think about* e.g. p. 24, 92, 105.

pect – the negative feeling of being a mere executor – they appear above all to be incapable of fixing the knowledge necessary for problem-solving, as opposed on the one hand to narratives, which have the advantage of safeguarding the context and the process of the inquiry itself, and on the other hand to understanding the internal mechanism of the machine.

The first point has already been made: in a way, the narrative gives the proto-history of the procedure, that of the reflection which identifies the generating cause of the breakdown and the remedy. In this sense, the procedure appears as the simple *trace* of a successful inquiry, but one that does not convey the essential knowledge useful for solving complex problems.

Regarding the second point – the need to understand the internal mechanism of the machine – its full clarification will only be completed with the elucidation of these core concepts, *machine* and *mechanism*, in chapter 8.

3.2 Practical reason *vs* procedural rationality

The position just reached is almost diametrically opposed to Simon's. It is not only that human action does not unconsciously follow procedures, but also that the knowledge from which it is derived cannot even be adequately described in terms of procedures. On the contrary, practical rationality seems to be repugnant to this type of artefacts and to have recourse to them only in very special circumstances, when it is a question of ensuring maximum attention to the situation and to one's decisions – either because one has to be wary of one's own automatisms, or because the situation is delicate and potentially risky, or when it is a question of obeying to an alien will or knowledge that does not explain its reasons.

How, then, could Simon have thought something so far removed from these obvious facts? And, to his credit, how then to explain the effectiveness of computers if it is not rules and procedures that make people effective?

One might be tempted to reply that this effectiveness is achieved thanks to the much greater precision with which computer rules are endowed, so that they compensate for the "holes" left by human rules. To use the hiking trail metaphor, computation rules would simply be markers that are much closer together than usual, guiding hikers 'step by step', so that they no longer need the slightest sense of direction. As noted in the introduction, such an interpretation of procedures as *more precise* methods is common.

Is such an explanation sufficient? The aim of this section is to observe *directly* situations of conflict between *rational* human practices and rigorous procedures (computational or not) in order to understand how their forms of rationality differ. The guiding thread here is that systematic procedures have a low tolerance for the inherent ambiguity of situations, unlike people, for whom ambiguity is often a resource, a source of increased flexibility. A few examples of this tension between the practical rationality of human agents and computer programs (3.2.1) are given. They show that it is not due to the rigidity of computer tools (although this may play a role) but to the requirement for systematicity and consistency that is at the very foundation of any software project (3.2.2). Humans do not follow rules in the same sense as computers: Daston's distinction between algorithmic and canonical rules, mentioned earlier, is helpful in making sense of this key observation (3.2.3).

3.2.1 The role of ambiguity in human action

The canonical example of *workplace studies* devoted to human-machine relations is Lucy Suchman's 1987 work, inspired by Garfinkel's ethno-methodology. It focuses (again!) on Xerox photocopiers. Suchman opposes head-on the thesis of computational cognitive psychology (and Herbert Simon's in particular) that action would be explained by plans and compositions of rules. This question has aready been dealt with in the previous section, and Suchman's argument is debatable.[6] The particular interest of Suchman's case study is rather that it deals directly with the interaction of people with a procedure book "in action", *i.e.* an interactive menu supposed to guide the user in programming their task such as, for example, making five bound copies of a bound document, double-sided, with enlargement.

Suchman shows, through the recording and meticulous analysis of interactions between the machine and the people (who were paired up, in order to encourage the oral explication of their reasoning and difficulties) that operational misunderstandings can very easily occur, because on the one hand the machine does not have access to the cognitive state of the people, and because on the other hand the people can misinterpret the machine's messages. For example, if the user misinterprets an instruction from the machine to close the cover of the copier (when it actually means closing the cover of the document), and the machine consequently asks their to re-open it, this request will be interpreted as an indication that they closed it incorrectly, or too roughly.

Such examples may make one smile. They date from the prehistory of human-machine interfaces, and designers have since made extraordinary progress on these issues. However, they illustrate a fact that remains fundamentally valid: a procedure, even an interactive one, can only interpret the situation in a predetermined way according to the information available to it, and does not have *the sense* of

6 See here the critical discussion in Schmidt 2011: 359.

what is happening. Even today, one can still be exasperated by the lack of appropriateness of interactive voice servers. Suchman's essential point is that people, during an interaction, have implicit resources for adjusting together the shared meaning they attach to the situation. The programmer, on the other hand, needs to make a bet on the expectations of the machine's future interlocutors, as such a negotiation of meaning is not available to them. They cannot check that the user has understood the message, and for their part the user has no opportunity to ask questions about the current state of the procedure (what Suchman calls a *meta-inquiry*), such as whether the machine's silence is a validation or a blockage. Thus there can be false alarms ("the procedure has gone off the rails") or, on the contrary, the installation of misunderstanding a long sequence of interactions (Suchman 1987: 120).

The interesting thing about this analysis – and this point must be stressed – is that it does not deal with the difficulties that are usually encountered in human-machine interaction. It does not, for example, concern the machine's inability to perceive the emotions of its human interlocutors. The difficulties relate to purely logical discrepancies concerning the interpretation to be given *to the procedures themselves.*

So there seems to be, in human-machine interactions, what could be called an essential *ambiguity* about the situation. This is not just an ambiguity about words and symbols, like the one that appeared in the example above, but an ambiguity about the goals to be achieved, the means available, the state of readiness of each of the protagonists. This ambiguity is not only, or even essentially, manifest in the human-machine dialogue and in their surface interactions, but in the chasm between their respective logics of action.

This chasm can be seen most clearly in *change requirements*, the IT term to describe requests from a principal to upgrade some software. These requests are an important part of the activity of the IT team serving an organisation. Most of them are benign and can be resolved in a few days; but some prove complex to satisfy, because they touch on sensitive points of the software's internal architecture. This does not mean that they are complex to formulate, and they can often appear to be common sense.

A change requirement can be due to a number of factors. It may be a question of correcting an initially faulty specification, if some ambiguity had infiltrated there, for example if the client had overlooked a need that seemed to them obvious, and of which they were sometimes not even aware.

It can also be a question of a change in the situation itself. A few examples, drawn from my personal experience in the ophthalmic lens industry, can make this point clear: 1) Ophthalmic lens production is partly made to order, as each complex lens is cut to suit the patient's eyesight. There are production parameters that optimise the thinness of the lens, but which may lead to other defects. Some customers (opticians) have different preferences in this respect. The computer system was able to record this preference, so that these parameters were activated or not depending on the customer who placed the order. When this software was implemented in Germany, the sales team there asked the IT team to allow the optician to specify *for each order* whether the optimisation parameters should be activated or not. German opticians, known for their meticulousness, wanted to be able to deviate *in specific cases* from the general rule they had set.

This was a legitimate request, but a complex one from an IT point of view. It required that the preference entered in the customer record should no longer be a fixed parameter passed automatically to the order, but only a *default value* which should be able to be modified when the order was taken (for example a box to be ticked or unticked). Such an evolution is not obvious, as changing the data model as well as the order form are tasks that have impacts on the entire information transmission chain. While the customer's request was perfectly justified and very simple to understand, it posed major problems for the IT department. This is a typical case of ambiguity: it is perfectly rational for a person to make *sometimes* an exception to their preference, a notion which, at the outset, implies a certain stability.

2) Quality problems had made the sales teams nervous. The (rare) event had occurred that the production anew of a faulty order was rejected a second time by the customer. The succession of two quality problems for one same order is disastrous in the eyes of a customer, so the production team wanted to be able to carry out systematic quality checks on these repeat productions: at the end of production, if the product was made locally, or on receiving it, if it was made abroad. This meant that these orders had to be explicitly identified by a special note on the production sheet.

This request, also perfectly understandable, was also very difficult to satisfy, as it still involved evolving the data model of a production order, and systematically referring orders with this flag to quality control, a task that was not provided for by the production software, since quality control was purely statistical. Here again, there is a situation of ambiguity, since, in the initial data model, it was simply not *foreseen* to describe a production order as a repeat-order or not.

3) In its order cancellation policy, the UK sales agency had set a deadline of one full day after the order was taken. This was because it was almost certain that production would not have started by then, since it had an average of two days leadtime in production. When the new software was introduced to automate production scheduling, the problem of orders placed on Friday evening arose. These orders could only be cancelled on Monday (a working day for the customer service department), while production worked also on Saturday. A significant proportion of cancellations could therefore arrive too late. The production team, when it sched-

uled production manually, used to delay orders from Friday until the next Monday. However, automatic scheduling did not have the capacity to define such a rule. For example, it was able to give priority to certain important customers, or certain types of product, but not to certain *days* of order taking over others.

Here again, there is an example of ambiguity. A parameter which is not significant for the decision can become significant in certain situations, in this case a difference in the organisation of weekend work between customer service and production. It is very simple for a person to follow this rule, but for software, once again its data model needs to be updated so that it can take this parameter into account.

These three examples are very similar. They are by no means representative of *all* the forms of change requirements with which software may be confronted, however they suffice for the purpose of this discussion, to make felt the *essential mismatch* that exists between a computer system and a human logic of action. The ambiguity that has been noted three times is not, as in the situation studied by Suchman, at the interface between man and machine, but in the *rules* themselves.

It is also notable that the ambiguities described are not linked, as Suchman seems to suggest, to the specific ability that people have to understand each other and to agree on the meaning that should be given to the situation. These issues *can* be solved by complementary programming.

3.2.2 Systematicity and adaptability

This is an important point: it is not a question of saying that there is a domain of human rationality that is inaccessible to rules and programming, but of pointing to the specific difference that there is, for humans and for programs, in adapting to ambiguity.

This difference may seem minor: it simply seems that, in the cases cited, resolving the ambiguity turned out to be *more difficult* for the program than for people, that there was a mismatch between the simplicity and common sense of the client's request and the technical difficulty experienced by the programmer in meeting it.

This difficulty is actually a significant one. In all three cases, it is due to the need to develop the program's *data model, i.e.* the program's representation of the situation. It is well known among software developers that this is the critical point of good or bad programming, as asserted by this adage of the computer scientist Fred Brooks (1982: 102): "Representation is the essence of programming". What is essential here is the notion of *model.* Any computational resolution of a problem must go through a global modelling of the situation and all its relevant dependencies. Each change in requirements can potentially require this modelling to be restarted from scratch, as the addition of a single parameter can call into question all the model's assumptions.

This method of integrating new rules into an existing system is specific to programming and, more generally, to the systematic approach to a problem. The term *integration* is itself significant: the point is to ensure that the addition of a new rule preserves the *integrity* of the new set. This is not a superfluous precaution, but *what programming means*: *ensuring* the automatic resolution of a general problem, through *all* the cases that suit its specification.

This is not at all how people act. If an operator is instructed to follow a new rule, they will simply try to comply, without updating any "mental model". They will adapt *locally*, so to speak, and not reconfigure a global system of behaviour. In the event that a conflict of rules arises and they become aware of it, they will ask their manager a question. It is often the case that they will not realise this and make an implicit choice. If the choice poses a problem, the manager will have to make a diagnosis, etc. Human action is thus naturally made up of *a posteriori* management of conflicts of orders and rules, which programming and automation come precisely to reform, by proposing an *a priori* management of all possible issues – which is clearly audible in the *pro-* of *programming*.

The relative ease with which people understand and integrate new rules, and even manage their conflict skilfully, is therefore in no way due to the superiority of human intelligence over that of the machine, which would be obtuse, rigid, clumsy, as one often hears. The flaw in this skill is its non-systematicity, *i.e.* the risk it takes of *mistakenly* solving the problem. Its modus operandi is that of exception management, *i.e.* the possibility of reconfiguring the problem during its resolution if the situation turns out to be novel.

This form of situational intelligence should not be contrasted with machine stupidity, which makes no sense, but with another form of human intelligence, which is planning. One is not inherently superior to the other. There is always a delicate trade-off between planning and *ad hoc* management. Leaving certain rules to the judgement of operators can be legitimate and advantageous in certain cases, particularly when decisions should be explainable. If one takes the problem of Saturday production in the UK subsidiary again, manual scheduling enables priorities to be managed more finely – for example, if the scheduler knows from experience that a particular customer never cancels orders, if there are emergencies, etc.

Nevertheless, plant managers are the first ones to ask for the automation of decisions along the production flow. They know that in a complex production system made up of hundreds of rules, only a few can be left to the judgement of each workstation. One has indeed to consider that several operators may occupy the same position, that they all need training, that new recruits arrive regularly, and so on.

Rigour should not be confused with rigidity. The rigour imposed by programming makes it more rigid in the short term and locally, *i.e.* from the point of view of the operator who can see that they could make a more intelligent decision *in this specific case*. On a large scale, however, in the context of a complex system, the rate of exceptions to be managed can quickly become unacceptable and suffocate the production flow itself. This is where automation comes into its own, as it can support the indefinite addition of specific rules while keeping the rate of exceptions under control. In these cases, rigorous planning is the condition for flexibility, not its opposite.

All these ideas point to a wide divide within technical rules themselves, between those that require interpretation and judgement – that aim only to *guide* and educate judgement – and those that suppress judgement and must only be obeyed. This crucial distinction is drawn by Lorraine Daston in her book *Rules*, which is the focus of the next discussion.

3.2.3 Rules as paradigms and as algorithms

According to Daston, as long as one goes back in Antiquity, the concept of *rule* (*kanon* in Greek, *regula* in Latin) split into three principal semantic clusters, as tools of measurement and calculation, models or paradigms, and laws (Daston 2022: 3). In the first cluster, there is for example the rules for adding numbers. In the second one, one finds typically rules of the arts and crafts, which are often embodied in masterworks or typical cases – e.g. Polykleitos's *Spear Bearer* (1st c. BCE), called the "canon" of artists by Pliny the Elder. In the third one are laws, regulations and social customs that organize, more or less forcefully, the collective behaviours of a community.

It is the two first clusters of meanings which are of interest for this discussion. The opposition between algorithmic rules and canonical rules was priorly made by Gilbert Ryle through his distinction between Procrustean and canonic rules (see above 1.2), which can be easily understood by the example of the game of chess:

> The player simply does not play chess if he does not observe the Procrustean rules of the game, but he does not play effectively if he does not also apply strategist maxims. (Ryle [1971] 2009a: 240)

Unlike Procrustean rules, then, which are in some sense constitutive of the very endeavour in question, canons are advice and maxims based on experience, which are relevant most of the time, but which elude any attempt at systematization – Ryle uses the term *codification* here, illuminatingly:

> Canonical rules [...] commonly resist codification. [...] They are learned by practice, not by training. They are taught by criticism rather than by rote, and we never finish learning them.

The strength of Daston's argument is to show that there is no mere homonymy between the two groups of rules. All aim to shape action, however they do so in different ways. Paradigms are flexible (one can copy them exactly, or just take them as inspiration) and only general: they do not claim that they should be followed in all cases, whatever the context. They can be transformed into prescriptive rules (of the *do-this* and *do-that* sort), starting to resemble algorithms, but in this translation they will become *thick*, *i.e.* rules with a lot of prescriptions and special cases and exceptions to attend. In contrast, algorithmic rules, at least in their *mechanical* flavour that emerges in the 19th century, aim at both rigid (mindless) application *and* universal validity, with no exception left unattended. An algorithm promises that, if its instructions are followed exactly, it will certainly produce a correct result, whatever the inputs, as long as they satisfy some preconditions stated in the algorithm. The transformation of paradigms into algorithms might seem a hopeless task in the practical world, because the resulting thickness would become unbearable: and this is what Daston observes about the practical rules of the early modern period:

> Rules were not free-standing: models, examples, tips, and observation propped them up and filled them out. This is not necessarily because the rules were vague or unspecific or approximate. Rather, it was because no universal formulation could anticipate all the particulars with which it would be confronted in practice. The gap between universals and particulars is hardly news; what is striking about the early modern rules are the provisions they included to bridge that gap. (Daston 2022: 60)

Algorithms, by contrast, are thin rules, *i.e.* have few exceptions. This is the condition of their mechanization: "There were no exceptions to the rules, and that is why machines could execute them" (Daston 2022: 141). How then could they become so prevalent in the practical world? According to Daston, the initial transformation appeared in the domain of computation itself, which rules until the 17th century were often stated as paradigms or thick prescriptions. It took a lot of work of gifted mathematicians to discipline computation and make it appear as a mindless task:

> [Human computers] received their problems and the methods to solve them pre-packaged, thanks to the labor of Maskelyne, Airy, Prony, and others who had divided and standardized procedures. There was no need for the induction from particulars to particulars so characteristic of pre-modern algorithm textbooks, and there was no scope for the classificatory generalization of algorithms practiced by earlier mathematicians. The problems had been pre-classified and the procedures specified to the last detail. No computer needed to ask what species of problem this was or what algorithms were needed to solve it. Nor were computers

> likely to encounter a wide variety of kinds of problems and algorithms. Their experience of algorithms contrasted sharply with that of pre-modern calculators. Their rules were genuinely thin. [...] They were thin not because they were independent of context but rather because their context had been carefully fixed. (Daston 2022: 120)

Computation showed the way for the standardization and division of labour across all disciplines. Standardization allowed to *fix the context* of work, and division of labour allowed to describe it in step-by-step tasks. This movement enabled mechanization of work, but was clearly prior to it:

> It was the division of labour, not the machines, that made the algorithms mechanical. (Daston 2022: 121)

Daston therefore argues that there is indeed a crucial break in technically organised work between traditional expertise and the systematic approach to problem-solving. This break is linked to the emergence and eventual dominance of algorithmic rules, and it long predates computers: it is characteristic of the division of work and thus of modern industry in general. It is this proposal that now needs to be explored.

3.3 Computing and industry

This section begins with the historical and conceptual proximity that exists between the fields of industry and computing, evident in the reflections of Babbage and Lovelace from the first half of the 19th century (3.3.1). The specificity of industrial thinking relative to mere technical thinking lies essentially in the systematicity of the analysis – the division of work and the standardisation of tasks, situations and skills – which makes it possible to objectively recompose a work process according to its own logic, and thus to *move* the human operator towards the periphery of a conceptually mastered device which is the machine (3.3.2). Information technology is a direct continuation of this mode of reflection. The discussion 3.3.3 aims to identify the benefits of effectiveness that systematic problem-solving brings, and lists five of them. The chapter concludes with a question on the very concept of systematicity (3.3.4).

3.3.1 Historical and conceptual proximity

The proximity between computing and industry is strikingly apparent in the comparison of their lexical field. *Process, procedure, task, step, instruction, execution, machine, system, operation, specification, rule, scheduling, planning, input, output, etc.* are central terms in each of these two domains. Admittedly, these are fairly general terms, but such lexical overlap is not found in other disciplines, for example law, medicine or even engineering. It can even be argued that, overall, computer science has a lexical field closer to the world of production than to that of mathematics or logic, with which it is often associated.

The terms mentioned have very different material meanings in each of the two fields of activity. A computer process is not an industrial process, and a computer instruction is not an instruction given by a foreman to an operator. Is this a metaphorical transfer from one field to the other, linked to the fact that machines are programmed precisely to *imitate* human tasks? But this imitative relationship is the very puzzle we aim to solve; we cannot be satisfied with such an answer. Moreover, a metaphor functions primarily by transposing a concrete term to another specifically distinct domain, and most of the terms mentioned here are abstract. It could happen that computing terms and industry terms are two different specifications of the same generic lexical field.

Actually, the transfer of concepts from industry to computer science has a long history, and it is not limited to semantic transfers. The first tool for representing programs, flow diagrams, was proposed by John von Neumann in 1946. This notation had become common for describing industrial processes in the 1920s; it is attributed to Franck Gilbreth, a production engineer influenced by Frederick Taylor's scientific management (Kiran 2017). It may be that von Neumann saw flow diagrams at work through his training in chemistry (Ensmenger 2016: 326). In any case, for computer historian Mark Priestley, von Neumann and Goldstine, in their 1947 report on programming, conceived of it on the model of their own experience in the organisation of work for large floors of human computers:

> [They] recognized that there was nothing in their description of the planning stage that was specific to automatic computing, and the details of their proposals reflected established practice in the organization of large-scale [human] computing [...] This continuity of practice should come as no surprise. All the developers of the automatic machines of the early 1940s had had extensive experience in carrying out or supervising large-scale manual computation. What was new was the need to be completely explicit about the operations that the machine would carry out. (Priestley 2017: 460)

This large-scale planning of computation, based on the principle of division of labour and quality control, was perfected throughout the 19th century (Grier 2005: 46 ff). Its model was the famous trigonometric table project carried out by Gaspard de Prony between 1791 and 1795 at the Paris Observatory. Tables of unrivalled accuracy were produced in record time thanks to the breakdown of computation into simple tasks within the reach of people with only very common arithmetic skills (wigmakers out of work after the French Revolution, according to legend). Now Prony himself claims to have thought of this organisation while reading the chapter of *The Wealth of Nations*, by Adam Smith, devoted to the division of labour (Durand-Richard 2010; Grattan-Guinness 1990; Peaucelle 2012).

The relationship between computing and industry is even more direct than that, since Charles Babbage states that he got the idea for his *Difference Engine*, considered to be the first conception of the computer, from reading Prony who had written:

> It soon occurred to me to apply to the connection of these Tables the division of labour, from which the Arts of Commerce draw such an advantageous benefit in order to unite to the perfection of labour the economy of expense and time. (cited by Babbage ([1832] 2010: 154))

Babbage thus asserts:

> The division of labour can be applied with equal success to mental as to mechanical operations. [...] A short account of its practical application [...] will show that the arrangements which ought to regulate the interior economy of a manufactory, are founded on principles of deeper root than may have been supposed, and are capable of being usefully employed in preparing the road to some of the sublimest investigations of the human mind. (Babbage [1832] 2010: 154)

What is intriguing here is the consequence that Babbage draws from his discovery of the applicability of the division of labour to the tasks of thought. One might have expected him to conclude on the importance of this principle. He draws the opposite implication, that it must itself rest “on deeper principles than have been supposed”. It probably suggests that the division of labour is based on a logical rather than an economic principle.

3.3.2 The notion of industry

In order to interpret this link between computing and industry, as well as Babbage’s suggestion, it is necessary to clarify what kind of activities are covered by the term *industry*, a notoriously difficult question. A recent historiographical update (*Artefact*, n°17, 2022) reviews the main characteristics discussed by historians and economists. Large-scale production, the existence of standards by which products can be recognised or judged, and the flow of production on global markets are characteristics broad enough to be recognised already in ancient and medieval productive activities (Pagès and Verna 2022). Dynamics of innovation, aimed at perfecting products and increasing productivity, can also be observed in these periods. From the 19th century onwards, industry is defined in opposition to craftsmanship, using all the above criteria, as well as complementary notions of mechanisation (as opposed to manual labour) and mechanical energy (as opposed to animal and human energy), division of labour (as opposed to the craftsman's unity of tasks), proletarianisation of the worker (as opposed to the craftsman's independence) (Caliste and Carnino 2022: 220).

One can see from the multiplicity and nature of these criteria that it is difficult to circumscribe precise classes of economic activity under the term *industry*. For example, it would be reductive to identify it with what statisticians call the secondary sector of the production of manufactured goods. Large areas of agriculture and services are now organised according to an *industrial logic*, such as vast cereal crops, motorways or call centres.

What, then, is meant by these expressions of *industrial logic*, or *industrial methods*, which have become commonplace today? Economist Gilles Le Blanc (2005: 29) proposes to circumscribe them through three criteria:

1. A logic of standardisation of products and processes, which enables the rational organisation of work, automation, scale effects.
2. A dynamic of innovation.
3. The importance of capital to finance the production tool and innovation, which assumes that the activity is always on a large scale.

This characterisation is not far removed from the apparently much more generic one proposed by historians Caliste and Carnino (2022: 232) who propose to define the industrial phenomenon "as the quantitative growth of a technical phenomenon to the point where a qualitative break occurs". What is to be understood by *qualitative* is very broad and may refer to the nature of the processes, the organisation of work, the accessible markets, and so on. Nevertheless, this makes it possible to respond to the objection that might be made to Le Blanc's definition, that innovation has always existed in craft communities. For Caliste and Carnino, industry is characterised by a break in the very mode of innovation, which is becoming systematic in companies' R&D departments. The advantage of this definition is that it places the *technical phenomenon* at the centre of industrial logic.

The fact that industry is driven by a dynamic of innovation, and that this dynamic itself is organised, therefore highlights the *reflexive* dimension that characterises the industrial approach to any technical activity. This approach is *technological* in the sense that this term had in the 18th century, *i.e.* related to the study of arts and crafts. It is essential for industry to constantly take a *critical* look at its products, markets, and processes in order to improve them. There is thus no industrial method or principle that has absolute value. Ure and Marx, for example, already note that the principle of division of labour is undermined by automation, which *integrates* tasks into a single production flow instead.

This requires to understand standardisation (the first criterion proposed by Le Blanc) as something else than the simple homogenisation of products or activities. Variety, or even customisation of products, is by no means incompatible with an industrial approach, as shown by the classic opposition in the 1930s between Ford and General Motors, the latter building its success thanks to the diversification of its range of automobiles. The essential act of standardisation is the *specification*, which makes it possible to determine *what someone is talking about* and to agree, *by anticipation*, on the problem they are trying to solve. Branding a product is to convey through names and images its promises and anticipations so as to facilitate the transaction; it is nothing else than *imaging* the specification.

The division of labour must be understood according to the same principle. What Ure and Marx criticise is the task-by-task organisation of labour, which is only one possible application of an *analytical* study of labour that resolves it into its elementary components. Yet such a study is just as necessary if labour is to be mechanised. What changes fundamentally with the English mercantilists, Adam Smith's precursors on the idea of division of labour, is that the increase in productivity with specialisation is no longer merely an anthropological given – a fact already noted by Plato – but "a lever or operator, [...] an economic instrument at the disposal of entrepreneurs" (Séris 1994a: 15). The essence of the division of labour lies in the reflective effort to optimise it. It is therefore first and foremost an *intellectual* operation carried out by the industrial engineer observing the work process and analysing it into its basic components. Their study considers this process as an *object* whose conditions – tools, gestures, force, precision, speed, etc. – can be explicated and therefore controlled. It is this particular point of view that marks the birth of *technology*, in the sense of the term given by Beckmann, this professor of cameral sciences at the end of the 18th century, when he mobilised the term to designate his method of describing crafts, no longer according to their field of activity, but through the analysis of their elementary functions:

> Planing, the carpenter's technique; polishing, the glass polisher's technique; beating the book, the bookbinder's technique; pressing canvas; calendering linen; all these operations, as different as they appear and as different as they are in reality, have but one aim, namely the smoothing of bodies [...] I would like to draw up a catalogue of all the aims pursued by craftsmen and artists in the course of their various operations, as well as a catalogue of all the means by which they know how to achieve each of these aims. I would call such a catalogue general technology. (Beckmann [1806] 2017: 67)[7]

7 Regarding the beginnings of technology, see the introduction to this work by Carnino and Hilaire-Pérez.

This idea of a functional taxonomy of technical gestures opens the way to the abstract consideration of work processes and thus to their general recomposition, not only as a function of the division of labour, but also as a function of their purpose. Therein lies the revolution brought about by machines according to Marx: they require the work process to be reconfigured according to its own logic:

> The size of the prime movers, of the transmitting mechanism and of the machines properly so called increased, the components of the machines became more complicated and diverse in form, they had to operate with stricter regularity, and they accordingly diverged more and more from the model which originally determined their construction under the handicraft system, and acquired an independent form restricted only by their mechanical task. (Marx [1867] 1976: 504)

It is indeed this lesson from Marx that Simondon will develop through his concept of *individuation*, and which Séris expounds as follows:

> From the point of view [...] of technical rationality [...], anyone who sees an industrial machine in operation [marvels] at the spectacle of the residue-free efficiency of an process entirely adequate to its purpose, better: entirely true to its essence and to its form. The machine gets rid of all useless accessories, strips the process of everything that is not necessary. Adapting from the outset to the required scale, it becomes a giant operating with hypertrophied organs (the crane, the mechanical shovel) or, on the contrary, a Lilliputian assembling and knotting the most tenuous silk threads with infinite delicacy and startling speed.
> The machine is the materialisation of the action to be accomplished, reduced to its pure form, to its simple silhouette or intelligible profile, to its rigour, to the demands of its rigorous execution. It standardises, regularises and rectifies gestures and tools, even more than it reduces and amplifies them.
> Freed from the "anthropomorphic" constraints of human labour, the machine is a collective worker, the model which the factory strives to resemble, when its ideal is to be a factory without workers. The point of view that triumphs with the machine is that of the organiser. [...] The design of the machine, far from copying the worker and their gestures, is able to reveal the pure operation. (Séris 1994b: 159)

This text has been quoted at length because it pinpoints the central characteristic of the concept of *industry*. Industry is linked to machines not because the latter would be a historically dated way of substituting more productive physical factors for human factors of production, but because machines make possible the free expression of the properly reflective movement that characterises industrial activity, manifest in *innovation*, which consists in positing the work process as an object (*standardisation*) and analysing it (*division of work*) in abstract terms endowed with a grammar (*technology*) in order to rearrange these in an optimal way (*automation*).

This characterisation does not of course aim at capturing the whole meaning of industrialisation, which also includes social and political aspects. It only highlights its essentially transformative dimension, which Carnino places at the heart of his definition. Industrial activities indeed refer to any type of human activity reconfigured in the context of a change of scale, whether productive or not, lucrative or not. Commerce and finance, public administration and health, as well as large-scale charity work, are all driven by industrial thinking, which is further illuminated and developed by computing. It is important to clarify what seems to be implied in Carnino's definition, namely that the industrial reconfiguration of activities on the occasion of a change of scale is the result of *acts of reflection* about them operated by certain agents – whether entrepreneurs, administrators or engineers, *i.e.* it proceeds from a *rational* intention of improvement. This is obvious if one thinks of the *crises* that the uncontrolled growth of an activity naturally gives rise to: shortages of labour, materials or know-how, distribution bottlenecks, and so on. Industrialisation is not the mechanical cause of growth, but the rational response to the challenges posed by growth.

All these explanations of the term *industry* can be summarised by the following three oppositions:

1. Industry is not related to the field of *production*, but to *operations* in general (as defined in the introduction to this chapter).
2. Industry is not related to machines in the common sense in which the term is used. Operations involving only human persons can be the subject of an industrial approach. However, it is characteristic of such an approach that people are then asked to behave *like machines*, *i.e.* to obey precise procedures. This point highlights the deep connection between the ideas of *mechanism* and *procedure*, which will be later discussed (see in particular section 8.2).
3. An industrial approach is not just technical, it is technological. Trades have always explicated and codified their ways of doing things. What is new about the industrial approach is that technical study, in the context of a change of scale that makes questions of controlling action crucial, becomes systematic in its search for standardisation *and* innovation, that is, both in the analytical reduction of the way of doing things to known elements *and* in its optimal reconfiguration.

3.3.3 A general industrial logic

In this context, it is easy to understand why computing is an industrial activity *par excellence*. Programming is an inquiry aimed at solving a problem through the configuration of a machine that intervenes in the problematic situation.[8] Its difference with the definition of industrial thinking just given lies only in its greater generality. In its traditional definition, an industrial problem concerns the optimisation of some production process, and leads to design the relationship between the machine and the other resources in the situation (such as the machine operator) essentially as the interweaving of their respective operations. On the contrary, the problems posed to programming, and the solutions it can propose, are in no way limited to the domain of production nor constrained by these restrictive forms. In this respect, programming appears to be a continuation and extension of the industrial approach, insofar as it is above all a posture of analysis and reconstruction of human practice.

From this point of view, it is not historical coincidences or metaphorical transfers that explain the proximity of computing and industry, but rather their common epistemic posture, to analyse thoroughly a dynamic process and to reconfigure it according to pure logic. Babbage had the intuition of this epistemic proximity between computation and industry, as did his famous associate Ada Lovelace when she defined *operation* as "any process which alters the mutual relation of two or more things, be this relation of what kind it may", and when she asserted that Babbage's machine "might act upon other things besides number, were objects found whose mutual fundamental relations could be expressed by those of the abstract science of operations, and which should be also susceptible of adaptations to the action of the operating notation and mechanism of the engine."[9]

It is now possible to gain a global perspective on the ways in which computing makes itself effective in industrial activities. The examples mentioned above (3.2.1) showed the mismatch between intuitive human logic and systematic (yet still human) logic. This is actually no different to the long-noted difference between craft and industry. All machines, whether computerised or not, create situations of ambiguity with human operators, because it is never clear whether they will be able to adapt to a situation that the operator considers "unexpected", or, to put it another way, whether they have the same notion of the unexpected as the operator. Furthermore, the machine can only be adapted to change through a complex engineering analysis, which is usually beyond the reach of operators.

In the mind of the industrialist, these deficiencies are offset by the expected gains of effectiveness allowed by the use of machines, an expectation that needs now to be substantiated. Effectiveness is indeed a concept that can be considered

8 I develop this idea in the companion book *Reasoning with Machines*.

9 Ada Lovelace, *Sketch to the Analytical Engine*, note A to the memoir of L.F. Menabrea (1843). Reproduced in Babbage 1989.

in both absolute and relative terms. Taken absolutely, a solution is effective if it solves a problem. Taken relatively, one solution may be more effective than another, in terms of certain measurable aspects, such as proximity to the objective, success rate or consumption of resources. It is this second sense that is the object of this discussion, where it is therefore measured as *advantage*.

The notion of advantage is tricky to handle, because it is relative (among other things) to a degree of contingency (circumstantial or structural advantage), to a norm (strength, profit, well-being, etc.), to a time horizon (short-term, medium-term), and to a certain focus (the workstation, the workshop, the company, the market). A median focus is here adopted, aiming at the effects of the introduction of computerised machines in a situation with a stable regular structure, and more specifically, the lasting advantages that contribute to the increased effectiveness of a work process relative to competing processes that do not benefit from this technological insertion, such as productivity or standardisation.[10]

There are five benefits brought by automation, most of which were already identified by Charles Babbage ([1832] 2010).

The first section of his *Treatise of Machines and Manufactures*, draws up a rather empirical list of the advantages procured by mechanisation. 1) It begins by mentioning the multiplication of the power of human labour, and the saving of time, in other words the gain in productivity in energy and time, respectively (Babbage [1832] 2010: 7–9). 2) However, he already notes that certain mechanical effects are incommensurable with human labour, such as the use of explosives to clear rocks: technology allows us to perform tasks that simply could not have been envisaged manually. He takes up this theme several times, notably in chapter 7 entitled *Exerting forces too great for human power, and performing operations too delicate for the human hand* (Babbage [1832] 2010: 35). In this chapter he explains that many power requirements often cannot be satisfied by an army of workers adding their forces. Their coordination becomes problematic at the time of the effort, so does the management of space if the force has to be concentrated in a specific point, and finally there are the ancillary difficulties of managing a large number of workers (recruiting them, feeding them, paying them). The title of the chapter also mentions the opposite case, that of operations requiring precision or consistency of gesture beyond the reach of a craftsman.

3) This theme is taken up again in chapter 9, which adds to it the similarity of products, what would be called today *stability* or *low variance* of the industrial

10 This median focus means that there is no intention here to explore the large-scale structural, economic, social and political changes brought about by industrial digital technology. From a social and political perspective, digitalisation changes the balance of power between workers and capitalists; see for example the Marxian analyses of Casilli ([2019] 2021).

process. But Babbage goes even further, visibly fascinated by a general property that machines have, of being able to guarantee the *identity of copies* of a series of manufactured objects, and devotes two of the longest chapters of his work to this theme:

> Nothing is more remarkable, and yet less unexpected, than the perfect identity of things manufactured by the same tool. If the top of a circular box is to be made to fit over the lower part, it may be done in the lathe by gradually advancing the tool of the sliding-rest; the proper degree of tightness between the box and its lid being found by trial. After this adjustment, if a thousand boxes are made, no additional care is required; the tool is always carried up to the stop, and each box will be equally adapted to every lid. (Babbage [1832] 2010: 48)
> [...] The two last-mentioned sources of excellence [precision and similarity] in the work produced by machinery depend on a principle which pervades a very large portion of all manufactures, and is one upon which the cheapness of the articles produced seems greatly to depend. The principle alluded to is that of copying, taken in its most extensive sense. Almost unlimited pains are, in some instances, bestowed on the original, from which a series of copies is to be produced; and the larger the number of these copies, the more care and pains can the manufacturer afford to lavish upon the original. (Babbage [1832] 2010: 51)

4) The fourth benefit of mechanisation is the constraint exerted by the machine on a third party action (human or otherwise). Babbage refers to this type of function in only one example, that of alarms that awaken the operator's attention, but it is not clear that he had in mind the regulation of work in an industrial environment (Babbage [1832] 2010: 43). Programmed clocks regulating factory, railway, etc. date from the end of the 19th century (De Mol and Bullynck 2021: 36). Computing has made it possible to develop this prescriptive function enormously for security benefits (access rights to resources) or compliance with rules, such as access badges or ATM machines. It should be noted that the rules to which users are to be constrained are not necessarily explicit or even visible to them. A computer system may present different options to different users without them being aware of this limitation nor its reasons.

This function is central to the argument developed in this chapter. Machines can force the person who depends on them (in one way or another) to act *like a machine*. In a factory, for example, an operator's cognitive ecology is entirely shaped by the machines and their own logic, so that it is the rhythm and needs of the machine that dictate what they must do. The usual constraints found in a cognitive ecology (natural objects, instruments and tools, signs and statements, the human environment) must leave, directly or indirectly, an important place to those dictated by the mechanical process. This is where the general idea of a person's *mechanical* behaviour can emerge, an adjective that will be used extensively by logicians to describe the notion of effective procedure (see below, 8.1.1).

5) The fifth benefit of mechanisation is the traceability of the evolution of the machine, of its interactions with the situation, and above all of the situation itself through these biases. Babbage was prescient about the importance of this function, since it occupies the whole of Chapter 8 of his *Treaty*, entitled *Registering operations*. It is easy to use a mechanical device to record the cycles of the machine, which makes it possible to measure production; or again, it is possible to record certain human operations for control purposes (Babbage gives the example of spirit stop-cocks), or finally to measure physical quantities, such as the distribution of gas, the quantity of liquid remaining in a cask, the distance travelled, and so on. One can see that recording operations can mean very different things: the operation of the machine itself, that of the operator, or the evolution of the objective situation (Babbage [1832] 2010: 39).

In summary, automation 1) provides productivity gains, 2) gives access to *new* operations, which would not be possible without it, 3) guarantees the exactness of the process, 4) allows to control agents or more generally people in the machine's environment, and 5) enables traceability of events. Automation's effectiveness goes therefore much beyond productivity, to which it is often reduced; an automation project may well be sought after, even if it increases the quantity of labour required; it is often a consequence of the general systematic logic of industrial work.

Four of these five benefits of automation (excepted the second one) can be well illustrated through the single example of a mechanical computing machine, the cash register, which spread in American retail shops from the 1870s-1880s onwards. Cash registers were invented in the United States in the 1870s by James Ritty. The company he founded, renamed *National Cash Register* (today NCR, still in business) was bought by John Patterson in 1884 who turned it into one of the most innovative, aggressive and emblematic companies of late 19th century American capitalism.[11] For the simplest models, cash registers consisted of a keyboard for entering transaction amounts, a cog for adding them up (or even printing them on an internal paper roll), and a cash drawer that opens only after a transaction has been recorded. Ritty called his machine *the Incorruptible Cashier*, a name that conveys quite clearly its initial function: to prevent theft at the cashier's desk by the employees manning it, since the shopkeeper could each evening compare the sums in the cash drawer with the recorded transactions.

However, these tools subsequently developed far beyond this initial function, as the historian Crandall summarises very well:

> The first cash registers were strongly promoted as a deterrent against theft. But they were also the instruments of a new analytical approach to business that led to the systematic use of in-

11 See Crandall 1988 for a history of this technology, as well as Friedman 1998 for the NCR company.

> formation to generate profits. [...]
> In 1902, you could obtain an audit trail of transactions, a count of customers, statistics on detailed amounts and cumulative totals. You could get all this globally, but also for each individual seller. You could even identify transactions for individual products. This made it possible to design special promotions, individualised employee incentive programs and collect data for quantitative market research. And, as if that wasn't enough, the cash register extended beyond its eponymous function (cash management) to credit management. [...]
> The cash register was a combination of a security system, an accounting machine and a management information system. All this was achieved using moving mechanical parts, gears, cams and levers housed in large casings. [...] Thus, the cash register was the mechanical force that first made it possible to exploit transactional information. (Crandall 1988: 3)

For these reasons, cash registers play an important role in the proto-history of computing, along with calculating machines and tabulating machines (Campbell-Kelly et al. [1996] 2014: 21). Crandall recognises three essential functions to cash registers:
1. Counting and calculating, which allows the cashier to make exact sums and to gain in productivity ("the accounting machine").
2. Constraining the cashier's action, for example by preventing them from opening the till drawer outside the context of a transaction ("the security system").
3. Keeping track of transactions carried out in greater or lesser detail, for various uses, control or decision-making ("the information system").

The first function illustrates very well the first and third identified advantages of effectiveness (productivity and exactness), and the two other correspond exactly to the fourth and fifth advantages. While the mechanical cash registers did not make possible any new operations for the shopkeeper, digitalization and connectivity have since allowed, for example, credit card payment.

3.3.4 The problem of systematicity

The conclusions of the studies in this chapter are clear: The way in which industrial and computational thinking solves problems differs from ordinary technical thinking, just as the latter differs from the immediate practical knowledge of the expert. If technical rules are devised by second-order thinking, which has immediate practical thinking as its object, then industry and computing are in effect third-order thinking; and *systematicity* is the name given here to this new modality of problem-solving, albeit intuitive and problematic at this stage.

In the previous chapter it was seen that, even if all the practical knowledge of a discipline could be recorded in a procedure book, its use would still be in need of rules. Its systematic expression in an algorithm would lead to the dissolution of this knowledge in a language alien to human practice, as the example of strategy games, characterised by "tangled" situations, has shown (see above 2.3.1).

This chapter has shown besides that such procedure books would in no way express the essence of technical knowledge, as it is usually understood in the workplace. This knowledge is materialised much more by *technical environments* than by rule books. These environments organise the cognitive ecology of agents and enable them to anchor their practical knowledge in patiently constructed habits. The few explicit rules and procedures that help human beings to accomplish their tasks are not sketches of computer procedures that simply need to be made more precise and rigorous, because they do not generally serve to "mechanise" human behaviour, but rather to coordinate it. This is the reason why confronting people with computer systems is often problematic. It highlights fundamentally different ways of assimilating change and adapting behaviour. Systematic, procedural logic can be obtuse, unable to take advantage of the ambiguity of situations, and above all cumbersome to reconfigure if the situation evolves in an unforeseen way.

On the other hand, this logic, apparently alien to humans' practical knowledge, has proven to be formidably effective in solving the problems it deals with. Daston has suggested that this logic was the same as that which presided over the industrialisation of human activities, long before the appearance of computers, that of a reflective attitude aimed at dividing work into standardised elementary tasks in order to recompose it in a systematic way, thus bearing within it the germ of that *science of operations* glimpsed by Charles Babbage and Ada Lovelace. Programming appears in this perspective as the continuation and generalisation of this analytical posture, capable of taking as its object not only productive work, but any type of problematic situation that can be resolved by regulated operations. Five forms of advantages in effectiveness have been thus identified.

It is therefore *another knowledge* that is made possible by the industrial and computational attitudes, foreign to humans' immediate practical knowledge, and yet constructed, at least initially, on the basis of its analysis and reconfiguration. Herbert Simon was therefore right when he foresaw *design* as an essential faculty of the modern practical man; but he went too far when he made it the source of *all* practical knowledge, encouraged in this by his own hypothesis of the computational nature of the mind. Establishing systems of rules to rigorously and exhaustively direct action is a highly *artificial* approach, if we may play on a different connotation of the term. It is only implemented in certain professional settings, particularly those where a significant change of scale is being experimented with.

What, then, is the nature of this *other knowledge*? The notion of *systematicity*, in the philosophy of knowledge, is usually associated with the theoretical organisation of knowledge and the sciences. Would this mean that there is a *theoretical* or *scientific* modality of problem-solving, to which industrial and computational thinking would belong, as opposed to the *practical* and *technical* modalities studied at the end of the previous chapter? Such a hypothesis is attractive: Does not the industrial approach find its culmination in what was called *scientific management* at the beginning of the 20th century? And is not programming the direct application of the mathematical science of computing?

4 *Technè* and method (Aristotle)

I noted in the introduction that problem-solving is not circumscribed to a definite domain of knowledge. From a certain point of view, it seems natural to relate it to practical knowledge, which deliberates in order to transform problematic situations into satisfactory ones. This is how Herbert Simon understands it when he makes it the essence of the *professions* (law, engineering, medicine, etc.).

It is also common to say that mathematics and science solve problems. However, problems in that case are *problems to know*, for example, if a proposition is true, or what is the solution to an equation, what causes the tides, and where will be the moons of Saturn one year from now. Procedures (like proofs, or computations) do not seem to transform anything, except our own knowledge.

Therefore, given the recent and rather loose sense of the term *problem-solving*, noted in the introduction, it is not obvious that these two classes of meanings have any deep connection with each other, beyond the simple analogy that both require reflective thinking.

Constituting *problem-solving* as a *unique* epistemological object across such diverse uses, and in particular across the distinction between theoretical and practical knowledge, is therefore not an obvious move. However, it is indeed towards this possibility of unification that computing seems to point, through its effectiveness in solving a large quantity of problems, whatever their level of complexity (low or high) and their domain (scientific, technical or other).

This question is made even more complex by another learning from the previous chapters. Programming, like industrial methods, seems to come under a very specific *modality* of problem-solving, which I have called *systematic*, as opposed to its *practical* and *technical* modalities. Yet the idea of *systematicity* seems rather characteristic, when it comes to knowledge, of a theoretical or scientific stance.

In a way then, the question now is about the very distinction between theoretical knowledge and practical knowledge, and about the role of systematicity in this distinction.

Theoretical and practical knowledge are usually distinguished in two ways. On the one hand, there is (theoretical) knowledge of *what is real* and (practical) knowledge of *what to do*. This distinction is similar to the one designated by Herbert Simon under the terms *is / ought* (see above 2.2.1). It is not obvious at all for a very large part of the philosophical tradition that problem-solving activities could be essentially the same across such a fundamental divide.

On the other hand, there is a practical *way* of knowing, and another one which is theoretical. The first one involves free reflection concerning the problem to be solved; in classical philosophy it is called, following Aristotle, *deliberation*; it may

https://doi.org/10.1515/9783111494500-004

refer to rules and cognitive artefacts prepared in advance to guide it, and in this case it is said to be technical. The second modality involves the mobilisation of a theory or doctrine to shed light on the problem to be solved, and this is why it is often presented as an *application* or as a *deduction*; it is often referred to as *systematic*.

These two meanings of the practical / theoretical distinction were introduced into philosophy by Aristotle, who closely bound them together. For the Greek philosopher, the *practical* modality of knowledge, deliberation, is relevant to deciding *what should be done*, whereas its systematic modality is the only way to determine scientifically *what is the case*. To seek to know *systematically what ought to be done*, even within a technical framework, would seem to Aristotle incongruous, if not impossible.

The three next chapters attempt to untie these knots. To understand the nature and apparent universality of systematic problem-solving, it is necessary to show how it can be applied to practical and technical problems, contrary to Aristotle's belief, and moreover, that the very notion of *problem* can be constituted as a category transversal to all fields of knowledge – about which Aristotle was also deeply sceptical.

This chapter exposes Aristotle's ideas and tries to make sense of them. To do so is not to indulge in the history of ideas for its own sake. In many ways, Aristotle's intuitions are still with us: engaging in systematic problem-solving in practical and technical matters seems like a dead end and a waste of time. The difficulties are not technical, but conceptual: how can anyone maintain the absurd belief that we could predict all possible cases in a concrete situation, let alone ensure that all its relevant terms have been properly identified and defined? In the history of crafts, attempts in this direction stumbled time and again with inadequate conceptual tools, such as lists of terms, decision trees or Lullian wheels. It was very late before systematic problem-solving was transferred from the abstract field of geometry, where it had its origin, to other disciplines. It is therefore important not to take this possibility for granted and to understand its difficulty from the point of view of the Aristotelian tradition – which in many ways is also that of today's common sense. This approach opens up the possibility of understanding this epistemological breakthrough and better appreciating its significance. It may also open up new perspectives on the nature of mathematics and modern science.

Difficulties begin as soon as one engages with the Aristotelian legacy. On closer examination, the explanations given by the Stagirite regarding theoretical and practical knowledge are no longer obvious to us. What does it mean, for example, that science (ἐπιστήμη) deals with "beings whose principles are necessary", while the productive and practical disciplines deal with those "which can be otherwise"?[1]

Aristotle also maintains, within this second domain, an equally clear separation between the productive disciplines, which produce artefacts or material results: carpentry, medicine, navigation, etc., and the practical disciplines, which deal with human actions, such as ethics and politics.

This rigid separation of theory and practice into separate fields of study no longer corresponds to our intuitive conception of rational knowledge. Firstly, it seems natural to assert that there are theoretical foundations for all technical knowledge, and conversely, to seek practical applications for all theoretical discoveries. Secondly, the emergence of psychology, sociology, economics and politics as scientific disciplines in their own right has shown the possibility of a theoretical study of human affairs, distinct from any practical precepts.

This challenge to the Aristotelian model extends in principle the field of theoretical knowledge to the whole of reality, and thus postulates a possible theoretical foundation for every technique and every practice. This thesis is the work of modern philosophy. Kant expresses it clearly when he proposes to reconsider the distinction between the two types of knowledge on entirely new bases. Indeed, if the rules of technique and the advices of prudence are by right only "corollaries" of science, practical knowledge, in order to have an autonomous foundation, must include only the teachings of the moral law:

> If one divides philosophy, insofar as it contains principles of rational cognition of things by means of concepts […] into theoretical and practical, as is customary, then one proceeds entirely correctly. But then the concepts that refer the principles of this rational cognition to its object must also be specifically distinct […] There are, however, only two sorts of concepts that allow an equal number of distinct principles of the possibility of their objects: namely the *concepts of nature* and the *concept of freedom* […]. Hitherto, however, a great misuse of these expressions for the division of the different principles, and with them also of philosophy, has pre- vailed: for that which is practical in accordance with concepts of nature has been taken to be the same as that which is practical in accordance with the concept of freedom […]. [But] all technically practical rules (i.e., those of art and skill in general, as well as those of prudence, as a skill in influencing human beings and their will), so far as their principles rest on concepts, must be counted only as corollaries of theoretical philosophy. […]
> Thus, as little as the solution of the problems of pure geometry belongs to a special part of it, or as little as the art of surveying deserves the name of a practical geometry, as a second part of geometry in general in contrast to pure geometry, even less should the mechanical or chemical art of experiments or observations be counted as a practical part of the doctrine of nature, or, finally, should domestic, agrarian and political economy, the art of social intercourse, the

1 *Nic. Eth.* VI.2, 1039a25. Consulted editions of Aristotle's texts can be found in the bibliography. Greek texts are taken from the *Thesaurus Linguae Grecae*.

> prescriptions of dietetics, the general doctrine of happiness itself or even the mastery of inclinations and the control of affects for the sake of the latter be counted as practical philosophy [...]; since all of these contain only rules of skill, which are thus only technically practical, for producing an effect that is possible in accordance with natural concepts of causes and effects which, since they belong to theoretical philosophy, are subject to these precepts as mere corollaries of it (of natural science), and thus cannot demand a place in a special philosophy which is called practical. (*Critique of the Power of Judgment*, Introduction, I, AA 05: 171, transl. Guyer and Matthews 2000)

This famous text has been quoted in full length because it operates a radical displacement of the domains of practical knowledge. Whereas Aristotle opposes both technical and practical disciplines to the theoretical sciences, Kant combines theoretical and technical knowledge into a single vast domain related to the conceptual knowledge of nature. What is more, he severs practical knowledge from prudence, which for Aristotle blends the ability to aim for the good and the skill to achieve it. And, to leave no ambiguity, both the "doctrine of happiness" itself, which for Aristotle is inseparable from the human aim of the good, and "the art of restraining the passions", so essential to virtue according to the Stagirite, are also part of this massive relocation of human knowledge into the "theoretical" domain, thus ridding the practical domain of any consideration other than that of the moral law.

The Aristotelian conception of *technè*[2] is still a very active topic in Aristotelian commentary. Not that the texts are obscure or contradictory; on the contrary, Aristotle often refers to the ideas of theoretical, technical and practical knowledge, in an overall very coherent way, and this across numerous texts (*Metaphysics, Ethics, Analytics, Of the Soul, Parts of Animals*, etc.) that refer to each other, explicitly or implicitly.

Rather, the difficulty encountered in reading the texts seems to be the very one raised above, which is to understand the nature and extent of the distinction Aristotle draws between theory and practice. The two major texts that set out the subject, *Met.* A.1 and *Nic. Eth.* VI, while not contradictory, have different perspectives and inflections. As James Allen says, "Aristotle finds it difficult to put forward what differences between theoretical and practical reason he considers fundamental" (Allen 2015: 68).[3] On reading the texts, indeed, there is no obvious major difference between science and *technè*, either in terms of their objects or their modes of reasoning. Thus, it is the Stagirite himself who seems to pave the way for a "reductive"

---

2 I do not translate τέχνη by *craft*, as is usual; the aim of this discussion is to understand the original meaning of this concept. *Technè* does not denote the domain of productive practices (crafts), but the intellectual virtue that makes them possible. I therefore write the Greek term *technè* as such.
3 This point is also noted by Johansen 2017: 121.

interpretation of *technè* as theory and thus, to the radical critique formulated by Kant.

The first two sections of this chapter propose a novel interpretation of the place that Aristotle assigns to *technè* in the realm of knowledge, and in particular of its autonomy from the theoretical sciences. This interpretation is based on Aristotle's analysis of practical deliberation (βούλευσις) as a form of *inquiry* (ζήτησις), which is opposed to science (ἐπιστήμη) as knowledge constituted and completed in its deductive presentation. *Technè* then appears as the positive knowledge that makes it possible to *guide* deliberation, and which therefore has to do with *methods*, taken as *generalisations of deliberations* (section 4.1), and their mode of truth is that of hypothetical necessity (section 4.2).

The next two sections explore the question of systematic thinking, which Aristotle also bequeathed to us. Aristotle points in the *Organon* to the possibility of exhaustive methods for solving problems, in particular in Book I of the *First Analytics* where he sets out the method of the *pons asinorum* (as tradition has called it). This idea seems to be closely related to the idea of a scientific system as outlined in Book II of the *Second Analytics* (section 4.3). Section 4.4 explores the reasons why Aristotle reserved systematic thinking to the scientific realm, to the exclusion of technical knowledge.

4.1 How *technè* achieves universality

This section is intended to interpret the intermediary position that Aristotle assigns to *technè* in the scale of knowledge set out in *Met.* A.1, between experience and science. It begins with the review of classical and recent interpretations, that tend to recognise in it an essentially Kantian position – namely that, if art imitates nature, then technical knowledge is merely a degraded form of scientific knowledge, which would merely "take it in reverse", *i.e.* start from the effect and work back to the cause (4.1.1). Another recent interpretation, which tends to equate it instead with experience, is then discussed (4.1.2). Against these two reductions, it is argued that *technè*, like prudence, is primarily concerned with decision-making *hic et nunc* and that there can therefore be perfectly rigorous practical reasoning that only links particular terms (4.1.3). Τέχνη, from this perspective, should be seen as the successful generalisation of certain deliberations (4.1.4).

4.1.1 The possibility of a reductionist interpretation

The main text that allows the reduction of *technè* to a subaltern form of science is the introduction to the *Metaphysics*, which describes it as an intermediate degree of knowledge, rising above experience and prefiguring science. It is even *already* science, if considered relatively to experience: "In general, a sign of knowledge or non-knowledge is the ability to teach, and this is why *technè*, more than experience, is considered science" (*Met.* A.1, 981b8).

This qualification is repeated in many other texts, for example: "...techniques (τέχναι), and productive sciences..." (*Met.* Θ.2, 1046b3) and "...no science, whether practical, productive or theoretical..." (*Met.* E.2, 1026b5).

Aristotle's use of the term ἐπιστήμη is "breathing", sometimes restricted, sometimes very broad, as in this last quotation.[4] This does not signal imprecision, on the contrary: as knowledge is a matter of ascending progress towards principles, it is given by degrees, and Aristotle uses the term ἐπιστήμη when he wants to represent a positive gradient, *i.e.* knowledge becoming firmer.

In the *Metaphysics*, it is more specifically the *theoretical* sciences that he describes as superior to the productive sciences (982a1). This is because they are concerned with going back to the universal and the principles of things (*Met.* A.2, 982a25-b3). Among the theoretical sciences, Aristotle cites mathematics (982a24), astronomy (982b15) – which is a part of mathematics – and finally theology (983a5). Strictly speaking, these are the only sciences whose objects are necessary and eternal beings, and which can therefore produce an absolutely certain discourse. But in his classification of the theoretical sciences of *Met.* E.1, Aristotle justifies at length the place of physics among theoretical sciences alongside mathematics and theology. Natural beings indeed, unlike artificial beings, carry within themselves the principle of their own movement, and the study of the causes of the natural development of things is therefore a theoretical undertaking.

It is the study of the same mechanisms and processes that also occupies technicians who wish to replicate or modify them to their advantage. Physicists and technicians therefore have, in a way, the same field of inquiry: "Surely as something is done, so it happens in nature; and as it happens in nature, so each thing is done, if nothing interferes." says *Phys.* II.8, 199a9. But their point of view is different: whereas technicians study causes in relation to the problem to be solved, scientists study causes from a universal point of view, and thus seek to trace them back to the first causes.[5] A favourite example of Aristotle's is medicine, whose necessary collaboration with science he clearly perceived:

4 On this subject, see Crubellier and Pellegrin 2002: 37–40; Moss 2014: 203 and 207, n. 33.

5 *Met.* α.1, 994b20–30; see also *APo.* I.13 and I.24.

> But it is also the office of the physicist, as regards health and disease, to see what their first principles are, for neither health nor disease can exist in beings deprived of life. Hence it may be said of most physicists, and, among physicians, of those who proceed in their *technè* in a more philosophical manner, that the former lead in their studies to truths of a medical nature, and that the latter, in questions of medicine, take as their starting point the principles coming from the science of nature. (*On sense and the sensible*, 436a17–436b1)

Therefore Le Blond ([1936] 1970: 328) may comment:

> There are not indeed two ways of producing, one according to nature and the other according to art, but both activities involve the same elements, and go through the same stages [...] This parallelism is explained, moreover, by the fact that art "imitates nature" (*Phys.* II.2, 194a21), and continues what it started.[6]

Practice, when it has access to scientific knowledge, can therefore reason with the same rigour, proceeding by "syllogisms" that go backwards through the chain of causes established by physics (the term *syllogism* is used in connection with technical-practical reasoning in *Nic. Eth.* VI.13, 1144a31 and in other texts such as *De anima* III.11, 434a10, or the *De Motu Animalium*. *Nic. Eth.* VI.13 justifies the specific expression of *practical syllogism*). The general idea is well expressed by the following text:

> The healthy man results [in fact] from the following reasoning: since health is such-and-such a thing, it is necessary, in order for health to be, for such-and-such a thing to happen, for example homogeneity, and for this, warmth. And the doctor continues to reason in this way until he arrives at what he can produce by himself. Then the movement from there to health is called production. (*Met.* Z.7 1032b7)

It is now clear how all these elements led some Aristotelian scholars to play down the difference between science and *technè*. If *technè* is already knowledge of causes, if it deals with natural things just as much as physics, if it also reasons by syllogisms, how then can it differ from theoretical science, if not by a simple difference in degree, or by its orientation towards production? Rodier (1926: 194), paraphrasing this passage, may therefore write:

> Knowing that the definition of health consists in such and such a proportion or harmony of the bodily elements; that this proportion presupposes the presence in the organism of such and such a quantity of heat and cold [etc.] is the fact of science.
> *It is easy to see that this science can serve as a foundation for practice. To achieve a goal we will only have to follow, in reverse order, the series of conditions determined by discursive reason.*

---

6 See on this subject Le Blond's study of the *schemes of industry* (1970: 326–346), which shows how they inform the Aristotelian conception of the four causes.

> [...] This ability to discover the means whose necessary relationships will serve us to produce a contingent end, is precisely prudence. Or rather, prudence is this aptitude applied to the realization of virtue. (my italics)

Rodier is not entirely accurate in his analysis, since medicine is for Aristotle a *technè*, not a form of prudence, hence his restatement ("... or rather...") at the end of the quotation. But *a fortiori*, his assertion that practical reasoning merely reverses the order of scientific proof resonates singularly with the Kantian thesis of art and skill as *corollaries* of theoretical knowledge.

In a recent interpretation, Jozef Müller, who aims against Rodier to reaffirm the autonomy of prudence from scientific knowledge, nevertheless takes up his argument, this time circumscribing it to *technè* alone. The key point for Müller is that only *technè*, which seeks to establish the universality of its methods, achieves such hypothetical necessity in its reasoning:

> In its idealised form, productive thought precedes by hypothetical necessity: if A is to be made, then, given what A is, B must be made, without any reference whatsoever to [the producer's] desires, preferences, beliefs or values. Unsurprisingly, Aristotle repeatedly associates reasoning by hypothetical necessity with production, [...and so] it resembles *theoria* more than *praxis*. (Müller 2018: 159)

Müller's interest in this interpretation is that it isolates the proper element of prudence (φρόνησις), that is, its capacity to determine the right end on which to fix desire.[7] For unlike productive thought, to which the object is *given*, and therefore indifferent – Aristotle often repeats that the physician is capable of healing as well as poisoning – practical thought is concerned with determining the object of desire. It is therefore particular to each person and to the given situation. For example, a person might think like this: *I want to make my life more comfortable, what should I do? – I'm going to extend my house*, and the architect could help them confirm this resolution, by providing a blueprint for this extension. However, this productive thought is logically consecutive and subordinate to the one that sets the desire and the end:

> [Practical] is reasoning about what to do vis-à-vis one's need or desire. Once it is concluded, it is followed by an action (done from one's decision) which includes productive reasoning [...] If decision is the principle of action, then productive thought can and in fact does occur as part of action. (Müller 2018: 159)

7 Against *Nic. Eth.* III.3's possible objection ("We deliberate not on ends, but on means"), see, in addition to Müller's own argument, Nussbaum [1986] 2001: 297.

Müller thus makes productive thought a species of skill (δεινότης), a term introduced by Aristotle towards the end of Book VI of the *Nicomachean Ethics*. He significantly translates this term as *expertise*, more general than *cleverness*, usually chosen by English translations. It is this "expertise" which, for Müller, is a modality of theoretical thought, and which must be contrasted with practical deliberation. Müller, it is true, does not assert this thesis explicitly, but he does suggest it. In addition to the passage quoted, he uses the term "theory" a second time in the context of the opposition of productive thought to practical thought (Müller 2018: 162). Thus he argues that Aristotle would associate hypothetical necessity only with productive reasoning (Müller 2018: 152, n. 9), and by no means to practical deliberation, which in his view establishes no necessary link between the desired end and the objective on which it fixes it (Müller 2018: 159).

This interpretation is not acceptable, since it introduces a split within practical deliberation that does not exist in the Aristotelian text. If one may admit, following the interpretation initially proposed by Wiggins (1975) of *Nic. Eth.* III.3, that deliberation does more than merely determine means, and that it also serves to fix desire on a determined object,[8] however Aristotle does not reject the consideration of means outside the deliberation of the prudent person, as Moss (2011) insists. Such a person is *also* skilful (*Nic. Eth.* VI.12, 1144a29), and thus a master of arranging means to their advantage. Moreover, *Nic. Eth.* III.3 compares deliberation to geometrical reasoning:

> When deliberating one seems to investigate and analyse the subject in question as it were a diagram. While it is clear that not all inquiry is deliberation – such as mathematics – yet all deliberation is inquiry.
> ὁ γὰρ βουλευόμενος ἔοικε ζητεῖν καὶ ἀναλύειν τὸν εἰρημένον τρόπον ὥσπερ διάγραμμα (φαίνεται δ' ἡ μὲν ζήτησις οὐ πᾶσα εἶναι βούλευσις, οἷον αἱ μαθηματικαί, ἡ δὲ βούλευσις πᾶσα ζήτησις). (*Nic. Eth.* III.3, 1112b20)

This analogy between deliberation and reasoning in mathematics suggests that hypothetical necessity holds in general for any deliberation; indeed, it is noted repeatedly by Aristotle as a remarkable fact, and notably in *Phys.* II.9, 200a15 in connection with productive reasonings, which shows that there is no need in this respect to distinguish them from deliberations in general.[9]

Müller's interpretation also stumbles over the other major text in the Aristotelian corpus concerning the division between theory and practice, namely the

8 On this passage and Wiggin's interpretation, see below 4.2.1.

9 See also *De Motu Animalium* 7, 701a6; *Eth. Eud.* II.11, 1127b28. In *Met.* Z.7 the example of practical reasoning is very detailed but does not explicitly mention mathematical reasoning.

development of *Nic. Eth.* VI concerning the faculties and virtues of the rational soul. In this text, Aristotle repeatedly insists on the *joint* opposition of practical and productive thought to theoretical thought. They both come under the same kind of thinking, deliberation (βούλευσις), which is the work of the "calculating" part (λογιστικόν) of the soul:

> Let it be assumed, then, that there are two reasonable parts of the soul, one by which we consider (θεωροῦμεν) those things whose principles cannot be otherwise, and the other those which can. Indeed when things differ generically, it is also generically that each part of the soul adapted to each differ, since it is on the basis of similarity and affinity that knowledge comes to them.
> Let us therefore call the one "epistemic" and the other "calculative". For to deliberate is the same thing as to calculate (λεγέσθω δὲ τούτων τὸ μὲν ἐπιστημονικὸν τὸ δὲ λογιστικόν; τὸ γὰρ βουλεύεσθαι καὶ λογίζεσθαι ταὐτόν). (*Nic. Eth.* VI.2, 1139a5–12)

In this text, Aristotle insists that the radical difference between theory and practice is based on the principles of being themselves, which is confirmed by the clarification that this is a "generic difference" (τὰ τῷ γένει ἕτερα), the highest difference in Aristotelian logic.

4.1.2 The interpretation from Lorenz and Morison

Such radicality has led Lorenz and Morison (2019) to put forward that Aristotle would here present two independent epistemologies, one rationalist, basing knowledge on deductive demonstration from first principles, and the other inspired by the medical empiricism of his time, basing it on observation and predictive inference. It is not difficult to see that such an interpretation is also, in its own way, reductionist: for it no longer bases the distinction between theoretical and practical knowledge on the difference there would be in knowing things as such and things as they can be "effected", but rather, within the very domain of theoretical knowledge, on the difference between two antithetical methods of establishing knowledge. Beyond the paradoxes that this interpretation gives rise to, for example that the same person could know the same fact twice,[10] this reading does violence to the many Aristotelian texts, some of which have already been cited, that promote a unified vision of knowledge: the *Analytics* describe science as entirely based on experience, and arriving at first principles only after a long journey, from the "most knowable to us" to the "most knowable in itself".

10 See Section 4 of their paper, which attempts to answer objections relating mainly to this fact.

Lorenz and Morison partly base their argument on the fact that Aristotle also uses the term "opiniative reason" (λογός δοξαστικόν) to refer to calculative reason.[11] They also identify practical knowledge as a whole with empirical knowledge in general. It is then no longer clear what is the difference between *technè* and experience; the example they give of technical knowledge indeed bears a strong resemblance to what Aristotle calls experience in *Met.* A.1. As an example of a technically accessible cause, they give the following inductive reason from Galen: "because it has been observed that, if something is released under such conditions, then a spasm occurs" (Lorenz and Morison 2019: 45). However, such a statement does not capture the cause according to Aristotle, but describes a general, ungrounded, experience.

More generally Aristotle defines *technè* precisely as rising above experience through the perception of causes. Aristotle uses the term δοξαστικόν to designate the practical rational part of the soul in contexts that have prudence and skill as their object; these indeed operate at the level of experience. But λογιστικόν is the generic term he uses to designate the practical faculty in general, and this includes *technè*.

The key term is therefore λογιστικόν. λογισμός means *count* or *calculation*. *Calculation* may be understood in two senses, which still resonate with the term today: the first, associated with the idea of number, gives mathematical computation, but the second, without the idea of number, means *reckoning, estimate of probability, forecast* and gives rise to expressions like *taking into account...., doing something with the thought that if..., as much as was possible according to the calculations of human reason...*, and so on. It is obviously in this second sense that Aristotle states that deliberating and calculating are synonymous, *i.e.* the calculation that takes place in everyday affairs, that assesses the value of things, the chance of an opportunity, and *that establishes plans of action*. Far from mechanically applying the rules of a mathematical calculation, it is rather the one which flexibly manoeuvres between particular facts to establish a result that is each time singular. This form of reflection is thus strongly opposed to the epistemic attitude which, as will be shown later, is concerned only with deductive demonstration. The "practical syllogism" therefore seems to conceal a more radical difference from the deductive one than a simple inversion of reasoning.

There is thus a tension between, on the one hand, the "vertical" gradation of *Met.* A.1, which presents *technè* as a first degree of science, and insists on their continuity, and, on the other hand, the *generic* distinction of the rational soul made in *Nic. Eth.* VI, which might be described as "horizontal", and which insists on the

11 For example, *Nic. Eth.* VI.5, 1140b26.

discontinuity between τέχνη and ἐπιστήμη. It is this tension that now needs to be further understood, taking into account the perspective and intention of each text.

4.1.3 The *à propos* of practical syllogisms

How then to understand that, according to *Met.* A.1 *technè is already science* (see above 4.1.1), while proceeding from a generically distinct rationality? Consistency between the two points of view could be ensured if can be shown that

1. the movement that makes it possible to rise from experience to *technè*, and that which makes it possible to rise to science (whether from *technè*, or from experience directly) are generically distinct;
2. the first movement can nevertheless inspire the second and be useful to it.

However, proposition 1 is not evident, since Aristotle seems to say that *technè* reaches the universal, like science, by grasping the cause: "[Technicians] see the cause (τὴν αἰτίαν ἴσασιν), and [experienced people] do not." (*Met.* A.1, 985a28) To resolve this difficulty, one must here return to the question of the practical syllogism, since to grasp the cause is to establish a syllogism between the terms that describe the phenomenon (*APo.* II.4, 91a15). Perhaps by understanding what distinguishes the practical syllogism from the scientific syllogism, one may better distinguish how the cause grasped by *technè* differs from that of science.

Now it must be immediately noted that, unlike the scientific syllogism, the practical syllogism is not necessarily a pathway to the universal. Even leaving aside syllogisms in the form of rule application (*rule-case syllogisms* in the Anglo-Saxon commentary),[12] the syllogisms which arise from practical deliberation, and in particular those of the skilful or prudent person, deal with *ad hoc* situations, which cannot easily be generalised. A typical deliberation indeed proceeds according to the method of analysis, described by Aristotle:

> [They] do not deliberate about ends, but about what contributes to the end [ἀλλὰ περὶ τῶν πρὸς τὰ τέλη] [....] They posit the end and look at how and by what means [πῶς καὶ διὰ τίνων] it may be attained. If several means seem able to attain it, they examine which is the easiest and most beautiful. But if there is only one means to achieve this end, [they investigate] how so, and the means to reach this means, until they arrive at the first cause, which is the last in the order of discovery. (*Nic. Eth.* III.3, 1112b12)

12 For example in the *De Motu, 7, 701a 13*: Every man must go for a walk, now I am a man, therefore I will go for a walk.

In this means-end heuristic, reasoning may remain confined to the concatenation of particular terms (an example is developed below). The cause referred to in the text is not necessarily a universal connection between general terms, but a means (διὰ τί) in what would be today called a *plan of action* rather than a syllogism.

The use by Aristotle of *syllogism* to mean plans of action designed by rational deliberation stresses the necessary concatenation of the consequences of actions; as shall be suggested later, an even better term than "plan of action" could be *procedure*, in today's language. However, the term *syllogism* suggests the convertibility between a procedure and a demonstration, but not the equivalence of their design. This point has been seen by Lorenz and Morison (2019: 447):

> The comparison between options, which was part of the deliberative process, is not captured by the practical syllogism. The latter presents the results of practical reflection (the choices made and the reasons why they were made) but it does not present the full picture of how these results were achieved.

If, then, action plans are the fruit of practical deliberation, why is it possible to generalize some them to reach technical knowledge, and impossible for others, especially those concerning the ethical and political domains?

Müller's interpretation, despite its limitations (see above 4.1.1), is on the right track: what allows productive reasoning to be universalized, he says, is that the particular and subjective circumstances of the agent *do not matter*. A productive problem is *anybody's problem*:

> The results [of productive thinking] – the recipes for producing or making something – are universal. They are not tied to any particular individual and are valid and true for anyone who wants to produce the intended products and is able to follow the instructions for doing so. (Müller 2018: 152)

This argument, as it stands, is not sufficient: a reasoning with a productive purpose is not necessarily universalisable because it is de-subjectivised. The merchant who determines the best route for their journey, taking into account a thousand specific commercial and logistical circumstances, is engaged in some reasoning that has no personal dimension (they would do the same if they were a shipowner's employee). It cannot be universalised for all that. Müller's argument must therefore be taken even more generally: it is not only reasoning involving personal desire that cannot be universalised, but also any reasoning that depends decisively on circumstances that cannot be subsumed under a general rule. There is certainly the circumstance of my own desire, which is malleable and capable of reconfiguring itself precisely as a function of situations,[13] but also any other circumstance that resists its reduction to a simple generic case. For example, in the case of the merchant, there will be the fact that the weather is now good for sailing, but also that pirates have recently been reported towards Aegina, that a ship with an experienced captain might be leaving soon, and so on. Not only is the *joint* assessment of these circumstances tricky to establish, but it is also important to know that it is precisely these ones that are relevant to consider, among the myriad of other facts available. This *à propos* of deliberation is precisely the point of experience. It is what essentially distinguishes it from mathematical reasoning, despite the formal analogy that brings them together. The demonstrative prowess of young men does not indeed make them wise for all that:

> [...] While young people may master geometry or mathematics and become wise in matters of this kind, it does not seem they can be prudent. The reason is that prudence involves a knowledge of particular things, which become familiar with experience. (*Nic. Eth.* VI.9, 1142a12)

It is not the reasoning or the type of necessity that distinguishes mathematics from deliberation, but their *problem space*, which is much larger in the second case, and which requires that we know our way around it, much more, that we first know how to circumscribe it appropriately in each situation.

In a passage from the *First Analytics* which aims to articulate a method for constructing all possible syllogisms between two terms, which will be called the *pons asinorum* by medieval scholars,[14] Aristotle states that this method "is the same in philosophy as in every kind of *technè* and study" (*APr.* I.30, 46a3). But he immediately adds that the elementary lists of terms on which this method operates must be supplied by experience; and it is precisely these lists that are difficult to establish once and for all in most practical deliberations.

So it is not the syllogistic form or the type of necessity that distinguishes practical deliberation from scientific reasoning, but their particularity. Allen rightly writes:

> The result of [the prudent person's] deliberations will be a decision justified by a syllogism. In comparison with a scientific syllogism, this syllogism will appear distinctly *ad hoc*. The considerations which will carry the decision will, to be sure, be general [...] but the reason why this consideration must prevail over others which, as generalities are equally applicable, is

13 "For example, Randy feels like having some salted nuts but he cannot find any at home. Since it is too late (Randy thinks) to go out and buy some, he decides to have some tea instead of the nuts" (Müller 2018: 159).

14 On the *pons asinorum*, see below 4.3.2.

> specific to the occasion, and will not be able to be formulated and re-applied as a general principle itself. (Allen 2015: 67)

It can be seen from this that a *technè* can emerge when, starting from multiple similar deliberations, it manages to describe their problem space in a uniform way, while simplifying and radically reducing the number of terms needed for this description, and above all guaranteeing their stability. According to Nussbaum ([1986] 2001: 94–96), in the Greek world of the 5th century BC, *technè* is defined exactly as the antithesis and antidote to *fortune* (τύχη), as "the deliberate application of human intelligence to a domain of the world, aimed at gaining some control over fortune there." The aim is to make things predictable, controllable, precise, and therefore assignable to a fixed general concept. Eliminating the variability of circumstances is the very object, and therefore the first condition, of *technè*. It allows to design operative prescriptions that produce the expected effect in a determined way, and it allows deliberation to reach the hypothetical necessity that characterises the practical syllogism.

Therefore, for a *technè* to exist, it is not enough that its end can be determined independently of any particular circumstance linked to the agent, as Müller argues. The means and any characteristic of the situation mobilised by the technical procedure must also be susceptible of a general description. Means whose quality could vary, or whose quantity could be limited, or dependent on a delicate or uncertain operation, or on a circumstance that could alter the course of the action, all these elements would ruin the technician's procedure, unless they were specified as such, and integrated as *constraints* into the objective itself. The technician's aim in their deliberation is not the end-result alone, but it also includes the means available and the circumstances of the action. Their thinking consists in reducing, as far as possible, the vast wealth of particular truths provided by experience to a small number of characters that they must *specify* in such a way as to make them unambiguous. In this closed field, where each term has its precise meaning, deliberation can begin to look very much like a mathematical inquiry; this is why only *technè*, but not prudence or skill in general, is a gateway to the universal and the beginning of science.

4.1.4 *Technè* and prudence

What appear on the horizon of the Aristotelian idea of *technè* are our modern concepts of *specification* and *standardisation*, as the modern engineer understands them. Specification achieves universality in the practical field by considering only hypothetical necessities: *If the result sought is such, the means available such, and* *the circumstances such (*specification*) then proceed this way (*method*)* says the technician. What prevents Aristotle to reach these modern concepts is, beyond the cultural and technological factors of his time, his own sensitivity to the contingency of certain trades such as navigation and strategy, where the art is precisely to know how to adapt to favourable conditions.[15] More profoundly, for Aristotle technical reasoning remains essentially tied to experience, because even once the situations relevant to the use of a method have been described, it is still necessary to know how to recognise these situations in the concrete case. This is why Aristotle writes that a technician, who had learned their art only theoretically, would fail in their practice and be less effective than a simple person of experience (*Met.* A.1, 981a20).

Techniques, then, essentially have as their proper domain of knowledge the objects resulting from the generalisation of certain deliberations, i.e. *methods and procedures*. Procedures are the general ways of producing a general result, whether it be making pottery, playing the flute, sailing, and so on. They guide the particular deliberation in such a way as to shorten it, or even render it unnecessary. *A minima*, they propose a list of ways to try out, perhaps more or less ordered, as the *Topics* do for dialectical discussion. At best, they prescribe definite rules, which the technician needs only apply to the present situation – an act that may have its own complexity – in order to obtain the desired result, *i.e.* the appropriate plan. "Art does not deliberate!", asserts *Phys.* II.8, 199b28 in a way apparently contradictory to all the developments of *Nic. Eth.* VI: art does not deliberate indeed, when the method is so well settled and assimilated that there is no longer any need for reflection to act, as with spelling and gymnastics (*Nic. Eth.* III.3, 1112b2–6).

This thesis allows to conciliate propositions 1 and 2 of 4.1.3. The generality that *technè* achieves is generically different from the universality that takes place in science, but it can nevertheless be a springboard for science. A procedure is not a demonstration, as already said, even if they are convertible one into the other, by inversion of the order of causes (see above 4.1.1). To see this even better, let us take one of Aristotle's favourite examples, the building of a house. A person of experience who has taken part in the construction of several houses will no doubt have learned to gauge the weight of a roof structure and to assess accordingly, for example, the necessary cross-section of the columns and walls intended to support it. They will also no doubt be able to adapt their quotes to some variations on the standard plan with which they are familiar. The architect, on the other hand – for example, the master builder of cathedrals, knowing the principles of constructive geometry and

15 Aristotle's sensitivity on this topic is undoubtedly superior to that of Plato, cf. Angier 2010: 38. See the development on this subject by Dunne ([1993] 1997: 254–258), mitigating M. Nussbaum's analysis of the antithesis *technè -tuchè*.

statics, will be able to determine methodically which stable structures to use, and what is the appropriate size of stones, *regardless of the configuration of the building to be constructed.* Proceeding with method, they can freely combine and vary the general means it provides, within the problem space it delimits. However, they know neither the essence nor the reason for it (gravity, lift, material tension, etc.), which are the subject of the physicist's research. Strictly speaking, the technician knows nothing of causes, but only of means (διὰ τί), which are, so to speak, *like candidates for the status of cause*, subject to the validation of the physicist. The example of cathedral architecture shows how distant an admirable art may be from the discovery of the true causes; it will be said, for example, that the circle or the square ensure the stability of constructions, assertions which are not false in a sense, but which do not touch the underlying causes. Rather, the circle and the square are tools of the method, and are the occasion for a provisional theory, a "bricolage" of thought, to use Claude Lévi-Strauss's famous expression (1985).

The dialectic between tool (ὄργανον) and cause (αἰτία) thus articulates the dual status of *technè*. Tool and cause are the syllogism's median term sought, respectively, in deliberative inquiry and in theoretical inquiry, and they are therefore entirely distinct from each other.[16] At the same time – which justifies proposition 1 – they are convertible into each other: any natural cause may be mobilised as a tool by the technician, and any tool is at the very least a hypothesis of causality for the scientist. *Technè*, as an inquiry of general methods, thus proceeds in a radically different way from science, while at the same time collaborating closely with it – providing it with clues as to the causes of things, and mobilising its results in return.

4.2 The truth of *technè*

It is now possible to return to Book VI of the *Nicomachean Ethics* in order to understand the radical originality of *technè* relative to other modes of knowledge, which grounds the distinction between theory and practice that Aristotle has bequeathed to us. It differs from prudence in that it only ever deals with ends determined elsewhere, *i.e.* with a *specification* (4.2.1). It also differs from science in that it never reaches the true causes, except by accident, but only the *means* of action (4.2.2). This section concludes with an overview of the status of *technè* in Aristotle, as a *middle ground* of knowledge, properly speaking, *i.e.* situated at its crossroads – between theory and practice on the one hand, between science and experience on the other. Aristotle considered this central position as revealing its inferior status (4.2.3).

16 *Nic. Eth.* III.3, 1112b29 asserts that the tool is indeed the object of deliberative inquiry.

4.2.1 *Technè*'s modality of truth

One purpose of *Nicomachean Ethics*'s Book VI is to characterise the relationship between reflection and ethical action. It must therefore first be distinguished both from pure reflection (which for Aristotle is theoretical) and from the reflection that determines action without an ethical character, that which is purely utilitarian. Chapters 1 to 5 carry out the bulk of this program, through the successive definition of science (ἐπιστήμη), τέχνη and prudence (φρόνησις). If this characterisation involves in part the description of their specific modes of reasoning, (deliberation *vs* demonstration), or of their "psychology" (relation to desire and sensation) it is the determination of their proper *mode of truth* that is at the centre of Aristotle's considerations.

First, in chapter 1 (text quoted in 4.1.1), when Aristotle distinguishes two parts in the rational soul, it is in implicit reference to the traditional idea in Greek philosophy that in truth, thought becomes equal to its object: to different objects correspond different parts of the soul.[17] Aristotle aims at the possible modalities of this relationship when he focuses, throughout Book VI, on the distinction between the objects of theoretical, technical and practical knowledge. It has already been seen (4.1.1) that the physician and the biologist both study the same class of phenomena, but from a different angle. It is also easy to imagine, alongside, a politician interested in curing an important patient, for example a general who could save the city. While the object of the three inquiries (the sick general) may therefore be the same, they do not approach it in the same way: which means, for Aristotle, that the same subject-matter is the occasion for three *formally* different objects: the biological causes of the disease for the biologist, its possible treatments for the physician, and its consequences for the city for the politician. What remains to be explained, however, is how these three relations to the same being are formally different – and not simply, for example, a mere difference in point of view.

Chapter 2 gives a first characterisation of this difference by the fact that, in the practical sphere, truth and error are correlated with correct desire, whereas they are absolutely valid in the theoretical sphere. This correlation is indeed a co-determination: true reason and correct desire presuppose and reinforce each other (*Nic. Eth.* VI.2, 1139a24). But because of this, this determination is insufficient, since the notion of correct desire is itself in question. Moreover, it does not yet make it possible to distinguish practical reflection from productive reflection, which aims only at relative goods.

17 On the genealogy of this idea and its treatment by Aristotle, one may consult J. Tricot's note on *Nic. Eth.* VI.2, 1139a10, in his edition (1987: 275, n. 5).

Chapter 3 takes a new start, this time explicitly from the notion of truth, as it is determined it in the treatise *On Interpretation*, *i.e.* as a specific property of affirmative or negative judgement (*Nic. Eth.* VI.3, 1139b14). The point here is to determine more precisely the *proper* disposition by which each of the two rational parts of the soul grasp truth, what Aristotle calls their "virtue". Such an approach makes it possible to distinguish explicitly the truth that takes place in productive reflection from that which takes place in practical reflection. Moreover, by focusing his research on the *dispositions proper* to each mode of reflection, Aristotle shows that he is concerned to find essential modes of truth, and not truth by accident. Opinion is for example immediately discarded. Prudence (φρόνησις) and *technè* are thus the two "virtues" of "calculating" reason, and on the "epistemic" side, Aristotle adds to science (ἐπιστήμη) proper the intellect (νοῦς) and wisdom (σοφία). Wisdom will be later presented as the conjunction of science and intellect, and intellect itself defined as the immediate grasp of principles. Only science, prudence and *technè*, then, are *discursive* dispositions to grasp the true, accompanied by *logos*, *i.e.* reason with a dual normative and explanatory value.[18]

These three ways of grasping truth are defined successively, with formulations that Aristotle seeks to make as homogeneous as possible in the three sections devoted to science (ἐπιστήμη) (VI.3), τέχνη (VI.4) and prudence (φρόνησις) (VI.5). Here are the central passages:

> Science is a demonstrative disposition, with the qualifications that we have detailed in the *Analytics*. One knows scientifically when one acquires a form of belief and familiarity with the principles; and if this is not superior to that of the conclusion, it is by accident that one has science.
> ἡ μὲν ἄρα ἐπιστήμη ἐστὶν ἕξις ἀποδεικτική, καὶ ὅσα ἄλλα προσδιοριζόμεθα ἐν τοῖς ἀναλυτικοῖς. ὅταν γάρ πως πιστεύῃ καὶ γνώριμοι αὐτῷ ὦσιν αἱ ἀρχαί, ἐπίσταται. εἰ γὰρ μὴ μᾶλλον τοῦ συμπεράσματος, κατὰ συμβεβηκὸς ἕξει τὴν ἐπιστήμην. (1139b31)
> *Technè* is a productive disposition accompanied with true reason. All *technè* have generation as its object, to exercise a *technè* is therefore to consider how to generate one of those things which may or may not be[...] *Technè* is therefore, as we have said, a productive disposition, accompanied by true reason, of contingent things. The lack of *technè*, on the contrary, is a productive disposition based on a false reason.
> ἡ τέχνη [ἐστὶν] ἕξις μετὰ λόγου ἀληθοῦς ποιητική. ἔστι δὲ τέχνη πᾶσα περὶ γένεσιν καὶ τὸ τεχνάζειν καὶ θεωρεῖν ὅπως ἂν γένηταί τι τῶν ἐνδεχομένων καὶ εἶναι μὴ εἶναι. [...] ἡ μὲν οὖν τέχνη, ὥσπερ εἴρηται, ἕξις τις μετὰ λόγου ἀληθοῦς ποιητική ἐστιν, ἡ δ' ἀτεχνία τοὐναντίον λόγου ψευδοῦς ποιητικὴ ἕξις, περὶ τὸ ἐνδεχόμενον ἄλλως ἔχειν. (1140a9)
> Prudence is a true pragmatic disposition, accompanied by reason, concerning the goods and

18 See on this subject Moss's clarification of the meaning of λογός in the *Nicomachean Ethics* (Moss 2014).

> evils of man.
> [ἡ φρόνησις] εἶναι ἕξιν ἀληθῆ μετὰ λόγου πρακτικὴν περὶ τὰ ἀνθρώπῳ ἀγαθὰ καὶ κακά. (1140b5)

These texts relate to the theory of affirmative and negative judgement, which Aristotle recalls in a preliminary note. The Aristotelian doctrine of truth and error (*Met.* Θ.10, 1051b1;b35, *On Interpretation*, 4–6) is built on this theory: if to judge affirmatively is to bind two concepts together, and to judge negatively to unbind them, then depending on whether this binding is in things or not, one may be in truth or error.

This perspective sheds light on the definition of prudence: it is a matter of judging the considered action in relation "to human goods and evils", *i.e.* to affirm that it is aligned with human goods, or on the contrary that it leads to evil. Whether such a judgement may be described as true or false is based on the fact that human goods and evils are determined by the very essence of what it is to be a human person. But they are also dependent on the subject's current situation, the options open to them, the means to be implemented that may have harmful consequences, events that may occur in the future, and so on. In general, therefore, no certainty is possible regarding these judgements. This is why, it could be argued, the adjective "true" here qualifies the disposition of the soul and not the reason that accompanies it, contrary to the definition of *technè*; Aristotle indicates by this that if the reasoning of prudence is fallible, its guarantee of truth proceeds first of all from the general attitude that the virtuous person adopts in their deliberation. This mode of truth contrasts singularly with that of *technè*, which is nevertheless the product of the same calculating reason. The major difference lies in the criterion by which the course of action is judged: in the technical field, the criterion is nothing else than the desired result, or "the specification", as Johansen notes:

> The true logos involved in craft will represent correct reasoning about the means that brings about a certain product. (Johansen 2017: 101)

If the technician correctly judges the adequacy of the means envisaged to the specifications, then they are in the truth, and if not, they are in error, in the ἀτεχνία, as Aristotle says. This term does not necessarily mean that their project will fail, but that it will not succeed thanks to the *logos*. It will be seen below (4.3.1) for example, that confessions extracted by torture are *atechnical* in relation to rhetoric, which is the art of persuasion; this does not mean, unfortunately, that torture is not effective. Aristotle's perspective is not to highlight the effectiveness of *technè* but to make explicit its proper mode of being true and false.

The mode of truth of technical deliberation differs radically from that aimed at by prudence. Even if prudence is also concerned with deliberating about means, it carries out an important task upstream, which is to *specify* what a good solution looks like, precisely what criteria it should satisfy. It carries out this specification work by taking into account the immediate and long-term aims of the person or community for which they are responsible, thus necessarily making choices in their deliberation. Thus Wiggins writes:

> The standard problem in a non-technical deliberation is quite different. In the non-technical case I shall characteristically have an extremely vague description of something I want – a good life, a satisfying profession, an interesting holiday, an amusing evening – and the problem is not to see what will be causally efficacious in bringing this about, but to see what really *qualifies* as an adequate and practically realizable specification of what would satisfy this want. Deliberation is still *zetesis*, a search, but it is not primarily a search for means. It is a search for the *best specification*. [...] When this specification is reached, means-end deliberation can start, but difficulties which turn up in this means-end deliberation may send me back a finite number of times to the problem of a better or more practicable specification of the end. (Wiggins 1975: 38)

Note here the central role of the concept of *specification* that was already used above to characterise *technè*. Whereas the technician does not start work until they have received a specification of the result they have to aim for and the resources at their disposal – for example the carpenter receives the specification of the rudder from the pilot (*Phys.* II.2, 194a36) – the prudent person is the one who *writes* the specification, and who writes it *freely*, with regard to what they think is the good for themselves or for the community they serve, and not under the compulsion of some other objective imposed by anyone else. This precision helps to distinguish them from an expert who would merely advise a master on their wishes, which would still be a technical activity. In prudent deliberation, as Wiggins remarks, a dialogue is established between the two phases of deliberation, which can lead to multiple iterations: for the prudent person also observes the collateral effects of the means that their reflection suggests, and may see in them prohibitive consequences, or on the contrary hitherto unnoticed opportunities. Practical reflection is thus constantly occupied with a free "back-of-forth" movement from the ends to their means, and from the means back to their consequences. The whole effort of prudence is to *integrate* all these elements into a holistic perspective that maximises good outcomes, taking into account chances and uncertainties. This is why, in a way, as Aristotle says, prudence is foremost the virtue of politicians, such as Pericles, who possessed an unparalleled integrative capacity for judgement (*Nic. Eth.* VI.5, 1140b7).

It is now possible to understand why Müller's interpretation already discussed (4.1.1) is reductive: by establishing a strong split between the two stages of deliberation, and identifying careful deliberation strictly with the first, it ignores cases where it is difficult, if not impossible, to distinguish between them, because the complexity of the situation requires a constant readjustment of ends in the light of the options considered. The whole purpose of *technè* is to achieve this separation between the customer and the producer, by defining a controlled space of production that also dictates the language in which requirements must be expressed. On the contrary, prudence is situated in an essentially *open* space of deliberation, where it must judge *what should be for the best*, taking into account all the relevant contingencies that it can detect in the present situation, and all the lessons learned from experience and past deliberations, whereas the second stands in a *enclosed* space, whose terms have precise and unchanging values, and which it must only try to arrange between them in the best possible way. The truth of the former aims at the indefinitely open horizon of the human good, while that of the latter is limited to the satisfaction of an order.

4.2.2 Technical knowledge and scientific knowledge

It is possible now to contrast these two modes of truth with truth *par excellence*, scientific truth. Here, the elements given by Aristotle are patchy; he merely states in the text quoted above that a condition of demonstrative knowledge is that the principles are better known than the conclusion, and refers the reader to the *Analytics*. This assertion indeed concludes the long development of *APo.* I.2 which aims to define scientific knowledge. Its starting point is the famous thesis of the Stagirite, that to know truly (scientifically) is to know the cause of a thing and why it cannot be otherwise than it is. This is why syllogism is essential to science, because it links a judgement to a principle that justifies it; and if this principle is established elsewhere, or already known, then the syllogism can be called a demonstration. Scientific truth aims to go deeper into the causes of things, into "what they are", their internal structure or principles of behaviour. It moves from what is more knowable to us (*i.e.* through perception) to what is more knowable "by nature" (*APo.* I.2, 71b33–72a5). Scientific truth lies entirely in this adequacy of thought with the causes and intimate structures of the thing. In order to be perfect, it requires both science (ἐπιστήμη) and the intellection (νοῦς) of principles, *i.e.* the theoretical wisdom that combines them both.

Therefore, although the scientist and the technician both know the cause of a thing – for example, that heat is the cause of digestion –, they nonetheless have absolutely different perspectives on it. The scientist will try to question it further and to understand its proper causes, whereas it will interest the technician only if it can be used to solve their problems. The results of science are nothing more for the technician than additional means for crafting their procedures. Johansen writes:

> We would expect the logos of the craft to feature theoretical information but only to the extent that it is instructive in production. A good way to think of the logos here is as a recipe. If you are to make ice-cream expertly, or teach another person the skill, you need to know about processes of emulsion, stabilising and freezing. [...] An Aristotelian craftsman may know that protein is a stabiliser and in this he differs from the experienced who Will just know that egg white has worked to smooth the mixture in the past. But the craftsman need not know the chemistry at work. And that is because she is just interested in making a particular product, ice-cream. (Johansen 2017: 125)

In this text Johansen expresses well why *technè* does not rise to the consideration of primary causes: it is because it stops at the satisfaction of its need. This does not prevent it from asking questions to science, or from looking further into the causes that may be useful to its enterprise; but it is not its original perspective, which is only to draw up effective plans to achieve the desired result.

The truth of *technè* lies in reasonings that correctly arrange means to ends. It must be judged only on the appropriateness of this arrangement. That the means come to be lacking in a given situation in no way alters the truth of the reasoning, no more than Pythagoras' theorem becomes false when it is applied to an equilateral triangle. It is essentially, as Aristotle often repeats (see for example *Phys.* II.9) *conditional necessity*, of the type "If you want this, then do that". It aims at reality as it might be, in contrast to the aim of science, which tries to understand things *as they are*. This explains Aristotle's choice in Book VI of the *Nicomachean Ethics* to dramatise the opposition between science "in the strict sense" (*Nic. Eth.* VI.3, 1139b19), as occupied only with necessary beings, and between *technè* and prudence as occupied with the contingent.

It could therefore be argued that it is prudence that should be brought closer to science, rather than *technè*. If the truth of prudence is indeed, as with *technè*, the adequacy of actions envisaged with their ends, the latter are not here determinate, but the general ends of the human person, their essential good, "objective" as one would say today. It is for this reason that the conclusion of Book VI of the *Nicomachean Ethics* finally distinguishes, among the five intellectual virtues initially introduced, only two supreme virtues, theoretical wisdom and prudence (*Nic. Eth.* VI.11, 1143b15). For the former aims at the real essences of things, and the latter aims at the real good of man. This proximity is signified at this point in the text by the strong expressions, with Platonic overtones, which describe prudence as "eye of the soul" (ὄμματι τῆς ψυχῆς), to whom only "the noblest good" appears (τὸ τέλος καὶ τὸ ἄριστον).

Of the five ways of being-true initially enumerated, therefore, only *technè* has really disappeared, since science and intellection are conjoined in wisdom. Is this exclusion merely due to a Greek prejudice against craftsmanship, from which Aristotle is not exempt (*Pol.* III.3)? Or is it rather related to its specific modality of being-true, which Aristotle judged inferior to the other two?

The second option is definitely the case (whatever the case of the first one). The adequacy of thought to the thing (science), the adequacy of the means envisaged to the objectives set (*technè*), the adequacy of the action envisaged to the good of man (prudence): these summary formulas of the respective modalities of truth show clearly that the middle one is not on the same level as the other two, which aim at objective being or good. *Technè* is concerned only with arranging certain causal processes with a view to certain ends, and moves within closed problem spaces, where it is ultimately concerned only with itself. Its truth is fundamentally the truth of conditional necessity "if… then".

The exclusion of this modality of truth is explicitly justified by Aristotle in *Nic. Eth.* VI.12 where he argues for the supremacy of wisdom and prudence over the other virtues. He then contrasts virtue with skill (δεινότης), which could be seen as a modality or close relative of *technè* (see above 4.1.1). Both indeed differ from prudence in that they are indifferent to ends, in that the physician is capable of healing as well as poisoning, and in that the skilful person can be prudent as well as cunning, depending on whether he possesses virtue or not. Aristotle remarks that one can be skillful and yet in error, for "vice perverts and falsifies (διαψεύδεσθαι) the grasp of the principles of action" (1144a34). The conditional truth of skill and *technè* is just that: true only if the end they aim at is good, but of that it knows nothing. Indifferent therefore both to the deeper causes of things and to the ultimate ends of the human person, *technè* is only capable of that half-truth provided by conditional necessity. It has no value in itself, but only some utility, which has value only when it stimulates science, or when it assists prudence.

4.2.3 The inferior status of *technè*

This interpretation thus makes it possible to explain the formal tension from which this discussion started, between the vertical and horizontal schemes of rational knowledge. It explains the originality of technical knowledge in relation to theoretical knowledge, while firmly distinguishing it from practical reflection, as well as from experience in general.

There is no doubt that these two schemes were compatible and consistent in the mind of the Stagirite. Deliberative thinking is generally the proper skill of the experienced person, for it presupposes good judgement of the circumstances. Some deliberations, which Aristotle calls productive, have their object determined by an empirically fixed concept, such as "health" or "a horse's saddle", rather than by particular ends such as "how shall I repay the money to my creditor?" – or by general ideas of good such as "happiness" or "honesty", the meaning of which will necessarily depend on each situation. When a method is established through productive deliberation, which mobilises means that are also described by general concepts, and when the connection between these means and the intended end is established causally, one then reaches technical knowledge. The truth of this type of knowledge is that of hypothetical necessity, *i.e.* the internal coherence between the specification of the problem and the proposed method of resolution. This mode of truth is inferior to "truth proper", *i.e.* oriented towards reality, which itself has two modes: that of theoretical wisdom, which lies in the adequacy of thought with things, and that of practical prudence, which lies in the right perception of the good to be achieved in the present situation. In particular, the means mobilised by technical methods – its "tools" – must be considered only as *indices* of causes, as pointers to a possible theory, and not as attested elements of that theory. They may be totally reconfigured, even discredited by scientific truth, whatever their practical successes.

It should be noted that this inferiority of *technè* contrasts with the central position it occupies in each of the two schemas, between science and prudence on the one hand, and between experience and theoretical science on the other. Such a situation might have evoked the idea of *technè* as a "crossroads" of knowledge, enabling trade between its different domains: it is by taking a technical form that empirical or practical knowledge can be studied theoretically, and vice versa as well. But for once in Aristotle, the median position is not a sign of pre-eminence. *Technè* is not only inferior to science in grasping causes, and to prudence in considering the good, but it is also inferior to experience in effectiveness, which is nonetheless its raison d'être. *Technè* thus appears as a minor form of rationality, limited in its height of vision, restricted in the horizon of its action, myopic in its judgement of circumstances.

This is where the essential tension lies within Aristotle's concept of *technè* , an autonomous domain of knowledge, and at the same time, a primitive and minor form of knowledge, which lacks the broad foundation of experience and the elevation of science. It is this tension that has so often led scholars, and in any case the philosophical tradition, to equate it sometimes with one, sometimes with the other, failing to recognise its distinctive feature of being a *knowledge of methods and procedures.*

4.3 Method and logic

It is now possible to return to the main problem of this chapter, and ask whether systematic procedures which were found to be characteristic of the industrial approach to work and computers in the previous chapter, is at all comprehensible, or even possible from an Aristotelian point of view. *Technè* aims to generalise particular deliberations that it abstracts from contingent situations, but can it go so far as to make them exact and exhaustive?

It is first shown why this possibility is implausible, taking the example of one particular technical discipline, rhetoric (4.3.1). There is, however, one place where Aristotle seeks to set up a methodic approach to inquiry: the *Organon*, where he sets out techniques for finding suitable arguments for solving given problems, notably in the *Topics* and *First Analytics* (4.3.2). The *pons asinorum*, a method for finding the middle terms of syllogisms, is of particular interest because of its distinctly procedural character, especially since Aristotle declares it relevant both "for philosophy [and] for techniques (τέχναι) and all disciplines". Yet the idea of *systematicity* is first associated with the Aristotelian doctrine of science expounded by the *Second Analytics*. The next discussion therefore attempts to show how these two ideas are linked, and why the approach to problem-solving proposed by the *pons asinorum* can best be envisaged only within the framework of a scientific system (4.3.3). This allows to finally reach a unified notion of inquiry, albeit one that is dramatically differentiated between science and *technè* (4.3.4).

4.3.1 The methods of inquiry

The main result of the former discussions is that techniques are the repositories of some kinds of practical knowledge, those which "generalize well", so to speak; and that their objects are procedures and methods. So far the term *method* has been used informally, in its modern sense. Now *method* is an important term for Aristotle; according to the catalogue of his works handed down by Diogenes Laertius, he is said to have written a treatise *on the Methods* in eight books, filed alongside the works of the *Organon*, and another treatise *on the Method*, which followed three treatises concerning *technè*, unfortunately all lost. Despite these gaps in the Aristotelian corpus, it is possible nevertheless to reason from an Aristotelian ground, and to explore the idea of method to understand whether technical knowledge may be systematic.

As Lennox (2011: 29) notes, μέθοδος is a difficult term to translate, as it colours with various nuances depending on the context. It appears above all in the introductions to Aristotle's treatises, during the reflective considerations that he makes about the approach that his study should take to the object in question. It is close to its current meaning in that it covers the precepts that Aristotle gives himself concerning inquiry. But it can also take the more general sense of *discipline, investigation, inquiry*, as in *Phys.* I.184a10: "looking at any study (*theoria*) and any μέθοδος", or the opening words of *Nic. Eth.* I.1, 1094a1: "Any τέχνη and any μέθοδος". It appears, in this second sense, interchangeable with σκέπτεσθαι and especially ζήτησις, which Aristotle uses very commonly, as well as ἱστόρια, less frequently, to refer reflectively to his own research, whether it concerns first philosophy, physics, ethical or logical disciplines. Σκέπτεσθαι and ζήτησις have their own nuances (*examination* / *inquiry* respectively) but are also translated, depending on the context, as *inquiry* or *investigation*. Also μέθοδος, ἱστόρια, ζήτησις and σκέπτεσθαι are easily interchanged in these reflective texts, without any significant nuance being detectable. All these terms designate knowledge in the process of being actively constituted, in opposition to the already constituted knowledge of science.

Inquiry (ζήτησις) is the term Aristotle uses to describe the search for the cause (*Met.* Z.17) and the middle term (*APo.* II.1–2;8), as the essence of scientific activity. It is also the term he uses in the passage already quoted from *Nic. Eth.* III.3 (4.1.1), when he states that deliberation is inquiry, but not all inquiry is deliberation, giving mathematics as a counterexample. Allen (2015) and Charles (2015) have both noted this parallel, comparing *Nic. Eth.* III.3 to *APr.* I.27–30 and *APo.* II.1–2;8 respectively.

If deliberation is an inquiry, then it is not knowledge, but knowledge in the process of being realised, the knowledge of what needs to be done in the current situation. The method is then situated at the reflective level that guides deliberation, just as it might guide a scientist seeking to explain a phenomenon. If the practical syllogism represents the completed procedure, the method is concerned with writing the procedure.

An idea of how Aristotle viewed the method of a technical discipline can be glimpsed in the *Rhetoric*. At the beginning of the treatise, Aristotle states that “the technical method concerns the means of persuasion (ἡ ἔντεχνος μέθοδος περὶ τὰς πίστεις ἐστίν)” (*Rh.* I.1, 1355a4) and further states:

> Among the means of persuasion, however, some are non-technical (ἄτεχνοί), and others technical (ἔντεχνοι). By non-technical I mean all the elements that are not provided by us, but existed beforehand, such as witnesses, confessions extracted by torture, written documents and the like. By technical I mean all the elements that it is possible for us to put in place by method and by our care (ἔντεχνα δὲ ὅσα διὰ τῆς μεθόδου καὶ δι’ ἡμῶν κατασκευασθῆναι δυνατόν). Thus one only has to use the former, but has to discover the latter (ὥστε δεῖ τούτων τοῖς μὲν χρήσασθαι, τὰ δὲ εὑρεῖν). (*Rh.* I.2, 1355b35)

What belongs to the domain of *techne* is therefore that for which one is responsible (*that for which one is the cause*, writes Aristotle further on in *Rh.* III.16, 1416b19), and which is conducted with method. Note the circularity of the definition, which characterises the method by its objects (the means of persuasion) and the objects by the method; this shows the closeness between the two notions.

Now what does Aristotle’s treatise contain? Certainly not a compilation of *practical syllogisms* that could take the form, in the case of rhetoric, of speech outlines that would simply have to be adapted to the circumstances (which is what many later treatises on rhetoric will do). Rather, it contains the distinction between the main purposes of rhetoric, an examination (often prescriptive) of the means and figures of speech available to the orator, an analysis of the circumstances they must take into account, such as the passions of their audience, the type of arguments their opponent is making and how to respond to them, and so on.

The rhetorical method therefore does not dictate speech-writing, but frames and guides it; it helps the orator to prepare their speech reflectively, by asking them questions: *"Are you sure what you are trying to achieve? Have you studied your audience? Have you thought of such and such an argumentative figure? Does your style meet such and such standards? etc."* It therefore consists in helping them to describe their problem space, *i.e.* to list all the terms they need to consider in order to find a solution – in this case a persuasive speech.

This result is consistent with the general idea about *technè* that emerged from the former sections of this chapter. A method for achieving one's ends is not in general absolutely determinative of action or infallible. The orator, the navigator, the carpenter, cannot rely on books of procedures to carry out their work, yet their techniques guide them in their thinking, can remind them of certain points of attention or highlight certain particular cases. This Aristotelian vision is exactly the one identified in the previous chapter (3.1 and 3.2) concerning human procedures, when they were contrasted with industrial and computing procedures.

4.3.2 The *pons asinorum*

The fact that there are no procedural patterns in the *Rhetoric* does not prove that they would be equally incongruous in any other technical discipline. The proximity of rhetoric to human affairs – ethics, politics, law – may explain this; but since the uncertainty inherent in each productive activity varies, they could be possible and legitimate for other techniques.

And indeed, clues to such schemes are found in the *Organon*, which is not only a sum of theoretical expositions on the form of proposition and reasoning, but also contains, at least for the *Topics* and *First Analytics*, numerous methodical considerations.

The *Topics* in particular is presented as a *technical* treatise dealing with dialectics, which has among its functions to train and educate in oral disputation.[19] Here, again, the construction of a method is the goal:

> The present treatise sets out to find a method that will enable us to reason deductively, from accepted ideas, about any problem.
> Ἡ μὲν πρόθεσις τῆς πραγματείας μέθοδον εὑρεῖν ἀφ' ἧς δυνησόμεθα συλλογίζεσθαι περὶ παντὸς τοῦ προτεθέντος προβλήματος ἐξ ἐνδόξων, (*Top.* I.1, 100a18)

19 *Top.* I.2, 101a25-b4. See the introduction, 13 ff, by P. Pellegrin to his edition (2015).

The method takes here a much more definite turn than in the *Rhetoric*. Book VIII, which deals with how to conduct the dialectical disputation, has a distinctly procedural tone. The bulk of the treatise consists of lists of places or *loci*, the purpose of which is to feed arguments into a common reasoning scheme, described by De Pater (1965). Brunschwig describes the *locus* as “a machine for making premises from a given conclusion”.[20]

This procedural approach is taken to its logical conclusion in *APr.* I.27–30, which constitutes a mini-treatise on the practice of syllogism. Crubellier (2014: 31–34) shows the remarkable progress this approach presents over the dialectical recipes of the *Topics*. Indeed, it sets out a method for finding certainly the middle term of a syllogism. This method, known in the Middle Ages by the name of *pons asinorum*[21] is represented, since John Philoponus, by the figure 4.1. In order to find the middle-term that links the predicate A to the subject E, six lists of terms are drawn up: B, C, D are the lists of terms respectively consequent, antecedent and incompatible with A, and F, G, H those which are consequent, antecedent and incompatible with E, in the same order. If the same term is found in two lists linked together in the diagram, then the syllogism of the figure shown on the line can be established. In addition, there is an order of exploration (CF and BG, then HB and DF, then GD and GC) which allows the most informative syllogism to be found first. Here are two examples of how the *pons asinorum* works, with lists of two terms

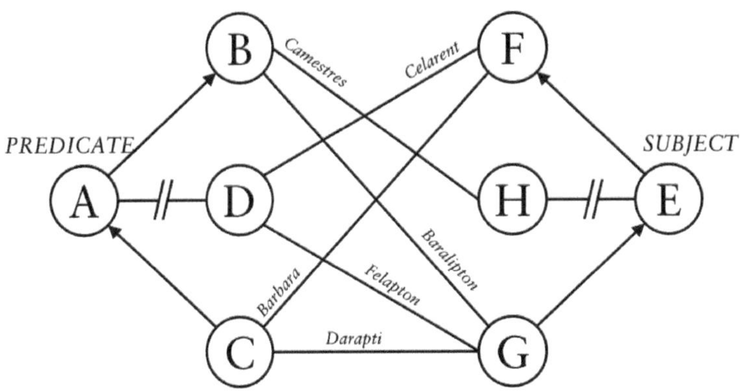

Figure 4.1: General diagram of the *pons asinorum*

20 Cited by P. Pellegrin in the introduction to his edition of the *Topics* (2015: 26).

21 See West and Thompson 1891 for a history of this term and the expositions of Hamblin 1976 and Crubellier in his edition of the *First Analytics* (2014: 31–34, 286–294).

each, represented directly in the graph next to the corresponding element in the diagram. The first one (4.2) allows to construct the following syllogism in *Barbara*:

> *Everything that consumes oxygen is alive.*
> *All humans consumes oxygen.*
> So, *all humans are alive.*

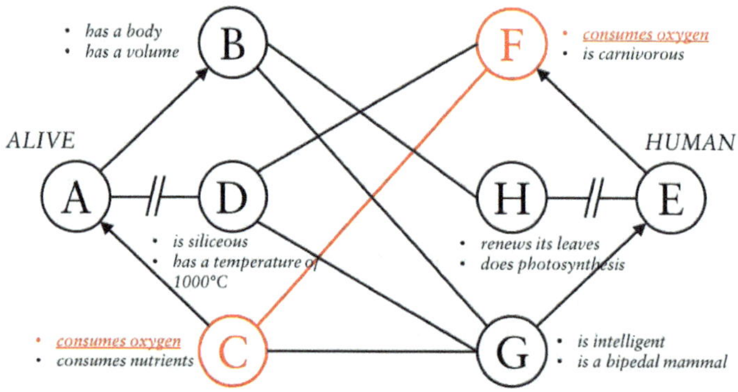

Figure 4.2: *pons asinorum*– example 1

The second example (4.3), distantly inspired by the *Logic of Port Royal*, allows to construct the following syllogism in *Camestres*:

> *Every saved man confesses their sins.*
> *No strong-minded man confesses their sins.*
> Therefore, *no strong-minded-man is saved.*

It may be noted that this syllogism is preferable to the syllogism in *Felapton* also possible:

> *No saved man is a libertine.*
> *Some strong-minded man is libertine.*
> So, *some strong-minded man is not saved.*

The order of inspection of the lists allows the best syllogisms to be found first.

This method has a properly procedural character, which allows Aristotle to conclude his exposition with satisfaction:

> We see, then, from what has just been said, not only that it is possible to find all syllogisms by this method, but also that it is impossible to find them by any other route. For it has been

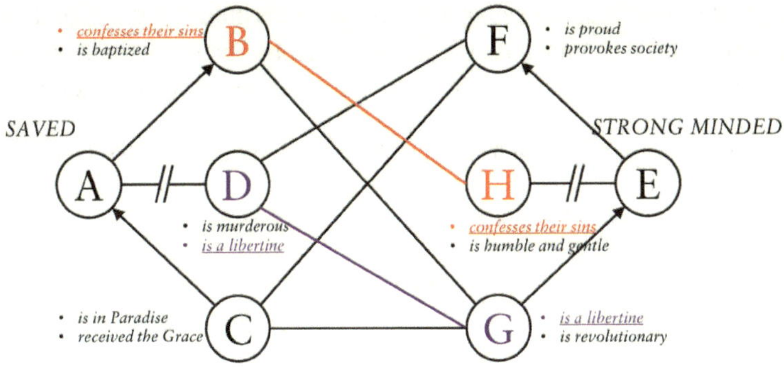

Figure 4.3: *pons asinorum*– example 2

> established that every syllogism is constituted by one of the figures previously discussed, and yet it is not possible to construct them in any other way than by means of the consequents and antecedents of each term. (*APr.* I.29, 45b35)

A little further on, he points out another benefit of this approach: it makes it possible to give all the possible relations between two terms *once and for all*, without the need to start the search over again each time, depending on whether one is seeking to affirm or deny a thesis, to establish universal facts or particular facts (*APr.* I.30, 46a10–17).

There is here the sketch of a systematic approach to inquiry: The *pons asinorum*, then, as long as it is provided with lists of terms exhaustively describing a field of knowledge, can proceed like a "theoretical machine", writes Crubellier, which structure and operations can be "described precisely" and which could be realised today "in the form of a very simple search engine associated with databases".[22] Above all, this "machine" leaves no stone unturned and identifies all the possibilities for solving the problem at hand.

The method is "theoretical", according to Crubellier, in the sense that it depends on lists of terms derived from experience which, as soon as the field investigated is sufficiently broad, become impossible in practice to constitute. He writes:

> Despite [Aristotle's] very assured tone, it is not clear how [this] condition […] could be achieved in practice, nor how it will be certain that it is actually achieved. (Crubellier 2014: 33, n. 2)[23]

22 Crubellier's edition of the *First Analytics* (2014: 32–33).

23 Crubellier expresses his doubt by saying that the *pons asinorum* "does not give a procedural method, *i.e.* a set of rules that it would suffice to follow carefully to be sure of success" (34), but he

Before coming to this question, this reliance on lists of terms indicates what kind of inquiry the *pons asinorum* is aimed at: it is *problem-solving*. The term πρόβλημα is indeed used frequently in this passage, and it has a precise meaning for Aristotle:

> In the technical sense developed in the *Topics* a problem takes the form of a double interrogation: "is X *p* or *non-p*?" and its solution consists in finding a syllogism that proves the position one defends. (Pellegrin, notes to his edition of the *Second Analytics* (2015: 416))

In its extremely restrictive form, the Aristotelian problem is only a distant antecedent of the present meaning of this term (see above 1.2). In particular, there seems to be no reference here to the other sense of the term developed by Greek geometers at the same time to designate questions regarding constructions of figures.[24] There is, however, the crucial idea that a problem determines a *problem space*, in which its solution (here: lists of terms) must be found. Within this space, the inquiry can be procedural.

So where could these lists of terms required by the *pons asinorum* come from? First of all, it should be noted that the requirement for exhaustiveness is relative to the context of the problem to be solved. The beginning of the *APr.* I.30, devoted to epistemological reflections on this method,[25] mentions two of them: that of the search for truth, *i.e.* the scientific context, and that of dialectical argumentation, *i.e.* the context of probable opinion – a classic opposition in Aristotle's thinking. Now the beginning of this text, which seems to be placed in this second context, simply says that one should "possess terms as much as possible". Dettel and Crubellier refer here to *Top.* I.14 which describes in a concrete way what seems to have been a common school practice, resembling that of the *florilegium* in the Middle Ages:

> Premises still need to be collected from books, and tables drawn up for each category of subject, with separate headings, for example "*the good*" or "*the animal*", and "good" to be taken in its entirety, beginning with the essence. Each time, the names of the proponents of these opinions will be noted [...] (*Top.* I.14, 105b13)

The *pons asinorum*, insofar as it is an improvement on the methods of dialectical argumentation described in the *Topics*, uses the same material. It is more effective and more exhaustive relative to these methods, not absolutely.

It is, however, in the scientific context of the *Analytics* that Aristotle expounds the *pons asinorum*, and it is to scientific inquiry that he seems to find it most suited.

---

is aiming here at the impossibility of reducing experience itself into procedures, and in no way at the central method of the *pons asinorum* which feeds on it and is straightforwardly procedural.

24 On this question, see our developments in Appendix B.1.

25 This is the title proposed by Michel Crubellier in his translation (2014: 136).

At the end of *APr.* I.30, he gives the example of astronomy, the science *par excellence*, to show the soundness of his method:

> We have explained [...] in what way the terms of syllogisms must be sought, so as to avoid our attention being focused on the totality of what is said in the terms of the problem; [...] What we must do is consider a smaller number of terms, well determined. [...] It is up to experience to provide the principles relating to each subject. I mean that, for example, it is astronomical experience that provides the principles of astronomical science, because it is only once celestial phenomena have been properly understood that the demonstrations of astronomy have been discovered. The same applies to any other art or science. Afterwards, [...] it is up to us to derive the demonstrations quickly. If, indeed, none of the true attributes belonging to things has been omitted from our study, we shall be able, in everything that admits of a proof, to discover and demonstrate it [...] (*APr.* I.30, 46a10–26)

For Aristotle, science is the best purveyor of lists of terms for the *pons asinorum*. Without science, the research for middle terms is condemned to draw up long lists of terms, without ever being sure of their completeness. Science provides certainty and efficiency. The *pons asinorum* will therefore be effective to the extent that science itself is completed.

4.3.3 Science and systematicity

It is therefore the characteristics of science that must be examined in order to understand at which conditions a systematic approach to problem-solving is possible. Although Aristotle does not use the term *system* to designate the form of scientific knowledge, it is indeed to him that Christian Wolff refers when he fixes its meaning in the sense familiar today:[26]

> It emerges from his *Organon* that, among the ancient philosophers, it was Aristotle who had a systematic mind. If we illuminate his reading with the notions necessary for its understanding, it turns out that the *Organon* is written in such a way that each truth is demonstrated by the others and, at the same time, the idea of a systematic mind emerges from the doctrines presented. (Wolff [1729] 2019: 40–41)

This view has been challenged by more recent commentators. As Jonathan Barnes reports:

26 More precisely, the reference to Aristotle concerning the idea of system can already be found before Wolff, who takes it up and subordinates it to that of Euclid (See M. Albrecht's introduction in his edition of Wolff [1729] 2019: 12–13). The history of this term is sketched below section 7.3.

> Aristotle's system – his "image of the world" – has been admired and praised for centuries. Some scholars, however, have challenged this view of Aristotle. They have denied that he was a systems builder. Suspicious themselves of the grandiose pretensions of systematic philosophy, they believed that Aristotle's virtues lay elsewhere. For them, Aristotle's philosophy is essentially "aporetic": it consists in posing particular enigmas or *aporiai*, and developing particular solutions to them. Aristotle's thought is provisional, flexible, changeable. [...] This anti-systematic interpretation of Aristotle's thought is widely accepted today. (Barnes 2000: 55)

Barnes may have in mind, among others, Pierre Aubenque's interpretation, who described Aristotle's *Metaphysics* as "the meticulous description of a failure" (Aubenque [1962] 1972: 487).[27] Against this perspective, Barnes maintains Wolff's view, particularly as regards his conception of science:

> The theory of science expounded in the *Second Analytics* cannot be dismissed as an irrelevant archaism [...] There are so many allusions and indications of systematization in the treatises that the solution of the *aporiai* cannot be regarded as the goal of Aristotle's scientific and philosophical inquiries; and – a point worth emphasizing – even fragmentary discussions of individual problems are given intellectual unity by the common conceptual framework within which they are examined and treated. Systematization is not achieved in treatises, but it is an ideal, always present in the background. (Barnes 2000: 56)

In this debate, it seems that the perspectives need to be reversed: the contemporary meanings of *system* are themselves inventions of the early modern period and, as said, they crystallize in part *with reference to* the theory of science expounded in the *Second Analytics*. This is what allows to use the term in a commentary on this text, while taking care that it does not introduce surreptitiously the meanings it conveys today.

Indeed, for Barnes, the *Second Analytics* is a "treatise about the axiomatic method – it is concerned not to develop a science but rather to examine the way in which a science should be developed." (Barnes 2000: 57) In other terms, it does not present a doctrine of scientific inquiry, but only precepts concerning the systematic exposition and didactics of science once its research has been completed.[28] Barnes, writing from the standpoint of the analytic philosophy of science, may have in mind the distinction between the *context of discovery* and the *context of justification*. And for logical positivists, logic is indeed only concerned with the completed form of the scientific system. In the so-called "classical" model of scientific explanation, due to Hempel (1965), the latter does indeed proceed by deducing laws from more fundamental laws, in particular by introducing intermediate terms that specify the

27 See the commentary of Crubellier and Pellegrin 2002: 329.

28 See Barnes 1969, as well as P. Pellegrin's introduction to his edition of the *Seconds Analytics* (2005: 9–10).

latter. Its form is therefore indeed the syllogism, as Aristotle asserts, particularly in its universal affirmative modality.

Pierre Pellegrin provides a more nuanced view of what systematicity might mean for Aristotle. He writes, in the introductory presentation of his translation of the *Second Analytics*:

> This eminent mode of knowledge [which is demonstrative science] takes the form of a deductive system founded on true principles, anterior to the conclusions drawn from them and to the causes of these conclusions, leading to an unshakeable conviction in the one who possesses it.[29]

Pellegrin argues that this last property is the essential one:

> The *Second Analytics* certainly contain the description of a system of knowledge that can be briefly characterised as a body of propositions put into a deductive and axiomatised form. [...] But this is undoubtedly not the essence of the project that Aristotle deploys in the *Second Analytics*. Indeed, his first intention is to define a certain *type of knowledge*, science understood as "demonstrative science" that "we possess by the fact of possessing a demonstration".[30]

In other words, it is indeed science (ἐπιστήμη) as an intellectual disposition or virtue that is essential here[31] *i.e.*, rather than the system of propositions itself: the ability to know *systematically* what things are, why they are the way they are, and so on. There is here an echo to Wolff's insistence on the systematic mind rather than the system, visible in the quotation given above.

Thus Pellegrin grants Barnes that it is certainly not scientific discovery that the *Second Analytics* are talking about, but he disputes that Aristotle had our contemporary formal considerations in mind. What he aims to make explicit is the "knowledge that a human subject can have of the necessary connections to be found in the world",[32] and which renders them capable of explaining phenomena or predicting them with assurance.

It may be possible to go a little further in this direction, and to rehabilitate the heuristic value of the *Second Analytics*. Anticipating somewhat the forthcoming reflections, it can be noted that two things may be understood by *scientific inquiry*. In the sense understood by Barnes and Pellegrin, it refers to inquiries aimed at establishing scientific theories themselves – what we would call basic research today. What we mean by *problem-solving*, on the other hand, refers to other, more widespread and less spectacular inquiring activities, which consist of explaining and

---

29 P. Pellegrin, edition of the *Second Analytics* (2005), back cover.

30 Introduction by P. Pellegrin to his edition of the *Second Analytics* (2005: 48).

31 Pellegrin takes up this idea from Brunschwig 1981.

32 Introduction by P. Pellegrin to his edition of the *First Analytics* (2005: 47).

predicting phenomena within the framework of theories that have already been established, for example: Why does this galaxy move at such a speed? What are those regular flows of light seen in a certain part of the sky? When will the next lunar eclipse take place?... Book I of the *Second Analytics* gives the main criteria and forms of a "good" explanation or problem-solving within the framework of an already constituted scientific theory, criteria by which it is essentially distinguished from the technical and practical problem-solving discussed above.

In particular, there must be a well-ordered sequence of connections between the premises and the conclusion, reflecting the underlying structure of the very things under discussion. Each connection must be necessary, *i.e.* minimal: it must refer exactly to the *actual* relation that exists between the connected terms; finally, the solution must be universal with respect to the conceivable cases of the problem at stake; for example, it must be invariable in time.[33] In this way, the scientist can respond exactly to the problems posed, with the minimum of necessary data, covering all the situations envisaged; they can justify and guarantee their response by rigorous sequences of reasons rooted ultimately in the correct observation of reality.

What gives scientific problem-solving these properties is that it takes place within a system whose elementary terms and laws have been identified and formulated in a minimal way to cover all possible problems. This more fundamental work of analysis is the subject-matter, among others, of *APo.* II's considerations on definitions. Indeed, whereas Book I is more oriented towards explanatory demonstration, *i.e.* the main form of the scientific problem, this second book is oriented towards the constitution of the scientific system itself. In particular, chapter 13 describes the requirement to constitute and organise *exhaustive* lists of descriptive terms covering the field in question. The Platonic method of division, of which Aristotle is otherwise critical, seems here be a placeholder whose merit is to give the idea of an exhaustive approach grounded in the observation of reality. Chapter 8 shows that there are terms that can be derived from other, more fundamental terms by demonstration. Once one has in effect the explanation of the *why* of an eclipse or of thunder, one has *at the same time* their definitions, which therefore need no longer figure among the elementary terms of science (*APo.* II.8, 93b4–14).

This brings us back to the *pons asinorum* and its possible application to astronomy. Aristotle, significantly, does not speak in the passage quoted above of *lists of terms*, but of the *principles* that make it possible to generate them exhaustively. A completed science is a discipline that provides *in advance* the terms necessary for

33 These properties are set out in particular in *APo.* I.2 and *APo.* I.4.

the application of the *pons asinorum*– or more generally, in today's terms, for the application of procedural methods for problem-solving.

In such a framework, a method such as the *pons asinorum* becomes conceivable *in principle*, because the space of possible solutions is bounded by the finite number of available elementary terms. At this point one can glimpse the essential relationship that exists between the notion of scientific systems and what I have called *systematic problem-solving* at the outset of this research. A systematic method must define a problem space which terms have a definite meaning and which can be explored in a definite manner so that the solution can be proven to be exhaustive or complete, *i.e.* correct for all cases of the problem. The difficulty lies therefore in this problem space itself: how to find it? how to prove that it is itself complete? That difficulty is lifted if an already warranted scientific system is available, from which relevant problem spaces can be carved out more easily and securely.[34]

One might object to the naivety of the *pons asinorum*, which is arguably one of the least impressive algorithms that the Greeks have bequeathed to us, especially if we compare it to mathematics. To this objection it may be first replied that this naivety derives directly from that of the whole syllogistic theory itself, which Aristotle presents as the universal form of scientific reasoning. Secondly, regarding practical matters, one may add that this naivety was shared (at least for a time) by Simon himself when he gave the boasting title of General Problem Solver (GPS) to an algorithm that was nothing more than a recursive version of the *pons asinorum*.[35] But beyond the inadequacy of the logical means available to Aristotle, the *pons asinorum*'s particular interest lies in its very simplicity, in the very simple claim it makes: that a problem sufficiently circumscribed by a scientific system could be analysed successfully by a method consisting in the exploration of this system in search of solutions. As Crubellier (already quoted) says, what interested Aristotle in the *pons asinorum* was that such systematic procedures should be possible in principle, *i.e.* that one could be sure of finding a solution by traversing in a definite way the space mapped by the scientific system. For then another approach to problem-solving becomes conceivable, the one promoted by Simon after the disappointment caused by the GPS, of developing "search strategies" to gradually reduce this complexity (see above 2.1.3); this is what many modern algorithms do, by progressively reducing the space of possible solutions to a given problem, and it is what the *pons asinorum* already does in a modest way, by classifying lists of terms into groups with strong incompatibilities between them as well as different priorities for consideration. Aristotle notes with satisfaction this property of his method, since it tells

---

34 The full development of this idea is the subject of chapter 7.

35 See above, 2.1.2, and appendix A.

how to "chase modes from syllogisms, without having to inspect all the predicates" (*APo.* I.30, 46a12).

4.3.4 The unicity of inquiry and its species

The former discussions provide a fresh perspective on the nature of inquiry (ζήτησις). It is indeed common to science, *technè* and practice.

Deliberation constitutes the *essential* part of knowledge in the ethical domain, where it is again required in each particular case. As Allen says (quoted above 4.1.3), ethical deliberation certainly determines a theorem, but this theorem is "specific to the occasion"; the reasoning will have to be reiterated each time, if only to ensure that it is similar to the situation that served as a model.

The case of techniques is intermediate, as they facilitate deliberation by means of a strong framework.

In the completed system of science, inquiry is doubly destined to disappear, firstly because the scientist will have reached their end, the contemplation of principles, and secondly because, if they sometimes still have to solve particular problems, the system of essential truths they possess will enable them to analyse them methodically. This explains perhaps why Aristotle, busy describing this final state of knowledge in the *Second Analytics*, says so little about scientific inquiry.

However, in the real and unfinished business of science, inquiry is very much present through the strong opposition that Aristotle institutes between the state of knowledge in actuality and the movement of acquiring knowledge. This is clearly seen in the formulation of *APo.* II.1 and its mirror construction:

> The things we seek are equal in number to those of which we have scientific knowledge.
> Τὰ ζητούμενά ἐστιν ἴσα τὸν ἀριθμὸν ὅσαπερ ἐπιστάμεθα. (*APo.* II.1, 89b22)

Thus, the distinction between epistemic and deliberative reason in *Nic. Eth.* VI.1 seems to hide another, equally fundamental one, which would oppose epistemic reason and *inquiring* (to translate "zetetic") reason. The latter would include the deliberative reason of practical and technical affairs, but also the reflection that is at work in scientific problems.

This unity of inquiry is alluded to in a discreet way in a text of the *De anima* that describes the passage from science in potentiality to science in actuality. This passage is described as a movement of the noetic faculty (τὸ νοητικὸν) which pursues the good and avoids the bad, and "for that which is without action, the true and the false, which is of the same kind as the good and the bad" (*De anima* III.7, 431b10). This elliptical text may be understood as an allusion to the scientific inquiry (ζήτησις) which, as "one calculates and deliberates future things from present things" (*De anima* III.7, 431b7) conducts its research by exploring leads with greater plausibility, and avoiding those with lesser. The procedural devices of *Topics* and *First Analytics* would thus aim at supporting this inquiring reason.

4.4 Contingency and technical systems

This section explores further the link between the notions of *system* and *problem-solving*. The system allows problem-solving to be exhaustive because it can be *traversed* entirely in search of a solution. In this feature lies the core idea of a *procedural method* that does not solve its problems through the penetration of their essence, which requires "quickness of wit" (ἀγχίνοιά), but by the external approach of searching a solution in a problem space (4.4.1). It is then shown how Aristotle could rightly be sceptical about the possibility of constituting technical systems that meet such requirements, by referring to his opposition between necessary and contingent principles (4.4.2). To conceive the possibility of systematic practical procedures requires thus to requestion his whole ontology (4.4.3).

4.4.1 Method *vs* quick-wit

The solution to a given problem is systematic, as said, not only if its correctness can be demonstrated, but also if it is exhaustive with respect to all possible cases. Such properties can only be demonstrated in the context of a system of reference that allows different options to be explored. In a surprising passage from *APo.* I.33, Aristotle argues that scientific knowledge is not distinguished from true opinion by the fact that the former would be founded by a demonstration, and the latter not, for it is possible

> that the man who opines will proceed by means of the middle terms until they reach the immediate premisses [...] For it is possible to opine on fact and the reason for it, *i.e.* the middle term. (*APo.* I.33, 89a14, transl. Trennedick and Forster 1960)

If such reasonings are not scientific demonstrations, according to Aristotle, it is because they do not grasp the concatenated propositions as necessary, and that in turn, because they do not recognise the definitions as essential, that is, they do not recognise *where* are the principles and where are the accidents, what is necessary and what is only contingent. In other words, doxastic reasoning lacks the "coordinate system" that a scientific theory provides, and which makes it possible to *locate* it, *i.e.* to make explicit how it relates to principles and thereby to consider different options and to compare their respective merit.

What allows this exhaustive exploration of the set of truths is an elementary property of the theory of demonstration that Aristotle presents in *APo.* I, namely that it has a *recursive* structure, as noted by Barnes (1981: 26, note). One may add that it shares this property with the Platonic dieresis it seeks to compete with. In the case of dieresis, a term is either prime or obtained from the division of other terms. In Aristotle's theory of demonstration, a true proposition is either a principle or a proposition demonstrated from other true propositions:

> A syllogism is a demonstration when it proceeds from true and first [premises], or from [premises] whose knowledge originates from true and first [premises].[36]

Barnes (1969: 147–150) has shown the difficulties encountered by Aristotelian logic in giving to this recursive structure an infinitary character, *i.e.* to show that a indefinite number of conclusions could be generated from a finite number of prime terms, a property that Aristotle nevertheless seems to assume.[37] However, whatever these difficulties, it seems that the essential point for Aristotle is that, in such a structure, exploration can be complete because it is enough to start from the first terms to reconstruct, with the help of the rules of derivation or division, the set of possible terms in order to examine them one by one.[38] It is this interest in establishing this possibility of complete exploration that might explain the many chapters that Aristotle devotes to establishing the "topological" properties of scientific knowledge in *Second Analytics*, such as the closure of demonstrations, the impossibility of circular or infinite demonstrations in any way, the possibility that there may be several demonstrations of the same proposition, the exclusion of randomness.[39]

These discussions do not imply an endorsement of Barnes's general thesis, nor that Aristotle was a precursor of the modern idea of recursive formal systems. It is sufficient for my purpose to show that the idea of a methodic analysis of problems, dependent on the availability of scientific systems, is consonant with the Stagirite's thinking about the ideal form of knowledge. The source of Aristotle's intuition seems

36 *Top.* I.1, 100a27. See also *APo.* I.2 whose formulations are less immediately recursive.
37 *APo.* I.32, 88b6. This passage is considered obscure by commentators and interpreted in various ways.
38 This is true of propositional logic, which is complete and decidable. In the case of more complex logics, exploration must take place at the underlying level of theorems, that of valid expressions.
39 See respectively: *APo.* I.7, *APo.* I.3, *APo.* I.19–23, *APo.* I.29, *APo.* I.30. See a similar reflection by Barnes, who speaks of the "structure" of an axiomatic system in his edition of *Second Analytics* (1994: xx).

to lie, as Pellegrin says, in his attempt to characterise the *state of mind* of the scientist, which distinguishes them essentially from that of the man of opinion. This attitude consists in examining problems not as they come, with their own provision of concepts and hypotheses, but *within* a predefined framework of reflection which describes them in a uniform manner, gives their exact "coordinates", and frames their resolution according to pre-established paths and rules.

Perhaps the innovation of Aristotle is best seen if it is compared with the *anamnesis* of Plato. Unlike the latter, Aristotle does not think that it is possible, faced with each question and each problem, to undertake their analysis by considering them in their own particularity and by going back, through the sheer power of conceptual analysis, to common notions accepted universally. Such a superhuman conception of intelligence does not seem to him to reflect the reality of knowledge, which progresses rather by accumulations local to each discipline. He thus interposes between particular experience and common notions the idea of "sciences", *i.e.* dispositions to apprehend a certain domain of things and their internal connections in their universality and necessity. These dispositions are not innate but must be patiently acquired and constructed, through discovery or learning.

It is thus possible to put forward the following hypothesis: Aristotle, it seems, perceived that the flaw of any dialectical method lies in the fact that, however sophisticated its internal argumentative procedures, its effectiveness ultimately depends on the quality of the data initially provided. The problem of the canonical lists of terms necessary for the proper functioning of the *pons asinorum* does not seem to have been naively ignored by Aristotle. On the contrary the text already quoted (2.3.3) from the *Top.* I.14 shows that he had a bitter and laborious experience of list-making. Through the theory of the syllogism he conceived, as Crubellier shows, a universal matrix of argumentation by which he foresaw the possibility of reducing these *ad hoc* lists to those, unique for each science, of the elementary terms from which all the others could be derived.

It seems, then, that Aristotle's intuition is more original and fundamental than the formal idea of a deductive axiomatics to which Barnes reduces the whole idea of a system. To aim to solve a problem methodically implies to consider it, at least in some aspect, *as just another* problem without worrying about its particularity. One might focus, for example, on the form of the question, on the terms it uses, on the means available, on the state of mind of the applicant, etc. and, on the basis of these characteristics *external* to the problem itself one would decide to act in such and such a generic way. The methods proposed by the *Topics* and the *pons asinorum* are characteristic of this approach. They do not address the subject by going straight to its content or by appealing to common sense, but they gradually encircle it with observations, auxiliary collections of facts, comparisons with pre-established patterns. In this sense they can appear rhetorical and dialectical, and deviate from the very spirit of dialectics as established by Plato, *i.e.* to aim directly at the thing itself in its essence.

The method of *diaeresis*, practised by Plato's Socrates, is each time a miracle of intelligence. Practised by less skillful students, it is no more than a vague recipe with uncontrollable results. Aristotle acknowledges this possible flaw in any method, that it can easily be deviated into repetitive, nonsensical formulas, or even hijacked for sophistical purposes. Alone, it cannot compensate for the inquirer's lack of intelligence or probity. However, he does not draw the conclusion that this methodical route should be rejected. On the contrary, he "doubles down" on the idea that any method, in order to be effective, must itself be inserted into a system of knowledge which constrains it, and which has been verified through prior inquiry. In such a system, one continues to approach a problem *from the outside*, but in a wholly different way than dialectics. Far from relying arbitrarily on accidental characteristics of the problem to be solved, the method here encloses it entirely within the system like a hunting drive draws its prey into an ever-narrowing circle of the forest, to extend the hunting metaphor of inquiry cherished by Aristotle.[40]

Behind the idea of the *pons asinorum*, and more generally of procedural methods, there is the very simple idea of a "systematic search" as it is meant in everyday language, when, for example, despairing of finding an object in the usual places, one decides to stop making assumptions about its location and to explore the rooms of one's home one by one, in their every nook and cranny. This method is not just a fall-back solution when quick-wit fails to find the middle term (*APo.* I.34, 89b10). The *pons asinorum*, pedestrian though it is, makes clear that the solution lies within a certain perimeter delimited by the system. It unmasks false intuitions, lists all possibilities, identifies the best of them, and often solves a whole class of similar problems in one fell swoop. The oppositions sketched in broad strokes in the previous chapter between human cognitive ecologies and industrial and computer procedures are in some ways already contained in the implicit and unresolved tension between the two ways of finding the middle term which Aristotle indicates less than forty pages apart in the *Analytics* (the *pons asinorum* and quickness of wit), without reaching a decision on the superiority of either one.

4.4.2 Contingency and necessity

This preliminary outline allows to better circumscribe why technical syllogisms cannot reach, in Aristotle's theory of knowledge, the same certainty as scientific

40 Barnes (1969: 147) notes that the term *hunting* (θήρα) appears four times in the *Analytics*.

ones. Crafts cannot constitute themselves in a system, for there can be neither principles nor demonstration of contingent things:

> There are certain things which, though true and real, can be otherwise. It is therefore clear that of such things there is no science [...] nor intellection (I call intellection a principle of science) or non-demonstrative science, which is grasping of an immediate premise. (*APo.* I.33, 88b33)

Therefore, even if the *pons asinorum* is open to crafts in principle, as to any other discipline, such a procedure will be of little benefit to them since they cannot hope, unlike the sciences, to enclose their concepts within a systematic totality. They will have to laboriously draw up, for each given problem, lists of terms describing the problem to be solved. As seen above with the example of rhetoric, *technè* may at best provide a framework for this preliminary activity of describing the problem, but it can in no way replace inquiry.

But this answer begs another question: Why did Aristotle think that techniques could not take the form of systems? Until now the oppositions established by Aristotle between science and *technè* were accepted at face value, namely that the former has to do with the real causes of things and their necessary principles, while the latter is content to identify the general means of action immersed in contingent situations.

These oppositions, which are intuitively acceptable enough, may justify why science and *technè* are related to two different faculties of the soul. However, they do not explain why, both aiming at the universal in their respective fields, they could not both also be organised in deductive systems around principles of reasoning and primitive concepts. The success of the "engineering sciences" in our time shows quite clearly, it seems, the falsity of such an assumption.

One needs to return to the two oppositions between cause and means on the one hand and between necessity and contingency on the other, in order to understand why Aristotle associates them (wrongly as it seems) with this third one between the systematic and the unsystematic mind. In particular, dependence on necessary and true principles seems quite accidental to our modern idea of a *system*. Its epistemic value resides rather, for us, in the validity of its internal connections, in its logical coherence, as Nicholas Rescher claims,[41] which may remain intact even when its initial principles (axioms, definitions) are requestioned. One speaks in this sense of a *hypothetico-deductive system*. Now this conditional mode of truth is precisely, as said in the previous section, that of *technè*: it therefore seems that the idea of *system* suits it very well.

---

41 On Rescher's views on systematicity, see chapter 7, introduction.

In *Met.* Δ.5, Aristotle considers four ways of understanding the term *necessity*, which he subsumes under the generic definition of "that which cannot be otherwise than it is" (*Met.* Δ.5, 1015a34). Logical necessity, which binds conclusions to premises, can be set aside since it is precisely what makes the unity of the deductive system, whose potential objects are now under investigation. In addition to the necessity of perfect and simple beings, which seems to refer to theology (a non-demonstrative science) Aristotle mentions two forms of conditional necessity: that of efficient causes, which constrain the natural movement of things, and that of final causes, which require means in order to be realised. The demonstrative sciences can therefore focus on these two forms.

Our modern instinct would naturally associate science with knowledge of efficient connections between things, and technology with knowledge of final connections. Since the latter must ultimately be related to human intentions, which are contingent, they would not be the object of science. However such an interpretation is only partially correct from an Aristotelian point of view. Aristotle, whose science of nature is through and through functionalist, sees rather in final connections the reign of true necessity, while efficient connections are the reign of randomness.[42] It is the internal dynamic of transformation of natural forms that explains external movements, which are violent *relative to another natural form* insofar as they come to constrain its own dynamic. Mechanical phenomena are in essence the locus of chance,[43] *i.e.* the accidental interference of otherwise perfectly explicable (final) causal series. However, when these interferences are regular because they concern necessary principles, such as certain meteorological phenomena which are due to the movements of the stars, they may become by extension objects of science. On the other hand, two categories of beings are excluded: first, irregular natural mechanical phenomena (such as an earthquake), and second, anything that derives from contingent human intentions.

These two forms of contingency are called *fortune* (τύχη) *insofar* as they disrupt human intentions. They limit the ambitions of techniques in different ways. Mechanical hazards limit their ambitions for reliability and precision. But it is the diversity and volatility of human intentions that essentially prevent technology from becoming a system.

If the contemporary point of view is abandoned, and the point of view of the traditional observer of crafts is recovered – which is the one still held by the early modern authors of *reductions in art*[44] and even by the encyclopaedists of the 17th

42 *Phys.* II.8. See the two studies by Lennox (1982, 1984) on the notion of chance in Aristotle.

43 The term here used in Greek is αὐτόματον. See Presti 2012: 26–32 for the different meanings of this term and their conceptual connection in ancient Greece.

44 On reductions in art, see below 7.3.2.

and 18th centuries – one may grasp the plausibility of such a position. Admittedly, a craftsman's workshop, such as a blacksmith's forge, can be impressive in terms of its organisation, know-how and procedures, and all of this may undoubtedly be described systematically on a local scale. But the specific ways of doing things in this workshop will depend on the traditions, the way the work is organised, the shape of the tools and the special quality of the materials used in the region, as well as on the diversity of the needs and habits of the customers, which will decisively shape the workshop's activities. In other workshops and other regions, one would find many similarities in the organisation and general processes of production, but no doubt just as many differences in detail, which would be compounded by the multiplication of the instances of observation. The universality of the description would be impoverished, and it would probably not be capable of organising the variations observed into a coherent taxonomy, since their factors could be combined in an undoubtedly arbitrary way – the human cognitive ecologies introduced in the previous chapter are precisely made up of multiple and reciprocal adjustments between human practices and their material and socio-cultural environment. Indifferent to causes and stopping, in their ascent towards the universal, at the first stable and efficient system of means they encounter, techniques are only capable of generality but never of universality.

Techniques are therefore incompatible in principle with the idea of *system* because they are too variable. They are not even of the type of "what happens most often" that impoverished degree of necessity that Aristotle nevertheless admits among the modalities of the science of nature.[45] They have only local stability, which allow mere generalisation, but cannot reach the universal. This is another sense in which the image of the ladder of knowledge, already commented on above (4.1.1), can be understood: technical knowledge systems, if there were to be any, would be local and modest – and undoubtedly of little use.

4.4.3 The problem of technical systems

The results of this chapter may be summarised as follows: Aristotle's *essential* distinction between theoretical knowledge and practical knowledge is justified by their *different* objects. The former concerns knowledge of *what is*, and the latter of *what is to be done*. It is from this essential distinction that all the others proceed, including those between science and *technè*, which can nevertheless both claim universality:

45 *APo.* I.30, 87b25, *Phys.* II.8, 198b35. See Barnes' comments in his edition of *Second Analytics* (1994: 192) and Mignucci 1981 on this much-discussed notion.

1. Practical problems and technical systems are radically contingent in that they proceed from *rational human intentions*, which *see* things not as they are, but as they *could be*. They are therefore not about natural beings (objects or situations) but about beings transformed or engineered – artefacts, infrastructures and controlled environments.
2. Technical knowledge concerns deliberations and methods, not facts; it may mobilise facts, but it is not reduced to them; it deals with the *how*, not the *what*.
3. The technical attitude does not aim to discover the root causes of things, but the most effective means of action; their criterion of validity is not truth, but success. In this respect, technicians are pragmatic and, while they can mobilise scientific knowledge, their horizon is limited by their interest.

This third distinction is closely related to the famous one between the active and the contemplative life, set out in *Nic. Eth.* X.7. The latter is considered superior to the former in that it aims to understand things themselves, in a disinterested way, being concerned only with grasping them according to their essence and to their causes. The deliberative mind, on the other hand, is interested in things only insofar as they serve its own ends. There is therefore a radical difference of intention in the act of knowledge which, according to Aristotle, justifies distinguishing them *generically*. It is important to note in this respect that, in the text of *Nic. Eth.* X.7, the epistemic virtues referred to by Aristotle to characterise the contemplative life are intellect (νοῦς, 1177a14, 21, b30), principally and theoretical wisdom (σοφία, 1177a24, 34), rather than science (ἐπιστήμη) which is not mentioned. The latter therefore certainly participates in the theoretic life, but only insofar as it provides access to first principles, grasped by the intellect. In other words, it is not the "labour" of inquiry (experiments, demonstrations, applications) that interests the theoretical life, but rather the grasp of what lies at the foundation of physical and mathematical reality – the nature of matter, energy, number, the Good, and so on.

Aquinas will summarise all these distinctions by saying that speculative and practical knowledge are distinguished according to their object (natural being or artefact), according to their modality (knowing what the thing is or how to do it), and according to their end (knowing for truth or for utility).[46]

To these three distinctions, this reading of Aristotle adds a fourth one, between the systematic form of science, which is possible because the process of searching for causes converges towards first principles, and the merely "rhapsodic" form of crafts, whose means have only a local scope. Any attempt to unify one craft in a truly

46 On the opposition between truth and utility, see below, 5.1.1.

universal system would disintegrate into an explosion of particular circumstances and cases.

How is it then that modern technology is possible? How is it that technical crafts could be expanded into gigantic systems of production, trade, finance and transport? How was it possible to begin to think systematically and on a large scale about human practices?

Some of the basic assumptions of Aristotle regarding the distinction between theoretical and practical reason must therefore be requestioned. It is not sufficient to say that knowledge about causes may be traded between science, which discovers them, and practice, which makes use of them in procedures. *Procedures themselves* are a common basic material for both science and practice. It is so because they are, as the discussions above made clear, generalizations of inquiry, and because inquiry is the fundamental common ground of all knowledge.

This remark opens up a new avenue for interpreting the emergence of modern technology, which would be not so much a consequence of the scientific revolution than a parallel movement, triggered by the same shift *in our approach to problem-solving itself.* What Aristotle would therefore not have seen (without, however, forbidding it) is that the systematic attitude is not the privilege of theoretical inquiry, but that it may also be suitable for certain practical deliberations. *It is a possible modality of inquiry in general.*

5 The rules of judgement and action

The purpose of this chapter is to understand the general nature of procedures as common to both theoretical and practical knowledge. This commonality must not be understood simply as some features or qualities that practical and theoretical procedures may share. Rather, the aim is to understand how procedures in general can become objects of theoretical knowledge, and how theoretical knowledge itself is in some profound way constituted by and essentially concerned with procedures. In other words, this chapter explores how all practices can become objects of theoretical study, and how theoretical study is itself a practice.

This understanding cannot be grasped within an Aristotelian framework. For the Stagirite, procedures are not only contingent on the intentions that motivate them, but they are not really objects at all, only prescriptions: How could there be true knowledge of these contingent prescriptions? And how could these prescriptions be constitutive of theoretical knowledge, which aims only at a necessary and immediate grasp of first principles?

This second difficulty was actually apparent in ancient geometry, and it deeply troubled the Greeks. It triggered a controversy reported by Proclus about the primacy that should be given to theorems or to problems in this discipline. A theorem states a general property of the geometric objects under consideration, which must be proved, whereas a problem requires one to construct, from given initial objects, an object related to them such as, for example, constructing the circumscribed triangle in the circle $\mathcal{C}$.[1] Pappus thus defines problems and theorems from their verbal roots:

> Those wishing to distinguish more systematically between the results sought in geometry, [...] think it fit to call that in which it is proposed (*proballetai*) to do or to construct something a "problem", and that in which, certain things having been assumed, one observes (*theoreitai*) what follows from them and everything that goes along with them a "theorem". (*Collection*, book III, 2023: 9)

The problem would thus be a *project*, and the theorem a *contemplation*: this distinction, which seems to transpose the theory / practice opposition to the very heart of mathematics, has still the status of evidence in the early modern period, as this definition from Furetière's *Dictionnaire* (1701) laconically expresses:

> Theorem, subst. masc. A truth demonstrated and determined, or a proposition which is only object of speculation, and in which the properties of things already made are considered. As

1 Knorr 1983: 127; Sidoli 2018: 405.

https://doi.org/10.1515/9783111494500-005

> opposed to problem, which adds practicality and construction. Geometry is distributed into theorems, and problems.

The ancient controversy stems from the fact that, while theorems seem to give a good account of a "theoretical experience", problems bring geometry closer to the productive arts, which is unacceptable to supporters of Platonic Ideas. Speusippus' school thus rejects problem-solving as an improper formulation of their mathematical content, in favour of theorems. Indeed,

> [Mathematics] deals with eternal things, among which there is no generation. Thus a problem has no place with them, since it proposes to generate or produce something that did not exist before.[2]

This position originated with Plato himself, who took offence at the use of action verbs in mathematics, as this passage from the *Republic* makes clear:

> No one with the slightest acquaintance with geometry would deny that it completely contradicts the ridiculous language geometricians are forced to use in dealing with it. They speak as if they were doing something and all their words refer to actions – squaring, adding, applying, and all the other terms they use – whereas in reality, I think, the whole study is pursued for the sake of knowledge [...] of what always is; not of something now becoming, now passing away. (*The Republic*, 527a-b, ed. and transl. Brann 1979: 187)

Plato suggests here that this unfortunate form of expression is only legitimate for the preparatory exercises for knowledge, which must however end in still contemplation. Human understanding must rely on imagination to represent geometrical figures as if they were produced in time, but pure intellection must do without it. Problems would therefore be pedagogical expressions of an objective content that could be reformulated into theorems. Proclus attributes to the school of Zenodotus the idea according to which constructions aim to demonstrate the existence of objects whose properties are studied by theorems. The problems would thus be "existence lemmas", preliminary and subordinate to the theorems.[3]

This chapter aims to show in which sense it is legitimate to consider procedures as part of theoretical knowledge. Together with the next chapter, it will lead to a restatement of Aristotle's distinction between practical and theoretical knowledge, opening up an understanding of the possibility of systematicity in all areas of human practice.

---

2 Proclus, *A commentary on the first book of Euclid's Elements*, transl. Morrow 1970: 77–78.

3 Proclus, *A commentary on the first book of Euclid's Elements*, transl. Morrow 1970: 78–80.

This inquiry begins with the analysis of Kant's critique of Aristotle's distinction. Kant brings all knowledge under the idea of *rule* as the elementary unit of discursive understanding and explains thus the convertibility between technical knowledge, which gives rules for action, and theoretical knowledge, which gives rules for judgment (section 5.1). Section 5.2 then examines two criticisms of this unification of knowledge. The first, by Elizabeth Anscombe, argues that it misses the *intentional character* of practical knowledge. The second is logical and pays attention to the difference between descriptions and prescriptions. This objection requires developing, in section 5.3, a deeper understanding of their convertibility, using Martin-Löf's constructive theory of types. The result of this chapter is to retain Kant's fundamental insight, but to enrich it with a view of sciences as intentional human practices, and to strengthen it with contemporary logical theories of rules.

5.1 The rules of judgement and action

This section examines challenges made to the Aristotelian generic distinction between deliberative and epistemic reason, arguing for the fundamental unity of knowledge. It begins with an outline of the major transformations of this distinction that took place in the Middle Ages, within a framework that was still Aristotelian: Firstly, the association of speculative knowledge with theology, second the recognition, by various means, of the complementarity, even unity, of theory and practice in any intellectual discipline, and finally the characterisation of practical knowledge as *ordinatio* or knowledge of the order of things. All these movements prepared the ground for Kant's strong denial of any generic distinction between theoretical knowledge (of *what is*) and practical knowledge (of *what to do*) (5.1.1).

I then study the Kantian thesis of the inclusion of technical-practical[4] imperatives in the domain of theoretical knowledge. This assertion is a consequence of Kant's thesis that understanding is the *faculty of rules* (5.1.2), and that technical prescriptions rely on the same rules that also regulate theoretical judgements (5.1.3).

4 Since Kant's use of *practical* often refers to the sole domain of morality, I will sometimes use the phrase *technical-practical*, as he does himself, to refer to the domain of human practices as it has been circumscribed in section 2.3. This domain includes technical prescriptions but is not limited to them.

5.1.1 The speculative unity of knowledge according to Thomas

The first transmissions of the distinction between theory and practice in the Latin tradition were made by Augustine and Boethius, who loaded it with Christian connotations.

Augustine, when he translates θεωρία as *contemplatio*, is primarily interested in *Nic. Eth.* X's contrast between active life and contemplative life. By associating this text with the Gospel story of Martha and Mary, he charges *contemplatio* with a religious meaning that only incidentally connotes the study of physics and mathematics. The neo-Platonic influence of Augustine also directs the term towards designating the perfect and completed act of the theoretical life, *i.e.* the direct vision of divine truths.

Boethius, wishing to translate θεωρία in the context of the distinction between theoretical and practical philosophy, is not satisfied with these too narrow connotations of *contemplatio* and chooses instead *speculatio* (Kerstiens 1958: 383). The term, originally meaning *to explore, to watch*, is used by Cicero to refer to physicists who "observe and explore nature" (*De Natura Deorum* (2003: 1: 83)) but also by Jerome to translate both the vision from a high point, in the literal sense, and the prophetic vision. Moreover *speculatio* retains the connotation of examination and inquiry; it therefore appears better suited to translate the rich meaning of the Aristotelian θεωρία, and becomes predominant in the Latin Middle Ages. *Speculatio* remains nevertheless resolutely oriented towards the search for God. A play on words on Paul's famous verse (1 Cor 13,12): "At present we see as through a mirror (*speculum*), but then we shall see face to face", allows Latin theologians to develop the idea that speculation on natural things has the profound purpose of contemplating God through the mirror of Nature. But this implies that the Aristotelian idea, according to which the theoretical life is concerned with knowledge "for its own sake", no longer really has any other object than God himself; mathematics and physics lose, strictly speaking, their status as "disinterested" knowledge.

As a result, the complementarity and even unity of practical and theoretical knowledge of things in the world becomes more acceptable, at least from three different perspectives.

First, the Arabic tradition delivers to the Latin Middle Ages the idea that one can distinguish, *within one and the same science*, a theoretical part and a practical part. This usage first appears for medicine, perhaps in the Alexandrian medical tradition of the 6th century (Cunningham 1982: 407–413). It is attested by the Arab physicians Hunayn Ibn Ishaq (801 – 873), Ali Ibn Abbas (930 – 994), Avicenna (980 – 1037) and spread through their translations in the Latin world. But it was in Al-Farabi's (872 – 950) treatise on the *Division of the Sciences* (2015), that this usage was applied broadly to several disciplines, and in particular to mathematical disciplines, making explicit a socio-cultural division rooted in antiquity (Asper 2009). Thus, theoretical arithmetic, which studies numbers "taken absolutely and insofar as they are abstracted by the mind from bodies and everything that is numerable" is distinguished from practical arithmetic "which studies numbers [of things we count] [...] such as *men, horses, dinars, dirhams*, and everything that can be counted and is the object of public transactions in markets and civil life" (Al-Farabi 2015: 125), which probably designates the various merchant systems of measurement and their rules of calculation and conversion that occupy a good part of medieval arithmetic textbooks. Similarly, theoretical geometry is distinguished from practical geometry "which studies the lines and planes in a body" whether that body be the wood of the carpenter, the iron of the blacksmith, the wall of the mason, or the field of a surveyor (Al-Farabi 2015: 160). It is interesting to note that Al-Farabi defines mechanics as the discipline responsible for the *realization* of mathematical conceptions in the material world, and algebra (by which he means what would be today called algorithmics) as "arithmetical mechanics" that gives the methods for "deducing numbers". However, this idea does not seem to have had any notable posterity.

Al-Farabi's treatise was translated into Latin in the 12th century. At the same time, Hugues of Saint-Victor (? – 1141) proposed a quadri-partition of knowledge into *theoretical, practical, mechanical and logical*.[5]

Theoretical philosophy – Hugues uses the Greek term again, but only in the adjectival form *theorica* – includes theology, the mathematics of the quadrivium, and physics. By *practice* he is aiming at the ethical domain (as Kant would later do). Under the generic term *mechanics*, he brings together techniques such as clothing, navigation, construction, agriculture, etc. *Mechanics* here refers only distantly to the use of machines, and Hugues associates it with the adjectives *servile* and *adulterine* (from a pun on *mechus*, adulterous, since technique creates artifices through the union of intelligence and nature) – pejorative terms that would constitute the bulk of its meaning until the early modern period.[6] However, Hugues has the merit of recognising techniques as an authentic field of knowledge. (Châtillon 1965: 546) He broadens decisively the horizon of knowledge, hitherto reduced to University teaching.

The classifications of Hugues and Al-Farabi eventually merged and formed the basis of the scholastic vision of techniques. The idea thus became commonplace from the 12th century that theoretical sciences could be of great use to the mechanical arts (Beaujouan 1957: 10).

5 *Didascalion de studio legendi*, book II, chapter 1, cited by Châtillon (1965).

6 On the meaning of *mechanical* in early the early modern period, see Gabbey 2004: 12.

Aquinas thus notes that some knowledge is purely speculative (e.g. mathematics), some purely practical (e.g. agriculture) and some “speculative in one aspect, and practical in another”, e.g. medicine (*Summa Theologica* I q. 14, a. 16). This new use of the *theory / practice* opposition, which may seem natural to us, significantly blurs the clarity of the Aristotelian scheme. It calls into question its ontological foundation, namely the distinction between necessary and contingent objects. Thomas explains this possibility by distinguishing three possible senses by which knowledge may be described as *speculative* or *practical*: according to its object, according to its modality, and according to its finality. This distinction allows him to extend significantly the domain of speculative knowledge:[7]

1. Knowledge is purely speculative when it relates to an object that cannot be produced by the knowing subject: thus natural objects, about which one can only speculate.
2. An artefact can also be observed speculatively, e.g. a house when one “thinks about its kind, differences and properties” in such a way as to define or classify it, following an analytical approach.
3. An architect might also think about how to build this house, but without intending to, asking “how it might be built, not in order to build it, but simply to know”. Knowledge here is “practical virtually”, but it is speculative “according to the end”, for it is only knowledge that is the objective here of reflection (*De Veritate*, q. 3, a. 3).

For Thomas, – and this is the second major shift made on the theory / practice pair, neither of these distinctions is absolute; only the last one is relevant:

> When we distinguish philosophy or the arts into theoretical or practical we must do so on the basis of their finality, naming theoretical that which is directed only to the knowledge of truth, and practical that which is directed to operation. (*Super Boetium De Trinitate* q. 5, a. 1)

Thomas thus considers that one can have speculative knowledge of the artefacts and methods of action, but that is not all: in a decisive move, Thomas completes his demonstration of the unity of knowledge by noting that the distinction of knowledge according to its finality, the only one he deems relevant, comes to it from outside. It is not even essential to it:

> An accidental element in the object that specifies a power does not diversify it, as we have already said. It is accidental to the coloured object that it is man, that it is large and small; also

7 I follow here closely the two parallel texts of the *de Veritate* , q. 3, a. 3. and the *Summa Theologica* I q. 14, a. 16.

> all these are grasped by the same power to see. But it is accidental to an object grasped by the intelligence whether it is ordered to action or not. And this is where the speculative intellect and the practical intellect differ. The speculative intellect is that which, when it apprehends something, does not order it to action, but only to the contemplation of truth. On the contrary the practical intellect orders to action what it apprehends (*Summa Theologica* I q. 79, a. 11)

Thomas's position is that the practical intellect "extends" the speculative intellect through consideration of the end, based on the text of *De anima* III.7. Aquinas sits here in the tradition of John Philoponus (6th century), according to whom the intellect becomes practical by adding, to the consideration of true and false, that of good and bad. *Good and bad* are to be understood here in their most general sense, which includes both their absolute (moral) and relative (utilitarian) valence. This interpretation presents practical knowledge as a *modality* that is added to simple knowledge which is essentially speculative.[8]

Finally, a third shift takes place if one considers the divine intellect: God does not know things insofar as he apprehends them, as if they were external to him, but insofar as he is their creator and cause. His understanding of things is therefore practical:

> God knows created things in the way that a craftsman knows the products of art, which knowledge is the cause of the products of art. The relation of this knowledge to the things known is thus the reverse of that which our knowledge maintains (*De Veritate*, q. 2, a. 8)

This idea, which goes back to Augustine (Crombie 1994: 2: 1088) can easily be traced back to the medieval tradition of the *Timæus*. Although technical knowledge refers here to final causes (there is still a long way from the idea of a "Mechanical God" that appears at the end of the 16th century), this idea nevertheless allows the emergence of an epistemological posture that runs through the early modern period, from Nicholas of Cusa to Descartes, and which the historian of science A. C. Crombie calls *hypothetical modelling*: namely that to know something is to know how it could be done (Crombie 1994: 2: 1081–1241).

What, then, is the structure of this practical knowledge? Its main characteristic according to Thomas lies in the notion of *ordinatio*, which I will translate here, in the precise context of this research, as *ordering*. It is first and foremost a theological concept, describing the fundamental orientation of all creation towards God, and underpinning the Thomistic conception of divine providence (Durand 2012). But it also has an epistemological interest, for *ordering* appears in three aspects: there is first that of finality (the order of use), then that of structure (the order of the

8 John Philoponus, *Commentary on the Treatise on the Soul* (1897: 554–555).

parts among themselves) and finally that of procedure (the order of generation) – themselves being ordered to one another as if by interlocking:

> One must note that an artisan, by conceiving in advance his artwork, considers primarily the end; only then does he consider the order of the reality that he intends to make to that end, and also the order of the parts to each other, like that the foundation will be under the walls, and the walls under the roof; and this order of the parts to each other is further ordered to the end of the house. Third, he must consider things that promote the attainment of the end, even such that he might remove things that can impede the end; whence he thinks through things that supplement the house by the addition of certain shrubs, windows, and the like, by which the house is suitable for habitation. [...] By reason of the order thought out in the thing to be accomplished, it is given the name 'plan' (*dispositionis*). For plan signifies a certain order, so the plan is said to be the ordination of the generation. (*Scriptum Super Sententiis* I d. 39, q. 2, a. 1, transl. Decaen 2025)

These epistemic schemes are very different from those taught by classical logic, centred around categories, predicables and universals. Duns Scotus takes up the same lesson as Thomas, and uses the concept of *ordinatio* to explain the difference that *Met.* A.1 makes between experience and technique:

> There are two ways to know practically: to know how to act (*scire operari*), and to know what is ordered to act (*scire ordinatum ad operari*). The first is properly practical knowledge; the second is called practical in the broad sense. And in this second way every man of art has practical knowledge, but not in the first. In the second way to know can be call 'speculative' relatively to the first. (*Questions on the Metaphysics of Aristotle*, I.5, transl. Etzkon and Wolter 1997 (modified))

This distinction allows Duns Scotus to explain to what extent practical knowledge is not reduced to knowledge of particular cases: if this is the case for know-how, assimilated to the simple skill of the man of experience, "the knowledge of art is by itself universal, and if it is knowledge of the singular, it is by way of consequence and by accident" – a remark that ties in with the previous analysis of Aristotelian *techné* as the universalization of a deliberation. The *ordinatio*, which is the act of putting actions in order *with an end in view* plays a central role in this new modality of access to the universal: it sets knowledge's objects to be means / ends structures, who can be applied to a diversity of situations.

This review shows that the essential conceptual shifts concerning the theory / practice pair had already been accomplished *within* the very scrupulous Latin reception of Aristotle: weakening of the ideal of disinterested knowledge (apart from the search for God), accidental character of the distinction between theoretical and practical knowledge, and primacy of the practical model of knowledge in the divine intellect. The value shift towards practical life characteristic of the Renaissance spirit is therefore deeply rooted in the Middle Ages. The role of the early moderns on this subject will be only to resolve the tensions already operated by these displacements.

Firstly, the association of theoretical activities with religious activity weakens their privileged position among human vocations. There has always been, within the Church and theological reflection itself, a current of mistrust for the pure intellectual attitude, tainted by the risk of "curiosity". Already in the 12th century, John of Salisbury asserts that speculation should only be sought when it bears fruit – which expressly contradicts the Aristotelian ideal. (Kerstiens 1958: 389) In the 14th century, Jean Gerson calls for a return to pure faith, freed from intellectual constructs concerning God, thus foreshadowing the humanist rejection of scholastic speculation (Ebbersmeyer 1995). The most virulent religious criticism of speculative activity, however, comes from Luther, who rejects the idea that natural reason can rise to an understanding of God. He can only be known through his own personal revelation, as God who acts in history.[9] Through this current emerges the idea that access to God does not involve speculative reflection, contrary to what Thomas asserted.

Alongside this religious critique of speculative activities, humanism revalues the active life, and the ideal of the good life. Here again, antecedents to this trend may be found as far back as the 12th century, with Bernard of Clairvaux, praising *praxis* and the organised and disciplined work of monks as the privileged path to God (Stock 1975). In the 14th century, Collucio Salutati asserts that the goal of human beings is not to know God through speculative knowledge. Referring to Cicero, he argues that the active life is the goal of the human being, thus inaugurating a tradition of "humanist pragmatism" (Ebbersmeyer 1995).

This shift in values partly reflects a societal change: established, educated craftsmen are aiming for recognition of the nobility of their arts and are rebelling against their characterisation as *servile arts*. They seek to show that their craft requires ingenuity and reflection just as much as the liberal arts. It is now the arts, both liberal and mechanical, that are at the centre of knowledge issues. The Renaissance period sees the multiplication "in almost incredible numbers" (Gilbert 1960: 69) of manuals devoted to methods and to the arts.[10]

This valorisation of practical knowledge becomes a standard of modern philosophy against the "ancients". The command made to students to visit the workshops of craftsmen is a recurrent *topos* of the early moderns, from La Ramée to Leibniz, via Rabelais, Descartes, and so on.

---

9 See Obermann's focus on Luther's relationship to philosophy (Oberman 2003).

10 I come back to these *reductions in art* in 7.3.2.

5.1.2 The concept as a rule of judgement

The three shifts outlined in the previous discussion – that all sciences have a practical as well as a theoretical dimension, that the distinctions between theory and practice (according to their object, modality and purpose) are relative, and that divine knowledge must be understood as productive of its objects – all these shifts take place within a predominantly Aristotelian tradition. They lead to the affirmation, against Aristotle himself, of the fundamental unity of knowledge. So when Kant asserts that technical knowledge is to be regarded as a mere corollary of the principles of theoretical philosophy (text quoted in 4), he seems to be settling the score of an opposition that has long since become obsolete.

The Kantian position nevertheless remains innovative in that it provides a foundation for this unity of theory and practice. However far removed mathematical theorems or the principles of physics may be from military tactics or business plans, they all fall under the same faculty of knowledge, which is understanding, and which determines both (theoretical) judgement and (practical) decision.

To see why, it is useful to start from the central proposition of Aquinas, who affirms the possibility of considering practical knowledge speculatively. This occurs, Thomas says, when the architect asks himself: "If I wanted to build the house, how should I go about it?" For the transformation of the practical consideration of an object into a purely theoretical one, it is enough simply to suspend the practical interest. The desired end must become, as it were, hypothetical. It is then that the rule-like character of practical knowledge is revealed: *if you want A, then do B.* It is precisely from this perspective that Kant states:

> All sciences have some practical part that consists of tasks to make some end possible for us, and of imperatives as to how it can be attained. these can therefore as such be called imperatives of SKILL. There is no question here whether the end is rational and good, but only what one must do in order to attain it. the prescriptions for the physician thoroughly to cure his man, and for a poisoner reliably to kill him, are of equal worth, in so far as each serves to effect its purpose perfectly. [...] They would, I believe, be labelled most suitably by saying that they are *rules* (*Regeln*) of skill[...] [or] *technical* imperatives (belonging to art). (*Foundations of the Metaphysics of Morals*, AA 04: 415–416)

One can note in this text the medieval theme concerning the "practical part" of the sciences (see above 5.1.1), and further on the Aristotelian example of the physician and the poisoner (*Met.* Θ.2, 1046b7). The Kantian doctrine of technical knowledge does not go much further than this. Kant devotes few developments to the topic, a sign that it must have appeared to him as non problematic. This is confirmed by the very few debates these ideas provoked – whereas the related topics regarding theoretical reason, the moral law, and even the pursuit of happiness inflamed post-

Kantianism. Kant and his contemporaries did not think that there was a break on this subject with the tradition going back to Scholasticism.

In the Kantian formulation, however, a decisive advance is made, which allows to perceive why technical prescriptions and theoretical judgments belong alike to the same domain of knowledge. Indeed, by thinking of concepts as *rules*, Kant opens the possibility of considering that *any rule for action can be also viewed as a rule for judgement.*

It is useful to summarize the thesis that the understanding is the power and source of rules before considering (in the next discussion) how it applies to practical rules. In this summary I rely on Longuenesse (1998)'s interpretation of Kant's theory of judgement, which has the merit of focusing on his reflection on rules. Kant's famous thesis is stated with some flourish at the start of the *Jäsche Logic*:

> Everything in nature, whether in the animate or in the inanimate world, takes place according to rules, although we do not always know these rules. Water falls according to laws of gravity, and in animals locomotion also takes place according to rules. The fish in the water, the bird in the air, moves according to rules. All nature, indeed, is nothing but a combination of phenomena which follow rules; and nowhere is there any irregularity. [...] Now, like all our faculties, the understanding, in particular, is governed in its actions by rules which we can investigate. Nay, the understanding is to be regarded as the source and faculty of conceiving rules in general. For just as the sensibility is the faculty of intuitions, so the understanding is the faculty of thinking, that is, of bringing the ideas of sense under rules. It desires, therefore, to seek for rules, and is satisfied when it has found them. (*Logic*, AA 09: 12, transl. Abbot 1963)

Concepts, as the product of the understanding, are therefore rules who articulate attributes (*Merkmale*) in one whole. Attributes are themselves partial concepts (*Logik*, VIII, AA 09: 59) *i.e.* rules themselves. These rules arise first from the synthesis of regularities found in experience, but they are then also rules for discursive thinking, as Longuenesse makes clear:

> The concept is a rule insofar as it is the consciousness of the unity of an act of sensible synthesis or the consciousness of the procedure for generating a sensible intuition. This first sense of *rule* anticipates what Kant, in the Schematism of the Pure Concepts of the Understanding, calls a *schema* (A 140 / B 180). But the concept is a rule also in another, discursive sense. It is a rule in that thinking an object under a concept provides a reason to predicate of this object the marks that define the concept. (Longuenesse 1998: 50)

The first point refers to the *determinative* judgement, which applies to the present situation the rules provided by the concept already formed (*Critique of the Power of Judgement*, AA 05: 179):

> If a matter is red and heavy, it is cinnabar.
> This has these characteristics.
> *So:* this is cinnabar.

The second point is illustrated by our ability to deduce a new rule from already existing rules:

> If *x* is an animal, then it is mortal.
> If *x* is a human, then it is animal.
> *Hence:* If *x* is a man, then he is mortal.

The propositions in this syllogism have been expressed hypothetically, rather than in the traditional categorical way (*All animals, etc.*), in order to make clear their status as rules. Kant is explicit about this:

> When I say 'A body is divisible', this means: 'some *x*, which I cognize through the predicates that together constitute the concept of body, I also think through the predicate of divisibility'. (Refl. 4634 (1772–1776), AA 17: 616)

This point, which shall be developed shortly, serves to emphasise that the "logical connection" of concepts to each other must always be grounded in reference to the object *x* to which it can be applied in experience. Discursivity, which allows us to organise our concepts into a coherent whole through the derivation of rules from one another, must always aim at their applicability to experience, under penalty of straying into the abstract and empty dialectics of pure reason. The rules stated in concepts make it possible to organise, through their meshing, the flow of sensible experience. They are also immediately *thought* to be objective, *i.e.* not due to my mere classificatory ingenuity, but as explanatory of the very regularity that I observe in experience.

The formation of concepts and rules is the object of the reflective judgement in which Longuenesse recognises the process "of reflection, comparison, abstraction", described in the *Logic* (§ 6, AA 09: 95). When we encounter phenomena that resist our immediate understanding, that is, for which we obviously cannot find a rule to which to subsume them, the understanding takes on a reflective attitude, aiming to formulate the rule adequate to this experience. This operation requires several phenomena to be *comparable* with each other so that, through *reflection*, the common rule can be identified (for example, the fact that pine, willow and lime all three have a trunk, which therefore becomes a character of the concept of tree under which we classify them) and that their differences are, on the contrary, *abstracted* from them.

This cognitive theory comes straight from the associationist tradition of Locke, Hume and Berkeley. However, according to Longuenesse (1998: 118–119), it differs from them in two essential features. On the one hand, against Hume and Berkeley, the concept of tree is not an image, a particular idea to which we confer a general value; it is indeed a general idea, as Locke intended; and on the other hand against Hume and Locke, we should not say that the general is not in things, but only in the way we think of them:

> Universals do indeed belong to the existence of things (they represent resemblances lending themselves to "rules of apprehension"), but are revealed in things only by the acts of "comparison, reflection, and abstraction" of the understanding. Although it is "made" as to its form, a concept represents something universal "present in itself" in the given object. (Longuenesse 1998: 120)

In other words, the simple empirical association of perceptions achieved by the reproductive imagination, and which is undoubtedly common to animals and us, is raised by the concept to the rank of *rule*, *i.e.* it is thought of as a property of reality itself, and therefore as necessary (Longuenesse 1998: 49).

Thus, in the Kantian doctrine, the concept, as a general representation, is framed by two acts: the reflective judgement that produces the concept, and the determinative judgement that subsumes the variety of experience under the concept. The concept is the result of a synthesis and makes subsequent syntheses possible. This structure clarifies the ambiguity that often hangs over Kantian assertions about judgement. First of all, the term designates two *acts* of different nature, the act of forming a concept and the act of applying it to experience or reasoning. But that is not all; it also designates the representations that result from these acts and that are expressed by propositions. Kant calls judgement both the act of forming the proposition "This is cinnabar", and this proposition itself. This second ambiguity also affects the other terms used by Kant to denote judgement: one hears the term *synthesis* rather as the act, and the term *relation* rather as the result of the act, although the other interpretation is equally possible in each case. Now this ambiguity is essential, because it denotes the fact that judging is an act that leads to an accomplished state (like "writing a letter") rather than a mere activity without result (like "running"). The judgement, in its propositional expression, states the rule elaborated by the reflective judgement and inscribed in the concept. It can be extracted again from the concept to form the major premise of a determinative judgement (*Prolegomena to any future Metaphysics*, §23, AA 04: 305).

5.1.3 Technical rules

An imperative of the will derives its prescriptive character from the fact that it is a judgement about the good or bad character of the state of affairs represented. Imperatives, Kant says, are nothing else than formulas that represent "an objective principle, insofar as it is necessitating for a will. [...] They say that to do or to omit something would be good, but they say it to a will that does not always do something" that is presented as good (*Groundwork of the Metaphysics of Morals*, AA 04: 413, transl. Gregor and Timmermann 2011). The same distinctions that apply to judgments also apply to imperatives. According to Kant's *Logic*, judgments can be categorical, when a predicate is judged as comprehended in a subject, hypothetical when it depends on another judgement being made, and disjunctive when it asserts that a subject containing a predicate also contain another from a definite set (*Logic*, AA 09: 105–109). Thus, an imperative is categorical if its principle is contained in the very object of the will, and on the contrary hypothetical if it depends on another imperative. This means that hypothetical imperatives "import" their validity (their "goodness") from the outside (*If* this *is good, then* that *is too*)[11].

Now there is no truly categorical imperative apart from the moral law – for nothing is intrinsically good except good will itself, as Kant claims at the beginning of the *Foundations of the Metaphysics of Morals* (AA 04: 393). For example, even the proposition "*Every merchant wishes to make a profit*", which seems only to make explicit what a merchant is, can be the object of a condition. For example, the merchant could happen to work for the State which has the goal to support local agriculture.

Hypothetical imperatives, on the contrary, infer the goodness of an object of the will from other goods which it allows to realize. They may have a categorical appearance, like "*Trade crops according to weather forecasts!*", whose corresponding judgement is "*All merchants in crops base their decisions on weather forecasts*". But these categorical forms, whether declarative or imperative, are only apparent ways of expressing a single judgement: *If you want to make a profit with crops, then you must consult the weather forecasts*. The State trader would on the contrary commit to long-term prices that guarantee farmers against the vagaries of the climate.

The rules which underlie hypothetical imperatives are final, *i.e.* they concern objects whose representation is an essential part of the causality of their cause (*Critique of the Power of Judgement*, § 63, AA 05: 367). This causality itself may be thought as efficient, *i.e.* when owning the house is conceived as cause of the rent, or final, *i.e.* when the rent is conceived as cause of building the house (AA 05: 373).

---

11 Disjunctive imperatives are (to my knowledge) only mentioned in a fragment (AA 29: 1017) and concern those which commend to take side, e.g. situations of war where one must not stay neutral.

These imperatives are therefore of two kinds, either technical - mechanical, when their means are material, or prudential - pragmatic, when their means are themselves representations. This second case includes, in particular, influencing other people's actions and representations to achieve one's end (AA 05: 172; see also AA 29: 1017). In a broad sense these ways can all be called mechanical, if mechanical is taken to mean any series of events occurring in time according to some natural causality. When the human will is considered as a natural cause upon which one can act, then it is considered as an *automaton spirituale* whose freedom "would at bottom be nothing better than the freedom of a turnspit" (*Critique of Practical Reason*, AA 05: 97, transl. Gregor 1997).

It is, of course, with regard to mechanical means in the strict sense that the identity of theoretical and technical rules can best be established. Kant often associates the rules of art with mechanical constraint,[12] and even takes it as a criterion for distinguishing useful arts from fine arts: an art reduced entirely to rules, in which skill would have no place, would not even be called such; it would be entirely *mechanical*, in the modern as well as in the old sense ("servile").[13] It is even a necessity that technical reasonings eventually resolve into some mechanical means:

> For where ends are conceived as grounds of the possibility of certain things, there one must also assume means the laws of the operation (*Wirkungsgesetz*) of which do not of themselves need anything that presupposes an end, which can thus be mechanical yet still be a cause subordinated to intentional effects. (AA 05: 414, transl. Guyer and Matthews 2000)

Judgement itself is the point of contact between theoretical and technical knowledge, *i.e.* it is judgement that identifies that the goals aimed at in a given situation can be met with natural means already present in the situation, directly or indirectly. This crucial role of judgement can be seen in the little treatise *On the common saying: That may be correct in theory, but it is of no use in practice* (AA 08: 275) which, while it deals mainly with political issues, is introduced by some general considerations on the opposition between theory and practice. There, in a famous passage, Kant mocks the gunman who would shrug their shoulders at the calculations of ballistics:

> No-one can pretend to be practically versed in a branch of knowledge and yet treat theory with scorn, without exposing the fact that he is an ignoramus in his subject [...]. All of us would merely ridicule the empirical engineer who criticised general mechanics or the artilleryman who criticised the mathematical theory of ballistics by declaring that, while the theory is ingeniously conceived, it is not valid in practice, since experience in applying it gives results

12 e.g. *Critique of the Power of Judgment*, § 43–44, AA 05: 303–304.

13 On these meanings of *mechanical*, see the previous discussion, 5.1.1.

> quite different from those predicted theoretically. For if mechanics were supplemented by the theory of friction and ballistics by the theory of air resistance, in other words if only more theory were added, these theoretical disciplines would harmonise very well with practice. (AA 08: 276, transl. Nisbet 1991)

The gunman is not entitled to reject the theory because its naive application is not validated by practice. On the contrary he must *judge*, according to the characteristics of the shot he is preparing, which parts of the theory to take into account, which approximations are acceptable or not, etc. If it is windy, for example, he won't be able to neglect to take air resistance into account.

> No matter how complete the theory may be, a middle term is required between theory and practice, providing a link and a transition from one to the other. For a concept of the understanding, which contains the general rule, must be supplemented by an act of judgement whereby the practitioner distinguishes instances where the rule applies from those where it does not. (AA 08: 275, , transl. Nisbet 1991)

The rule is what makes the link between theory and practice, for the proper act of the practitioner is to judge what to do in a given situation according to the rule provided by the understanding. Indeed, the text continues:

> And since rules cannot in turn be provided on every occasion to direct the judgement in subsuming each instance under the previous rule (for this would involve an infinite regress), theoreticians will be found who can never in all their lives become practical, since they lack judgement. There are, for example, doctors or lawyers who did well during their schooling but who do not know how to act when asked to give advice. (AA 08: 275, , transl. Nisbet 1991)

This last example echoes the Aristotelian assertion of the superiority of the man of experience over the "theoretical" technician who does not know how to apply their knowledge. It is therefore judgement and experience that form the basis of technical-practical knowledge: knowing which rule to mobilise, in such and such a circumstance, knowing when to apply or not to apply the rule. But the positive knowledge that is mobilised, *i.e.* the rule itself, must be considered to be entirely a matter for the understanding.

5.2 The objective and modal distinctions of knowledge

Because imperatives and judgements are based on the same rules, the three characters by which Thomas distinguished practical from theoretical knowledge appear to be quite relative: whether the object of the rule is natural or artificial, whether the rule is mobilised to form a judgement or a decision, and finally whether the aim of the judgement or decision is disinterested knowledge or not – all this changes nothing about the rule itself, and therefore nothing about knowledge either.

This section examines some challenges to this equivalence between practical and theoretical knowledge, focusing on the first two Thomistic differences (the discussion of the third one is postponed to section 7.1). These challenges are deep ones; together, they help to make more sense of Kant's claim on the unity of knowledge, within a framework of knowledge as inquiry that will be progressively built, with the help of constructive logic and Dewey's philosophy of knowledge. Some of Aristotle's core intuitions will be recovered in this process.

The first discussion returns to the idea that contemplation cannot take as objects the contingent goals of practical knowledge, as it aims only at the necessary principles of things. The discussion draws on the reflections of Elizabeth Anscombe, who argues that Kant and modern philosophy misunderstood the meaning of this distinction (5.2.1). Replying to Anscombe's argument requires to understand in which sense sciences can aim at *pure observations* (5.2.2).

The second discussion considers the *modal* distinction, according to which the forms of technical and theoretical knowledge are radically distinct, one having to do with prescriptions, and the other with facts. A recent analysis of philosophers of technology Zwart, Franssen and Kroes develops the logical background of this argument, which takes root in dynamic logic and state transition systems (5.2.3). The meaning of the duality between propositions about *states* and about *transformations of states* is problematised in the last discussion of the section (5.2.4).

5.2.1 The intentional order

To begin with the first distinction, Aquinas and Kant object to Aristotle that it is quite possible to consider contingent objects – artefacts – speculatively, that is, as natural objects. Thus, one can study a house from the point of view of the laws of mechanics just as any natural structure.

To this argument, Aristotle may reply that in studying the house in this way, one is not studying it *as the object of deliberation* (e.g. whether it is comfortable for a family of six, or convenient for home-working), but as a certain structure with such and such resistance to wind, or as a certain mass of stones, and so on. What the theoretical mind cannot do is reason about intentions themselves. As it was noted in the former chapter, this kind of reasoning has specific properties, because it concerns not only the search for solutions and their evaluation, but also the selection of their criteria, *i.e.* the very specification of *what is to be wanted*, including the resources available (money, materials, construction schedule, etc.) and the constraints (so many people to accomodate, garage, garden, energy efficiency, etc.).

The rationalist response to this objection, the one Simon makes (see above 2.1.3), is that if the problem specification must itself be the result of *rational* reflection (and not, for example, the expression of preference or intuition), then it must in turn be thematised as a problem. To arbitrate between several constraints in the problem space is to relate them to more distant utilities and thus to embed them into a larger "meta-problem space", and so on.

Elizabeth Anscombe, in her famous Aristotelian-infused work, *Intention*, takes the exact opposite view of this type of argument, and asserts the irreducibility of the practical syllogism to the theoretical syllogism:

> The role of 'wanting' in the practical syllogism is quite different from that of a premise. It is that whatever is described in the proposition that is the starting-point of the argument must be wanted in order for the reasoning to lead to any action. Then the form 'I want a Jersey cow, they have good ones in the Hereford market, so I'll go there' was formally misconceived: the practical reasoning should just be given in the form 'They have Jersey cows in the Hereford market, so I'll go there'. (Anscombe [1957] 2000: 66)

By beheading the practical syllogism of its major premise, Anscombe wants to show that will is not a hypothesis of reasoning, but its very motivation. We would never have thought to notice that there were Jersey cows in Hereford if our *interest* had not directed our attention to them in the first place. Anscombe thus refers explicitly to the notion of order, the centrality of which she has clearly perceived in the ancient and medieval conceptions to which she wants to return:

> If Aristotle's account [of practical reasoning] were supposed to describe actual mental processes, it would in general be quite absurd. The interest of the account is that it describes an *order* which is there whenever actions are done with intentions. (Anscombe [1957] 2000: 80, my italics)

Anscombe uses the term *intention* to refer to something that is below the psychological processes of desire and will. Intention is not indeed limited to these higher forms of "trying to get" something; it is common to all animals. Intention is not a phenomenon, but a way of looking at phenomena. It is what we are aiming at when we ask the question "Why did he do that?", or what physiologists are looking for when they study animal behaviour; they naturally distinguish between the movements of a dog reacting to a sudden noise, and those of a dog looking for food. In other words, for Anscombe, in these examples intention is not an object of description or reasoning, but their very reason. To use Kant's vocabulary against himself, it might be qualified as a condition of possibility of practical reason, and thus as part of the *transcendental* domain. Kant's error was to believe that the will was a natural object. This caused him to miss the autonomy of the practical domain (in the usual, non-Kantian sense), and to repatriate it to the theoretical domain. Anscombe's criticism of modern philosophy applies in particular to Kant:

> Can it be that there is something that modern philosophy has blankly misunderstood: namely what ancient and medieval philosophers meant by *practical knowledge*? Certainly in modern philosophy we have an incorrigibly contemplative conception of knowledge. Knowledge must be something that is judged as such by being in accordance with the facts. The facts, reality, are prior, and dictate what is to be said, if it is knowledge. And this is the explanation of the utter darkness in which we found ourselves. (Anscombe [1957] 2000: 57)

In order to make clearer this idea that intention governs an order of knowledge ("The term intentional refers to a *form* of description of events" (Anscombe [1957] 2000: 84)) irreducible to that of observational knowledge, a more telling example than that of the Hereford cows given by Anscombe may be summoned to illustrate practical reason. The anecdote says that Democritus, in order to prove to his fellow citizens the superiority of science over practical pursuits, had successfully speculated on the price of olive oil thanks to his knowledge of meteorological phenomena. Having shown that science could make you rich, he then distributed his profits to his fellow citizens.[14]

The deliberation followed by Democritus, as expounded by Pliny, can be explained by the following syllogisms, whose minors are omitted for the sake of brevity, except for the last one (penultimate proposition):

- The weather is bad when the Pleiades rise (6–10 May).
- Olive trees will bud poorly.
- Oil crops will fail.
- Oil will be scarce.
- Oil will be more expensive than it is today.
- Anyone who buys oil now will be able to resell it at a profit.
- I want to get rich.
- I am buying oil now.

The first five sequences, taken in isolation, are entirely theoretical syllogisms. They could be related to sciences: meteorology, agronomy and economics. Only the last sequence is a practical syllogism in the traditional sense. But that is not what practical intelligence is about, or at least not the essential part. The main thing was to organise, under the aegis of the main intention (to get rich), all the theoretical syllogisms so as to arrive at the major of the practical syllogism ("Anyone who buys oil

14 Pliny, *Natural History* 1961: 18: 68: 273. This anecdote is commented in the *Comparison between the Philosophies of Plato and Aristotle*, by Italian humanist Mazzoni ([1597] 2010) on the very topic of the theory /practice distinction.

now, etc."). The essential thing resided, as it were, in the "quickness of wit", to use Aristotle's expression, which *saw* all the middle terms between this practical finality and the fact originally observed (the rising of the Pleiades), extracting them in turn from their particular sciences in order to determine action. *This* practical wit is the central point in Anscombe's key idea, that intention is not to be seen as the major of the practical syllogism, but as its descriptive order.

The other interest of this example is that it clearly shows that practical reasoning is not limited to the sequential scheme of ends and means. Again, the key is not in determining the order *of actions* to be performed, but in the formation of the antecedent judgement that will determine the single action (buying olive oil) realising the intention. Anscombe does speak of a new *descriptive* order that is established by the intention, not a *prescriptive* order.

Any rule whose *raison d'être* necessarily refers to a human intention would therefore be practical. For example, a take-off check-list does not tell the aviator *how* to take off, but *if they can* do so. It looks like this:

> Auxiliary fuel pump — Off
> Flight controls — Free and correct
> Instruments and radios — Checked and set
> Landing gear position lights — Checked
> Altimeter — Set
> Directional gyro — Set
> Fuel gauges — Checked [...]
> Seat belts/ shoulder harnesses — Fastened
> Parking brake — Off [...][15]

Note that this check-list refers to other check-lists. For example, the fuel gauge check itself has two main items to check: 1) that the gauge is working properly, that it is displaying the actual amount of fuel present in the tanks, 2) that there is enough fuel for the flight we are undertaking. Checking the first point, that the gauge is working properly, is itself a delicate task, and there are several procedures designed to ensure this. Such a series of judgments, which does not contain any trace of an imperative, has however as its justification the intention to take off. In the same way, all material artefacts are practical objects, because the functional part of their definition necessarily refers to a human intention.

It might be objected to this vast expansion of the domain of intention that a gearwheel, for example, is defined more by its properties of physical interaction with other parts, than by reference to a human intention. However, the raison d'être

15 https://www.aopa.org/training-and-safety/students/presolo/skills/before-takeoff-checklist (last visited March 28, 2025).

of these engineered interactions must ultimately be referred to human intentions. Thus Anscombe writes:

> [An action] involving encounters with artefacts, like switching on or off, can of course be effected by an inanimate object; but the description only exists because we make switches to be switched on and off. (Anscombe [1957] 2000: 85)

In other words, even if "on and off" are not performed by an agent to whom we attribute intentionality, the fact remains that our use of these terms to describe the behaviour of a machine refers to an intentionality *in the second degree*, the one that led us to make these switches and this machine in the first place. Describing them "in action" therefore belongs to the intentional order, as does describing all the artefacts and systems of artefacts we have constructed, and within which we move. This reasoning could be iterated to artefacts of "third intention", such as spare parts of machines, or conversely systems of machines, and so on. Another example taken by Anscombe is that of the telephone, a systemic artefact *par excellence*, since beyond the individual device we use, it presupposes a whole underlying telecommunications infrastructure, but also the electrical network, civil works, and so on.

Such a position concerning technology is close to the one developed by the philosopher of technology Houkes:

> Technology is related to human, intentional actions. Most technical artefacts and processes do not occur naturally, but need to be designed and manufactured. [...] Even artefacts that function more or less automatically, such as fire alarms or assembly-line robots, require monitoring and maintenance. Because technology is intimately action-related, it makes sense to assume that technological knowledge is related to designing, using and other actions as well. (Houkes 2009: 339)

Artefacts are motivated by human intentions, as aids and instruments of action. But in doing so, they motivate second-order actions, such as *designing, making, using* artefacts, and which may themselves give rise to higher-order actions and artefacts. A central concept for philosophers Vermaas and Houkes is that of *plans of use*, which describe the possible uses of an artefact. These uses may go beyond the original intention of the designer, as when a bottle is used as a flower vase. Superimposed networks of intentions develop around artefacts, which are accompanied by a wide range of prescriptive nuances (permitted uses, traditional uses, recommendations, conditions of use, guarantees, prohibitions, etc.) whose complexity goes far beyond that of simple linear procedures for achieving an objective. A whole socio-technical world is thus set up by human intention "above" the physical world; it remains unknown to the scientist looking only for natural causes or laws in it.[16]

It is now possible to see how intentional descriptive frames can be superimposed without conflict on simple descriptive ones. A carpenter who examines tall pines in a forest in order to select a mast for a ship *inserts* these natural beings in an intentional order without prohibiting the botanist from examining them for entirely different reasons. In the same way, rules that are theoretical in their substance, such as the one concerning the imperfect budding of olive trees in a late spring, can be mobilised by practical reasoning and thus be seen instrumentally. There is thus a fundamental epistemic dualism, which distinguishes the description of things that derives from simple observation, and that which is established by intention:

> Thus, in any operation, we really can speak of two kinds of knowledges – the account that one could give of what one was doing, without adverting to observation; and the account of exactly what is happening at a given moment (say) to the material one is working on. The one is practical, the other speculative. (Anscombe [1957] 2000: 88–89)

5.2.2 Pure descriptions?

But is such an epistemological dualism tenable? Take again the example of the machine that pushes the switches for us. Anscombe may not have chosen the example of binary actions characteristic of electronic components by chance: in a sense, switching on and off is the only thing computers can do. A program of any kind, as long as it solves a problem, is part of an intentional world in which that problem is posed and solved. What about a program that is automatically written from its specification? Anscombe's answer would be the same: this automatic writing would not take place unless, at some level of description, there was an original human intention and a problem posed by a human being.

If computer programs are thus systems of practical rules, what about logical rules, which are rules of rewriting and mathematical demonstrations?... Are such objects "facts" or "reality", which "are prior [to knowledge] and dictate what must be said", to use Anscombe's words? Such a posture seems to fall within the realm of mathematical Platonism. There is actually no hard distinction between formal procedures and formal demonstrations, and the whole world of mathematics seems

16 Houkes 2009: 341. See also Houkes and Vermaas 2004; Houkes, Vermaas et al. 2002; Vermaas and Houkes 2006.

ready to tumble into Anscombe's intentional worlds, taking with it the whole of the mathematical sciences of nature.

Anscombe captures correctly what Aristotle probably meant in the difficult text of *Nic. Eth.* VI.1 when he contrasted beings whose principles are necessary and those "which can be otherwise". Intentional order makes the mast of a ship appear in the middle of a forest, a machine in an assemblage of metals, a logistical network on a geographical map, a program in a configuration of electronic circuits. All these "forms" are contingent in the sense that their principle lies in the way some practical inquiry is looking at them. In contrast, the forest "itself", metals "themselves", geographical locations, electrical currents – all these "forms" are a matter of simple observation.

The problem, however, is to understand what could be *pure observation*, devoid of any other intention than the things "themselves". The following text testifies the struggle of Anscombe with the notion of *observation*:

> But isn't the role of all our observation-knowledge in knowing what we are doing like the role of the eyes in producing successful writing? [...] Our observation is merely an aid, as the eyes are an aid in writing. Someone without eyes may go on writing with a pen that has no more ink in it; or may not realise he is going over the edge of the paper on to the table [...]; here is where the eyes are useful; but the essential thing he does, namely to write such-and–such, is done without the eyes. So without the eyes he knows what he writes; but the eyes help to assure him that what he writes actually gets legibly written. (Anscombe [1957] 2000: 53)[17]

But if observation is only a moment in intention – and on this point Anscombe may be right, as it will be suggested in the following discussions – then how solid are her distinctions? Isn't she saying that there is really *only one form of knowledge* , the intentional one, of which science would be a part? One could interpret Anscombe's critique of theoretical knowledge as reduced to its mere content (axioms, theorems, etc.), in order to better bring out its essential and dynamic moments *i.e.*, upstream of the theory itself, the moment of its research and elaboration, and downstream of it, the moment of its use and application to problem-solving – both moments of an obviously intentional character.

Anscombe's critique of Kant's critique of Aristotle is therefore itself caught in a quandary. It is true that sciences are human practices, in MacIntyre's sense, and in that regard as intentional as anything else. But why then do they appear as a fundamentally different kind of activity, so much so that Anscombe takes their status for granted? The answer may well lie in their *systematic attitude*, which is the very term of this inquiry.

---

17 A fundamental text on this question is *The Intentionality of sensation* (1981), which should be studied in detail to extend these remarks.

5.2.3 The modal interpretation

Aristotle's second distinction between theoretical and practical knowledge is simply that they deal with different modalities, one with prescriptions, and the other with facts. Prescriptions may mobilise facts, but they are not reduced to them; they deal with the *how*, not with the *what*.

Such a distinction already no longer holds for Aquinas, who considers this to be a minor modality of the distinction between theoretical and practical knowledge: The architect can consider *theoretically* the ways of building houses, *i.e.* they can reduce the method to reasoning about facts. However, the fact that this reflective attitude is possible does not mean that theory and practice refer to the same *process* of knowledge or, to use Aristotle's terms, the same intellectual virtue. That science can profitably study technical constructions and derive general propositions from them does not mean that the knowledge that guided this construction consisted of these propositions from the outset. Similarly, the fact that science uses technical modes of reasoning to explain phenomena does not mean that its results are themselves methods.

This counter-argument is quite vivid in contemporary philosophy of technology. It is sometimes related to the Rylean distinction between *know-that* and *know-how* (see above 1.2) and often takes the form of the grammatical remark that theoretical knowledge is expressed by indicative propositions, whereas practical knowledge by imperative ones. It may be remembered that it was also on this argument that Simon grounded the independence of the sciences of the artificial, borrowing from Hume his *is / ought* distinction (see above 2.2.1). Today it often takes the form of the opposition of the *descriptive* and the *prescriptive*, as here in Walter Vincenti's influential book on the engineer's knowledge:

> Descriptive knowledge is [...] knowledge of truth or fact; it is judged in terms of veracity or correctness. Prescriptive knowledge is knowledge of procedure or operation; it is judged in terms of effectiveness, of degree of success or failure. (Vincenti 1990: 197)

Zwart, commenting on Vincenti's definition, states:

> Prescriptive knowledge consists of advice or prescriptions for which engineering actions to take to achieve some technical goal. Being prescriptive, it is normative means-end action knowledge, it has empirical content, and it can be explicated in words. More precisely, [...] prescriptive engineering knowledge is identified with all concrete engineering action-knowledge that can be paraphrased as technical norms of this (canonical) form:
> (*) *If you want to achieve technical goal B and you are in context C, you should perform engineering action A*. (Zwart 2020: 113)

The distinction between prescriptive and descriptive knowledge partially overlaps with the distinction already drawn by Anscombe (see above 5.2.1), as prescriptions may be seen as deriving from intentions. This difference is not what is at issue here.

To see the difference clearly, it is useful to draw on the formalisation of technical prescriptions proposed by Zwart, Franssen and Kroes (2018), which shows that a double transformation is required to transform a technical prescription of the form:

(1) *If you want goal B* and the situation is C*, then perform action A*.*
If you want to make the hut habitable and it's cold, then heat it up.

It seems that the "theoretical" rule at its foundation is as follows:

(2) *If C, and A, then B.*
If the hut is cold, and one heats it, then it is habitable.

At first glance, the difference between the two propositions is that the technical prescription 1 envelops the assertion of a will, and its consequence a command. Deontic logic captures the intentional aspect of this prescription. Zwart uses a logic called DDeL (*Dynamic Deontic Logic*) for this purpose, which formalises the idea of a *transfer* from the intention or obligation to see the goal *B** realised, to that of realising the action *A**.

It suffices for the purpose of this discussion to state the following result, which Zwart calls the *technical norm*:

(3) *If the situation is C*, then achieving A* implies that B* happens.*
If the hut is cold, then heating it will make it habitable.

It is this technical standard that enables the *transmission of intentions* from ends to means, in the felicitous phrase of Zwart, Franssen and Kroes (2018: 22). This notion seems to capture perfectly the Kantian intuition, that the rule of action is nothing else than a rule of judgement, which can be used indifferently for theoretical or practical syllogisms.

However, Zwart, Franssen and Kroes's analysis does not stop there. Indeed, they show that there is still a gap between the formulations 2 and 3. The technical standard cannot be logically converted to (2) because if, as they rightly note, (Zwart, Franssen and Kroes 2018: 7) for if *C∗* and *B∗* can easily be converted into propositions *C* and *B* that describe states of affairs (if *B∗* is an object, the proposition *B* is "*B∗* exists") such is not the case with *A∗* which describes an action to be performed, hence a transition between state C and state B. If the predicate "realise *A∗*" was to be into a proposition, it would need to be set to a certain time ("*A∗* has been realised" or "$A*$ will be realised"), which are possible instances of the predicate "realise A*", but not its general meaning.

Zwart's solution is to model the proposition 3 in propositional dynamic logic (*PDL: Propositional Dynamic Logic*), which defines actions A* as transitions between states *S* of the system under consideration, described by classical propositions. Partially adopting the PDL notation, formula 3 is written:

(4) $S \models [\alpha] B$

Where *B* represents the proposition "$B*$ exists", *S* an initial state that satisfies the context proposition *C*, and $[\alpha]$ the fact that performing the action α (*i.e.* $A*$ in the previous notation) necessarily leads to a state that satisfies *B*. Intuitively, α actions should be conceived in PDL as paths or segments of paths that lead from one state to another. They can therefore be composed sequentially or be trees of paths.

In other words, actions are not objects of the same order as the facts they manage to bring about. Rather, they function as modality operators, requiring or making possible certain facts, as Zwart puts it very well:

> Whereas the logic of descriptive knowledge is *first* (or perhaps higher) *order* with static truth-value distributions, the logic of prescriptive knowledge must be captured in some form of *dynamic logic* where actions may change the truth-value of a sentence from one state of the world to another. ([...] In these logics, the fact-value gap is unbridgeable. (Zwart 2020: 114)

It seems (though this remains to be confirmed) that such irreducibility is also seen in the impossibility, in the logic of knowledge, of simply converting propositions about know-how into propositions about descriptive knowledge. Saying *"I know how to do X"* does not translate into *"I know that [I can do X]"*, because this second proposition does not say that one knows *how* to do it, contrary to what is understood by the first. Know-how is not the combination of a knowledge operator and a competence operator (Y. Wang 2018: 4421; Herzig 2015: 740).

There is thus a logic of action whose difference from that of propositions does not lie in the expression of wishes and wills (which is the object of deontic logic), but in the fact that action always amounts *to modifying the truth values of certain propositions*, since it aims to change the state of the world. This remark sheds even more light on Anscombe's claim that, in the practical syllogism, the intention does not play the role of a deontic major (*"I want a Jersey cow"*) but that it structures the representation of the world. It is now possible to specify that this representation includes both facts and possibilities for *changing* the facts, which are called *actions. To see things practically is to see these possibilities for action that a theoretical gaze can only ascertain after the fact.*

The distinction that emerges here is quite elementary; it proposes to distinguish *states* and *actions* as two radically distinct objects of judgement, the first being about facts, relationships and constraints that hold in the situation, and the second about transformations of those facts and relations under those constraints. Could this modal distinction be at the heart of the distinction of the theoretical / practical knowledge divide, and could it mean that it is irreducible? It must be first noted that this duality between states and actions is not to be interpreted physically. For example, an organism can be seen as in a state of rest, but this state is in fact the equilibrium of many metabolic *actions*; therefore it should be be modelled logically as a transformation keeping stable some observables of interest, but not as a static state. Conversely, an industrial process in action can be described as a certain *state* of the factory, if the problem at hand is to reflect on its potential improvement.

Furthermore, PDL is a logic inspired by the so-called Hoare logic, conceived by the computer scientist Tony Hoare as a framework for analysing the formal correctness of computer programs (Harel, Kozen and Tiuryn 2000; Pratt 1976). In Hoare logic, one writes directly:

(5) $C\,\{\alpha\}\,B$

to mean that if precondition C is satisfied before the execution of action α, then postcondition B will be satisfied on its completion (Hoare 1969: 577). This logic therefore does not originally aim to account for human action, but more generally to describe the rules of evolution over time of any system from its elementary states and changes. In particular, it is aimed centrally at logical machines operating in logical time, actuated by the succession of judgements made by the logician about their object. PDL makes it possible to explain why, for example, if we follow a particular sorting algorithm, we can be sure of obtaining eventually a well-ordered series of data.

The troubling result here is that the situation is the judgement itself, and the action the carrying out of an inference. The duality between state and action seems inscribed in the very logic of inquiry; it may even seem fairly obvious since one needs, in order to solve a problem, to distinguish its specification (what to do) from its resolution (how to do it). It is now the meaning of this *logical* duality that must be interpreted and understood.

5.2.4 The relational reading of rules

The logician Yuri Gurevich considers the distinction between state rules and transformation rules to be basic. He claims that *state transition systems* are not just particular mathematical structures, convenient for studying discrete dynamical systems, but a rather fundamental concept, capturing the essence of a core mathematical activity, computation. Computation itself should be understood in the widest possible sense, and not limited to its traditional scope, that of natural integers and countable sets. It should be understood as encompassing any procedure that allows the construction and evaluation of new functions from given ones.

A state, in Gurevich's *abstract state machines*, is as general as a first-order structure as described by Tarksi, *i.e.* a *domain* of mathematical elements, functions and predicates (Gurevich proposes that every element be described as a function, even if it is 0-ary) and an *interpretation* mapping variables (basically symbols, in finite number) onto elements of this domain. As Gurevich writes:

> The justification for postulating that structures or algebras are appropriate for capturing algorithmic states is the enormous experience of mathematicians who have faithfully and transparently presented every kind of static mathematical reality as a first-order structure. (Dershowitz and Gurevich 2008: 315)

In particular first-order structures are the standard logical frame where mathematical proofs can be expressed. Abstract states machines (ASM), by contrast, are sets of rules for moving from one state to another by giving new values to variables. Rules all have the form:

(6) **if** p **then** $t := u$

where p is a predicate (or a function evaluating to true or false), and t and u potentially complex functional terms of the form $f(t_1, t_2, ..., t_k)$ with $t_1, t_2, ..., t_k$ being terms themselves.[18] The application of this rule "updates" a given state by changing the interpretation of t to that of u if p is true in this state. An ASM *evolves* autonomously over some logical time until it reaches a state where no updates are possible any more.

In Gurevich's interpretation, mathematical procedures transform or construct states, whereas mathematical proofs operate within a single state. They are thus fundamentally different objects. Gurevich does not attempt to model proofs themselves. True, they could be modelled as syntactic transformations of structures which domains would be the well-formed-formulas of a theory, but this formalist move would situate proofs at a different abstraction level than procedures.

What is strange about this interpretation is that state transition systems are quite mundane mathematical structures (basically, graphs), and it seems dubious that they could represent a fundamental structure of inquiry.

---

18 For more detail, see Gurevich 2000: 8 and Dershowitz and Gurevich 2008: 321.

An alternative approach is to state that there is only one kind of rule, and that the difference between state rules and state transition rules is only one of psychological understanding or subjective interpretation. It is subject to the same kind of ambiguity than the one characteristic of the notion of *function* in mathematics[19] which is susceptible to a dual interpretation, dynamic and relational. Take for example a simple functional expression, inspired by λ-calculus (Barendregt and Barendsen [1984] 2000):

Listing 5.1: Illustration of the principles of λ-calculus

```
first of (sort(append ('dog','rabbit') (sort('mouse','cat'))))
→   first of (sort(append ('dog','rabbit') ('cat','mouse'))))
→   first of (sort('dog','rabbit','cat','mouse'))
→   first of ('cat','dog','mouse','rabbit')
→   'cat'.
```

In this expression, `first of` refers to the operation of selecting the first term of a list, append to the operation of concatenating two lists together, and `sort` to sort the terms of a list alphabetically. Each operates on the list indicated by the parentheses that follow it.

This expression can be interpreted in two ways. One way is to see it as a sequence of dynamic transformations due to the execution of instructions that the operators would represent: the arrow → here represents the unfolding of a process. The process in question here is not, however, that of material phenomena, observed in the world. It is nothing else than the progress of the judgement itself, which advances, line by line, to the final solution. It invites to see judgement itself as an act that modifies the subject's knowledge.

The other way is to see it as a sequence of statements, *i.e.* judgements of equality between two successive terms: the arrow → here takes on the meaning of the copula *is*. The succession is merely psychological, it is that of the mind taking time to go from one equality to another, which all stay however the same and unchanged in this process. This is the heart of Plato's criticism of those who believe that geometers perform *intellectual constructions* whereas they truly only observe relationships (see chapter 5, introduction).

The algebraisation of geometry from the 17th century onwards sanctioned this view. Construction problems were widely interpreted in the last century as existence theorems, particularly following the influential paper by Zeuthen (1896). Thus the axiomatisations of Euclidean geometry by Hilbert and Tarski feature only theorems. In Tarski's theory, the only objects given are points, together with their re-

19 See on this subject the reflection of Granström (2011).

lations of betweenness (between three points) and congruence (between two pairs of points) (Tarski 1959: 17). Complex geometric objects (lines, triangles, etc.) which would be sets defined using these fundamental relations, are not necessary for the expression of the content of geometry because, as Tarksi notes, they are always defined from a finite number of points, and can therefore be directly expressed by their relations.

Such a representation of geometry amounts to turning all its propositions into theorems, since any request to construct a complex object is represented there as that of a relation between points. For example, Euclid's proposition I.1, which asks to construct an equilateral triangle, is expressed in Tarski's geometry as follows:

$$\forall a, b, \exists c : \neg Col(a, b, c) \wedge ab \equiv ac \equiv bc$$

In this proposition, a, b, c are points, and $\neg Col(a, b, c)$ means that they are distinct and not aligned (the collinearity relation $Col(a, b, c)$ is defined from the elementary betweenness relation (Tarski 1959: 17; Beeson 2015: 1205)). Finally, $\equiv$ denotes the elementary relation of congruence between segments.

This reading, however, tends to cast the very notion of problem-solving in mathematics as merely psychological. Not only would geometric construction problems become unintelligible, but so would all theorem-proving problems. A proof, in itself, would offer the mathematician's mind only a chained sequence of relations from which a summary one (the theorem) could be inferred in a series of parallel inferences. But these acts of inference themselves, which are collectively called *the act of proof*, would be non-mathematical *per se*, and only the psychological correlate of these relations.

5.3 Type theory as a logic of problems

This section examines an alternative to the relational, set-theoretic interpretation of mathematical problems. It is anchored in the intuitionist logic of Martin-Löf, also called the *constructive theory of types*. It begins with the presentation of the four forms of judgement identified by this theory (5.3.1). The equivalence between problem and proposition is then discussed (5.3.2). For Martin-Löf, his logic extends far beyond the realm of knowledge and applies to any human practice: it is a logic of action (5.3.3). A reconstruction of Euclidean geometry according to these principles (detailed in the appendix B) shows that acts of *constructing figures* and acts of *reading figures* must be distinguished. Any complex domain of human practice makes (at least) such a distinction; it is this distinction that provides the logical justification for state transition systems (5.3.4).

5.3.1 Problem forms

To better analyse the relationship between proof and problem, it is useful to consider a few examples. A simple problem is to find the double of a number. Natural integers may be defined recursively by positing the element zero and defining any other integer as the successor of another:

$\mathbf{0} \in \mathbb{N}$
$\mathbf{succ}(n) \in \mathbb{N}\ [n \in \mathbb{N}]$

The **double** function may then be defined recursively as follows:

$\mathbf{double(0)} = \mathbf{0}$
$\mathbf{double(succ(}n\mathbf{))} = \mathbf{succ(succ(double(}n\mathbf{)))}$

This definition says that, to calculate the double of an integer, add 2 (expressed as **succ(succ())**) to the double of the previous integer, and for **0**, take **0**. Consider the problem of finding *x* such that $x = \mathbf{double}(5)$. Here, the solution is very simple: just apply the rules defining **double()**:

$\mathbf{double}(5)) = \mathbf{double}(4) + 2 = \mathbf{double}(3) + 2 + 2 = \text{etc.}$

Here one may say that the proposition P1(*x*) defined by $x = \mathbf{double}(5)$ is the *specification* of *x*. The calculation here is the proof that 10 meets this specification. If the answer had been given intuitively, the calculation would have served as *verification*.

Establishing the truth of a proposition may be done in a similar way. For example, the parity property of an integer $E(x) \equiv \mathbf{even}(x)$ may be defined by the following two rules:

$\text{ev0} : \mathbf{even}(0)$
$\text{evn} : \mathbf{even}(n) \rightarrow \mathbf{even(succ(succ(}n\mathbf{)))}$

E(10) may be seen as a problem similar to P1(5), for which an intuitive answer can be given and verified by applying the rule evn five times and finally ev0. It could also be calculated directly.

Now here is a second problem, slightly more complicated than the previous one: find *x* such that $18 = \mathbf{double}(x)$. Let us call P2(*x*) this specification. Here the calculation of the function **double()** is of no help to solve this problem; it can only be used to check the solution. The solution must be found by another means, either intuitively or by another calculation. In the case where the problem is $17 = \mathbf{double}(x)$, the nature of the problem changes: it must be *shown* that there is no solution, for example by enumerating the doubles of all the integers until they exceed 17.

Here is a third problem: find, for any $y \in \mathbb{N}$, x such that $y = \mathbf{double}(x)$. The specification, which may be called P3(y, x), is a function of y, and so a general solution must be found, a function $g(y)$ such that $y = \mathbf{double}(g(y))$. In this type of problem there is generally no foolproof method for deducing the solution. This must be found and then verified. For example the partial function **half**(y) defined recursively, for even integers only, is a candidate:

$$\mathbf{half(0) = 0}$$
$$\mathbf{half(succ(succ(}n\mathbf{))) = succ(half(}n\mathbf{)))}$$

This definition says that to calculate the half of a number, add 1 to the half of the number twice before (when it exists) and for **0**, take **0**. However it must be *proven* that this function does conform to the specification, as there is nothing in this definition to indicate this. In other words, it must be proven that:

$$\forall y, \mathbf{even}(y) \rightarrow \mathrm{P3}(y, \mathbf{half}(y))$$

For the sake of brevity, this proof is not given here.

Finally, (and this is the fourth form of problem) one may want to establish a general property for integers, like this one:

$$\mathrm{P4} : \forall x \in \mathbb{N}, \mathbf{even(double(}x\mathbf{))}$$

Here, the question is whether the double of any integer is even: this is the reciprocal of the previous proposition. It may also be seen as asserting that every solution of the calculation problem **double**(n) for n any, is also a solution of the problem **even**(x). Its proof also consists of universal verification.

These four examples have points in common. For problems P1 to P3 it was necessary *to find* the solution, even though in P1 the method for doing so was given. For P2 and P3, the search was heuristic, and so the solution had to be verified *a posteriori*. This is not a pleasant situation, and P3 in fact aims to reduce P2 to a P1-type search (since P2 will be solved by simply computing the **half**() function). Problem P4 only requires verification. However, as with P3, and unlike P2, this verification is parametric, *i.e.* a universal method for carrying out this verification must be found.

This last point suggests that P4 should also be seen as a problem similar to the first three, of finding or constructing a requested element. Only, in problems P3 and P4, one has to find the verification method itself!

This is where type theory comes in, as it attempts to provide a logical basis for this possibility. As it is well known, the verification methods required by P3 and P4 proceed by recursion. This is a quick example for P4:

Proof. First it must be proven that the property looked for is true for **0**, *i.e.* that **even(double(0))** holds. The proof is simple (indications on the right side of the equation point to the rule used to make the deduction):

even(0)	ev0
⇒**even(double(0))**	Def. **double()**

Next it must be proven that if the property is true for *n*: **even(double(***n***))**, then it is also true for **succ(***n***)**: **even(double(succ(***n***)))**. One has:

even(double(*n***))**	Hyp. of recurrence
⇒**even(succ(succ(double(***n***))))**	evn
⇒**even(double(succ(***n***)))**	Def. **double()**

□

Type theory holds that this demonstration by recursion is itself an object that has been found. Its verification is a calculation, similar to the one in problem P1. Simply, the rules of calculation here are not those of **double()**, but those of ℕ itself, *i.e.* the rules of reasoning by recursion that characterise it.

Such a proposition brings to a conclusion the Hilbertian idea that it is possible to study mathematical proofs mathematically. This peculiarity, which is part of the foundational project of type theory, is well exposed in the following text:

> We note that a set-theoretic foundation has two "layers": the deductive system of first-order logic, and, formulated inside this system, the axioms of a particular theory, such as ZFC.[20] Thus, set theory is not only about sets, but rather about the interplay between sets (the objects of the second layer) and propositions (the objects of the first layer). By contrast, type theory is its own deductive system: it need not be formulated inside any superstructure, such as first-order logic. Instead of the two basic notions of set theory, sets and propositions, type theory has one basic notion: types. Propositions (statements which we can prove, disprove, assume, negate, and so on) are identified with particular types [...] Thus, the mathematical activity of proving a theorem is identified with a special case of the mathematical activity of constructing an object—in this case, an inhabitant of a type that represents a proposition. (Univalent Foundations Program 2013: 17)

20 Zermelo-Fraenkel's theory of sets.

From all this analysis it follows that there are three possible situations in front of a problem, and four forms of judgement.

1. There are problems for which it is only needed to check that a given solution is correct, *i.e.* that it conforms to the rules of the problem. These are calculation problems. Problem P1 falls into this class, since calculating the solution according to a given method is similar to a verification. The corresponding type-theoretic form of judgement is simply $a = b$ where a and b denote two calculations leading to the same result, e.g. **succ(succ(succ**(1))) = **double**(2).
2. There are problems (such as P2 or P3) for which a solution must be found, whether it is a precise object or a parametric object, *i.e.* a method for calculating an object according to the parameters of the problem. If A is the problem, the corresponding type-theoretic form of judgement is $a \in A$, where a is an object or a method.
3. Finally, there are "meta-problems", where the point is to check that two problems are equivalent, *i.e.* have the same solutions, or are simply related, like P4: $\forall x, y$, P3$(x, y) \Rightarrow E(y)$ where $E(y)$ denotes **even**(y). The corresponding type-theoretic form of judgement is $A = B$, where A and B are two problems. When these problems are parametric, a verification method must be found.
4. This implies that there must be a fourth form of judgement, which defines the rules by which a parametric verification can proceed. If the parameters are finite in number, the rule is simple: check all possible cases one after the other. However, it cannot be applied to infinite sets. In the case of natural numbers, it is replaced by the rule of demonstration by recursion, as just seen. Each type of parameter thus has its own rules by which it can be used to define new problems, whether these involve objects of the same type or of a different type. Of course, defining these rules presupposes that the set of parameters in question has also been defined, *i.e.* that we have a method for traversing all its elements (this is **0** and **succ**(n) for natural numbers). This fourth form of judgement is simply A set, "A is a set", and it consists of the stipulation of all these rules.

This presentation gives the four forms of judgement recognised by the constructive theory of types as developed by Martin-Löf (1984: 5–10) although in an order different from that usually followed, It is now possible to examine the way in which this theory helps to conceive of the identity of the rules of judgement and of the transformation of judgement in mathematics.

5.3.2 Type theory

This identity appears very well in table 5.1 from Martin-Löf 1984: 5.[21] It asserts, in the intuitionist tradition, the identity of problems and propositions, or constructions and proofs:

Table 5.1: Martin-Löf's table of judgements

	A **set**	$a \in A$
1	A is a set	a is an element of the set A
2	A is a proposition	a is a proof (construction) of the proposition A
3	A is an intention (expectation)	a is a method of fulfilling (realising) the intention (expectation) A
4	A is a problem (task)	a is a method of solving the problem (doing the task) A

Each column sets out one of the forms of judgement identified in the previous discussion, namely the fourth (which is generally presented first) and the second. Each line gives a different expression of these judgements, which Martin-Löf posits as equivalent. In the first column, the notions of set, proposition and problem are brought together, and the latter notion is also represented by the terms *intention, expectation, task*. In the second, it is the notions of element, proof (or construction) and method for *satisfying, solving, realising, doing* that are brought together.

Martin-Löf notes that the "analogy" between lines 1 and 2 refers to the correspondence established by Curry and Feys, and generalised by Howard (Martin-Löf 1984: 6).[22] This connection is quite apparent in the first two examples: the property **even**(x) can be seen as the predicate of a proposition, e.g. **even**(6), or as the set of all x satisfying it. For primitive sets the corresponding proposition must define what it means to be a member of that set. For example, for the natural numbers, $\mathbb{N}$ can be seen as the ordered succession of elements starting from an initial element **0**.

The concepts introduced in lines 3 and 4 of the table correspond to the same fundamental idea (according to Martin-Löf), introduced concurrently by Heyting (first formulation) in 1931 and by Kolmogorov (second formulation) in 1932 to interpret intuitionistic logic (Heyting 1931; Kolmogorov 1932). Martin-Löf also brings Kolmogorov's interpretation closer to programming:

21 I have added the numberings and omitted a last column, which is not useful to this discussion.

22 On the links between Howard and Martin-Löf, see Wadler 2015: 10. On the Curry-Howard correspondence, see Bourdeau and Mosconi 2022: 680.

> "a is a method..." can be read as "a is a program...". Since programming languages have a formal notation for the program a, but not for A, we complete the sentence with "... which meets the specification A". In Kolmogorov's interpretation, the word problem refers to something to be done and the word program to how to do it. (Martin-Löf 1984: 6)

It remains to establish the link between 1–2 and 3–4, between proposition and problem. On the one hand, it is easy to see why a proposition can be assimilated to a problem, because every problem must describe the object being sought, and this description is a proposition, or a verification that can be made of any object presented, leading to a judgement of conformity. But on the other hand, a general proposition, stating a property of a class of objects, can itself be taken as a problem, because it is then a question of showing, as already seen, the link between two specifications, for an integer for example, one of being even, the other of being the double of another integer.

Therefore, the relationship between judgement and the transformation of judgement is nothing else than that between problem and proof. Each of the three forms of judgement that solves a problem requires a proof, which is an act of knowledge – whether it is a calculation, the construction of a new method of calculation or some reasoning that compares methods. In the first case, rules are applied (Kant's *determining* judgement, see above 5.1.2), in the second they are elaborated (the *reflective* judgment), and in the third they are organised one relatively to the other (the activity of reason).

Proving and constructing are first and foremost *acts*. They must not be confused either with their result (the theorem, the constructed object), or with the description of these acts, which indicates how to reproduce them, but can in no way substitute them.[23] Martin-Löf writes:

> The word act, of course, is the most general word that we use for all of our acts, and, similarly, the word object has that complete generality. Thus we have acts of conjecturing and doubting and wishing and fearing and so on, and we have the objects of those acts, that is, objects of conjecture and doubt and wish and fear and so on. One particular kind of acts, and, correlatively, kind of objects that we direct ourselves towards in those acts, are acts of knowing and objects of knowledge, respectively. My answer to the questions, *What is a judgement?* and, *What is a proof of a judgement?* is simply that a proof of a judgement is an act of knowing and that the judgement which it proves is the object of that act of knowing, that is, an object of knowledge. (Martin-Löf 1987: 417)

Type theory thus makes it possible to show that the act and the object of judgment *require* to have the same rules, for the act has meaning only in relation to its object;

23 See for example Troelstra 1969: 3; Sundholm 1997.

but conversely this object without the act that fulfills it would be empty and entirely problematic. This idea is well expressed in the following text, which compares the two forms of judgment presented in the table above:

> To know a judgement is the same as to possess a proof of it, and to know a judgement of the particular form A is true is the same as to know how, or be able, to verify the proposition A [...] whereas knowledge of a judgement of the first form is knowledge of a problem, expectation, or intention, which is knowledge what to do, simply. [...] So the difference between these two kinds of knowledge is the difference between knowledge what to do and knowledge how to do it. And, of course, there can be no question of knowing how to do something before you know what it is that is to be done. (Martin-Löf [1983] 1996: 25–26)

In the examples in the previous development, know-what denotes the property $P1(x) : x = \mathbf{double}(5)$ or $P2(x) : 18 = \mathbf{double}(x)$, the parametric problem $P3(y, x)$ or the assertion P4. Know-how designates the verification that 10 is a solution in the first case, 9 in the second, and the function $\mathbf{half}(y)$ in the third. In the last case, the idea is to check the details of the connection between the notions of parity and doubling of an integer.

5.3.3 The scope of type theory

Martin-Löf explicitly defends the idea that his logic of problems is not limited to acts and objects of knowledge, but could hold for any kind of constructive act:

> The notions of act and object that have turned out to be crucial in the analysis of the notion of judgement or knowledge, ambiguous as it is between the act of judging or knowing and that which is judged or known, that is, the object of knowledge, those two notions are not at all limited to acts of knowing, because the concept of act can be understood in the widest possible sense, comprising in the traditional division both thoughts, words and deeds, that is to say, mental acts, verbal acts and concrete acts, like baking a cake or scrubbing a floor or something like that, concrete acts or real acts, whatever you prefer. [...]
> It turns out that several of the notions that I have already discussed can be lifted to the level of acts and objects in general. This is so, in particular, for the notions of logical possibility, actuality and real possibility: they certainly make sense for any objects whatsoever, not only for objects of knowledge. Indeed, an object is logically possible if it merely has been laid down what is meant by doing it, that is, by performing an act with that object, and an object is actual if it has been done, that is, if an act with that object has been performed, and an object is really possible if it can be done, that is, if an act with that object can be performed. (Martin-Löf 1991: 145–146)

Martin-Löf himself thus prompts his reader to extend his theses beyond mathematics to the whole field of knowledge and even beyond, to the whole of human action.

The title of one of his articles, “A path from logic to metaphysics”, clearly indicates his ambition, and he often defines logic as “general theory of knowledge”.[24]

However, he has not published any work dealing with these issues. Mäenpää mentions a lecture by Martin-Löf discussing the epistemology of the natural sciences (Mäenpää 1993: 35). In accessible works, his reflections take the form of examples of empirical evidence, which recur in several texts. Here are two of them, taken from (Martin-Löf 1994).

The first example is knowing whether the sun is shining. In this case, Martin-Löf says, all you have to do is stand outside, or look out of the window. Seeing the sky is the elementary act that allows one to know whether the sun is shining. This criterion is partly arbitrary, since we could simply rely on weather reports. However, if we want to find the most immediate evidence of an unclouded day, seeing the sky is undoubtedly the most elementary experience. We may do it ourselves: if we are inside a house, the procedure is to get out, or go to a window. It may also be carried out by a trusted third party, another person, or a machine. In all these cases, we have to devise and check procedures which, when carried out, provoke the elementary experience “seeing the sky”. Knowing that the sun is shining involves performing, in one way or another, this act of evidence.

Martin-Löf's second example seems very similar to the first. It involves knowing whether the temperature is 10°C. In this case, the elementary act corresponding to this knowledge is to read a thermometer. The point of this second example is to show that immediate knowledge cannot be reduced to immediate perceptual experiences of the phenomenon in question. Martin-Löf is not referring here to the experience of being cold or not, but to a *measurement* of temperature. Any measurement implies a measurement protocol, and usually an instrument, in this case the thermometer. We could certainly ask to audit this thermometer to make sure it is working properly. In that case shouldn't we count as an elementary act the *couple* ⟨audit the thermometer, read the thermometer⟩? Such refinement is certainly possible. In my opinion, the naivety of Martin-Löf's examples is feigned. After all, the half-learned person asking for the thermometer to be checked could be told to check the checking procedure as well, to measure the risks of the system going wrong between the time of the audit and the time of the actual measurement, and so on. The essential point of this second example therefore seems to be that a measurement is already a *constructed* object, and that its demonstration is always situated in a complex context which involves the manipulation of more elementary objects according to a procedure. Experience, apart from simple cases such as the first example above, presupposes such a *set-up* – Martin-Löf (1994: 90) uses the term “complex” – within

24 For example in Martin-Löf 1987: 419, [1983] 1996: 20.

which an act of *empirical acquisition* of knowledge may be conducted. Mäenpää (1993: 36) develops pragmatic remarks along these lines.

Within this device, which is assumed to be correct (the thermometer, the procedure, my own mental capacities), saying that it is 25°C when the thermometer reads 25°C is, according to Martin-Löf, an analytic act, because the measurement of the thermometer precisely describes a canonical proof of this proposition, it constitutes its very meaning. The assertion of the proposition is synthetic only if it is made *outside the presence of the thermometer*, because the proof is missing, and "you have to go beyond everything contained in the judgment to make it obvious to you" (Martin-Löf 1994: 88).

The example of the thermometer shows that in a given theory, what is defined as an elementary object does not in general have immediate evidence. What is determined as "immediate experience" must not therefore be understood in the sense of a sensualist doctrine of knowledge, which would reduce it to ultimate *qualia*. Rather, it designates what is posited, or postulated, as the immediate acts and objects of a system of knowledge, from which the latter can be deployed in discursive procedures. Thus the immediate objects of quantum physics, its famous observables, such as the spin of the electron, are associated with extremely complex measurement procedures, from a pragmatic point of view. Yet it is these procedures that must be considered immediate from the physicist's point of view, even if they have to rely on a sophisticated epistemic infrastructure – mechanical, electromagnetic and computer knowledge and tools, etc. – to carry out these acts. In another text, Martin-Löf notes that the terms *evidence* and *demonstration* are metaphorical of vision, but that they could just as easily be replaced by those of *justification* and *warrant*, where it is the legal metaphor that holds sway. (Martin-Löf 1995: 188–189)

There are certainly avenues of reflection here concerning epistemology that could be developed, and which seem to point towards an operational approach to experimentation that may bring Bridgman to mind (Chang 2021), *i.e.* which defines the content of the experiment on the basis of its protocol for carrying it out.

5.3.4 Complex action and interdependent types

How well does constructive type theory explain the duality of problems and theorems in Euclidean geometry? Appendix B outlines one of the attempts at a constructive re-expression of geometry due to Mäenpää and Von Plato, and detail one of the examples they propose, concerning the construction of a parallelogram. The basic goal of Mäenpää and Van Plato is to unify theorems and problems in a single form of judgment, which they call *problem*. Mäenpää (1993: 90) gives the following form:

$$\exists(y: B(x))C(x,y) : \text{problem} \quad (x: A)$$

where

$x: A$	is	the *given*,
$y: B(x)$	is	the *thing sought*,
$C(x,y)$	is	the *condition*.

The theorems, Mäenpää says, are therefore special cases of problems where the element $y : B(x)$ is void, so the problem can be simply written:

$$C(x) : \text{theorem} \quad (x : A)$$

This blunt identification of theorems to special cases of problems falls short on several points. In particular, it does not capture adequately the problematic nature of the *properties* by which geometrical figures are specified, and especially the magnitudes (distances, angles, areas). The detailed study of a theorem (I.13, in B.2.2) shows that the discourse on magnitudes proceeds through acts of "reading" the figure, which makes it possible to identify relations concerning them. These relations can then be transformed thanks to the rules of inference constituted by the common notions. Then, the study of a sequence of theorems deriving from I.13 shows that their purpose is to make licit more and more sophisticated acts of reading, in the same way that problems learn to construct ever more complex objects.

This result seems to militate in favour of "theoretical" acts, composed according to their own logic, and independent from the acts of construction, and that the duality between theorems and problems needs to be maintained.

It must be noted that there is no real symmetry between constructions and observations, because magnitudes are, so to speak, "phantom objects". Indeed, the acts that would properly access them, *measurements*, are *not* admitted in geometry; magnitudes are talked about and compared one relatively to the other, but never actually carried out. Does that mean that they are abstract and formal relations between figures, such as in Tarski's axiomatisation? However, in this axiomatics, most theorems concerning magnitudes actually lose their original meaning, since it is no longer possible, for example, to speak of their *sum*.

But then, where do these "acts of reading" mentioned above come from, and what is their legitimacy? Here one must return to the postulates, which are too often associated, particularly the first three, with acts of construction alone. In fact they also provide *rights* to make elementary inferences about magnitudes. Thus, the circle is defined as the set of points equidistant from the centre (definition 15), and it is postulated *furthermore* that we know how to construct it (postulate 3). Having constructed a circle, and marked two points on it, it is therefore licit to judge that their respective distances from the centre are equal, *by definition*. From the extension of a segment *AB* from *B* to a point *C* (postulate 2) we are entitled to infer that *B* lies between *A* and *C*, or that the distance *AC* is greater than *AB*. Postulate 4, which simply states that all right angles are equal, makes it possible to infer the congruence of such angles even if they are not adjacent and hence, thanks to algebraic reasoning from common notions, of all other angles. This postulate is necessary because the definition of a right angle (def. I.10) is itself relative: the angle of one line laid on another is right when it is equal to its adjacent angle. Postulate 4 therefore ensures that there is only one way of drawing such a line on the same side of the base. As for the additivity of adjacent angles, which is crucial for Theorem I.13 discussed in appendix B, it is undoubtedly one of those rare implicit postulates found in the Elements.

These remarks are not meant to be exhaustive. They are merely intended to show that the elementary acts of construction and observation are mutually dependent. Postulates and definitions together make it possible to define *elementary couplings of meanings* which tell the geometer which figures he must be able to construct and which relations he must be able to read into them. From these elementary couplings, geometry develops increasingly sophisticated couplings, showing how to construct ever more complex figures and how to infer ever more subtle relationships from them. The alternation of problems and theorems is therefore an essential part of the geometer's thinking.

If there is therefore a primacy of theorems over problems, as tradition suggests, it is because they provide the logical framework for geometry. Propositions *frame* constructions, setting their specifications but also validating their possibility.

A construction program, taken on its own, does not say nor guarantee anything. It does not indicate what the properties of the final figure will be. It is not even certain that it will succeed. For example, an instruction might ask one to take the intersection of two circles, even though the distance separating their centres would be greater than the sum of their radii. Geometry begins when the aim is to make certain judgements about constructed figures, and to ensure that constructions always succeed. It formulates the conditions for carrying out certain acts (in particular, the construction of intersections), describes the properties of elementary constructions (the line and the circle), and the rules for inferring certain properties from others.

In this perspective, geometry could be seen as a Hoare logic (see above 5.2.3) which goal would be to ensure that the proposed constructions are correct. It enables the geometer to construct objects with complex properties (chiefly, magnitudes) with a guarantee of exactness, free from the approximations of empirical measurement.

There is indeed an essential difference between geometry and arithmetic, from a constructive point of view. Natural numbers can be defined independently, *i.e.* without any constraint imposed by another type. Indeed, in many respects, they are the foundation of all the others. This is not the case in geometry: figure and magnitude are mutually dependent types, whose relationships, described by the set of definitions / postulates, are encapsulated in the elementary figures of the line and the circle. The elementary constructions that can be made and the relations of magnitudes that can be read into them are inseparably linked – for example, that a single straight line passes through two points, that the angles of parallels alternating with a transversal are equal, etc.

This complexity of geometry, where several types must cooperate to capture an aspect of reality – to account for the rules by which actions can succeed in the world – explain the pragmatic primacy of state transition systems for modelling any complex practice.

They model indeed successfully possible relationship between two kinds of actions, where one appears as "static" (the realm of judgement about states of the world) and the other "dynamic" (the realm of action modifying those states of the world and thus those judgements themselves). That these notions of *static* and *dynamic* are relative, and basically metaphorical, is amply shown by any example where the state of the world is not naively modelled as a "still picture" of static things, but as the regular operation of processes – for example industrial processes which themselves require ongoing actions, judgements and decisions on the part of the agents involved. Changing this state of affairs, e.g. introducing a new industrial norm, such as changing the speed or order of a process, requires these agents to modify all their behaviours – not only their actions, but their decision-making procedures themselves in order to adapt to this new situation and recreate a new dynamic equilibrium.

Computer programming, in particular, has much more to do with this geometric complexity than with the arithmetical simplicity with which it is often associated. A program, insofar as it acts on memory space, is a construction whose correctness depends on the structure of that space. Any program of any complexity has several data structures and uses several types of data. This is a notable difference from algorithms, which often only manipulate a single type of data. This fact also explains the relevance of computer science concepts such as *object-oriented programming* or *abstract data types* that closely associate data structures with operations associated with them.

The question about the unity of action and judgement is thus enriched by the consideration of geometry. It appears to be more complex than a uniform interpretation of type theory would initially suggest. An answer in two moments is thus required: 1) At the level of a given domain (or *type*), action and judgement do appear as two sides of the same coin. They are both governed by the same rules, as can be seen clearly in the case of arithmetic. Judging according to a rule is a form of action that transforms a context of knowledge. Conversely, any action is taken with the intention of transforming a situation, which implies that its effectiveness can be judged. So computing numbers is no less an action than stacking bricks. And stacking bricks in accordance with a plan requires perhaps as much judgement as drawing up the plan itself. Type theory interprets judgement as the assertion that a problem is solved in accordance with its specification, or that a particular method systematically solves a parametric problem, or that two problems are equivalent. The judgement always relates to an action, which itself always ends in a judgement.

2) Most complex human activities, however, involve the interaction of several domains of action, each with its own types of outcome. In particular, a common – but by no means absolute – dichotomy takes place between the domains considered to produce interesting effects in the world and those that verify or control these effects; and *in these cases*, different rules may seem to govern the domain of action and the domain of judgement; but this is only one form of that division of intellectual labour that develops from the early modern period onwards.[25] There is thus no distinction in essence between act and judgement, between the realm of actiosn and the realm of contemplation, between practical and theoretical knowledge; but an indefinite multiplicity of interdependent actions that, together, build up the consistency of a given practice.

25 This point is developed further in chapter 7.

6 The theory of inquiry (Dewey)

If the faculty that Kant calls *understanding* is essentially about knowing what to judge about a situation, or what to do about it (ends being given), then its activity is essentially what Martin-Löf calls *problem-solving*. The theory of types gives the unified and generic logical form in which such *knowledge of rules* could be expressed. The rules of a concept are, on the one hand, the different ways of accessing its objects under certain conditions, and, on the other hand, those by which they can themselves be conditions of other objects. Thus, within a theoretical framework whose elementary concepts have been defined, one can express any form of complex problem as a question regarding the accessibility of a complex object, using a composition of elementary acts.

In this framework, as Kant maintained, there is no essential distinction to be made between theories of knowledge and theories of know-how. Martin-Löf also says that it is indifferent *to the form* of the theory whether the objects treated are equations or cakes, and whether their acts of construction are inferences or cooking gestures (5.3.3). Geometry shows that any complex human practice deals with many different kinds of objects, and that "intellectual" acts often serve as validation or specification of "practical" ones. The division between theory and practice appears to reflect only some division of labour, which is variable according to the domain of inquiry; it is in no ways absolute.

It may be disputed, however, that such a unification really demonstrates the fundamental kinship of all problem-solving activities across human practices. Martin-Löf's constructive logic describes problem-solving already *theorised* in a formal language. It may be argued that it makes no more claim to the kinship of the objects it subsumes than a differential equation that applies to both mechanical and electrical phenomena (e.g. the second-order linear differential equation) proves the kinship of mass and electricity. Once theorised and mathematised, all phenomena look the same.

The argument is not rhetorical. Would practical procedures, expressed in the theory of types, bear any resemblance to the practical activities they would seek to describe? Nothing is less certain. In fact, in this formal translation, they would be subjected to the very process of systematisation that this research seeks to understand as a possibility.

Indeed, in Martin-Löf's logic, types (problems) and their elements (procedures) must be given by the mathematician themselves. Nothing is told about how they arrived at this result, about the problem-solving activity itself, *i.e.* the deliberation. For example, when the function **half**(y) is put forward as a solution to problem P3(y, x) (see above 5.3.1), this function is certainly correct, but no clue is given about how

https://doi.org/10.1515/9783111494500-006

it was found. The whole complexity of human deliberation lies *precisely* in getting the problems right and defining the paths to solve them, not in applying known procedures.

So it could be that, by describing practical knowledge too directly with the help of type theory, a "theoretical" vision of knowledge is inadvertently promoted. This theory certainly presents procedures that rigorously express how to effectively solve a problem, but this quite hieratic presentation of the steps to be followed no longer retains any trace of the living inquiry that, perhaps with difficulty, brought it to light. In other words, one may fail to see how this systematic resolution relates to the reflective attitude of deliberation tending towards a problem to be solved.

But doesn't Martin-Löf insist that the actions denoted by the constructive rules of types are in fact those that the inquiry must follow to solve the problem? Aren't these acts conceived as *productive* or *constructive* in the same way that deliberation constructs procedures? Interestingly, logician Göran Sundholm, close to Martin-Löf, distinguishes between 1) the proof as a constructed object, 2) the act of proving, and 3) the traces of that act. The proof-object is what is constructed, the theorem that contains its own demonstration, by the mental act of the mathematician who carries out this construction. But the traces left by the mathematician – the written proof, or the diagrams on the blackboard – are rather indications that enable other mathematicians to reproduce this act, which may be elliptical or, on the contrary, give additional explanations. Sundholm (1993: 61–62) quotes an analogy Martin-Löf is said to have given: in a ski tour, the arrival point of the tour is the object, the tour itself is the act of proof, and the traces of the skis, or the markings one would have left on the path, are useful traces for subsequent hikers.

However, these tracks are those of a hiker *who already knows the way*, as Sundholm makes no mention of false trails, hesitations, U-turns that they would have had to carefully erase later on in order to determine an optimal path for subsequent skiers, and spare them their own wanderings.

In another paper, Sundholm (1998: 32) speaks, concerning the traces of a mathematical demonstration, of a publication in an academic journal, but also of traces of chalk on a blackboard, of used coffee cups, and of scribbled notes. One hopes that here Sundholm evokes the trace of a *research*, of long hours spent turning the problem upside down in every direction, but this is not the case. This example is not commented on and is merely added to other examples, such as an academic publication, which in no way evoke the hesitation of research. These traces are those left by mathematicians in full possession of their faculties, who carry out the protocol contained in the proof object without hesitation and without any false steps.

These texts cannot tell much more than what Sundholm aims at: the analysis of the notion of proof, in order to distinguish it from both a subjectivist and a formalist interpretation. His intention is not to reflect on the processes of knowledge and the elaboration of proofs, nor to characterise the creativity of the mathematician, as he explicitly states (Sundholm 1993: 68). However, it seems significant that, in all these meticulous analyses, no place is given to the act, not of *executing* the proof (or of finding it by a quasi-somnanbulic intuition), but of patiently seeking and constructing it, with multiple back-and-forths.

Sundholm's argument is situated within the framework of Brouwer's "ideal mathematician" , but the problem with this framework is that it is not easy to understand *who* ultimately carries out the proof-object, this ideal method that mathematical inquiry constructs. This paradox has already been noted by Maronne (2017): the accounts of mathematicians indicate possible acts (proofs, constructions, computations) that one should do to solve a problem, without ever giving the appearance of actually carrying them out, since this "does not serve to cultivate or entertain the mind, but only to exercise the patience for of some laborious computer."[1]

What is required now is to leave the domain of logic and formal calculation and to observe deliberation in progress, problems as they present themselves, and theories only in so far as they are actually mobilised to solve them. Only once this grounding in the *experience of knowledge* has been acquired will it be possible to find an adequate place there for type theory, and for the logic of problems that it proposes; only then will it be possible to understand what *systematic* problem-solving is, relative to its merely *practical* or *technical* ways.

What is characteristic of any difficult deliberation is its hesitations, its wanderings, its successful or abortive reconceptualisations. Is one to conclude from the observation of these apparently disordered movements that deliberation proceeds by simple association of ideas, and that it may therefore only be the subject of psychological studies, as the logical positivists suggested? Or should one assume, along with early cognitivists like Herbert Simon, that it executes elementary algorithms whose complexity of behaviour results simply from that of its external and internal environment (its memory)? The problem with both of these solutions is that they fail to do justice to the central characteristic of inquiry, namely that it is a search and a tension towards..., or an intention, in Anscombe's terms. They immediately adopt an external perspective that describes it as a natural process of some kind, or as a mechanism.

It has already been noted in the introduction that John Dewey probably played a central role in the emergence of the concept of problem-solving as a generic and elementary activity of knowledge. This is no coincidence. Dewey asserts that the tension towards the resolution of problematic situations is an elementary fact of

1 Letter from Descartes to Elisabeth, 17 November 1643, AT 4: 42, ed. and transl. Shapiro 2007: 77, modified.

life, both natural and intentional, and that all reflective thinking can be defined as a problem-solving activity, or one of restoring a balance in belief.[2] His *Logic* of 1938, a late work that systematizes his reflections on the nature of knowledge developed very early on, thus defines logic as a *theory of inquiry*.[3].

This fundamental thesis allows to show the unity of the notion of *problematic situation* across the whole spectrum of human experience: If inquiry is the development of an elementary fact of life – the search for equilibrium with one's environment, then all problems, from the simplest to the most theoretically subtle, may be seen as the discursive explication of an elementary tension that must first be felt before it may be posited in language. Some of these tensions are caused by the resolution of other tensions, and therefore by discursivity itself, such as, for example, the difficulty we sometimes have in bringing together two descriptions of the same situation. However, all these tensions, which *compose* with each other and can be *inferred* from one another, are ultimately anchored in the same and unique and dynamic relationship of the human person to their environment. Composition and inference between problems can only be seen in beings endowed with language. Language indeed enables a person to *mark* their habits and their natural tendencies as objects and to consider them in symbolic planes, where they can follow their consequences at leisure (section 6.1).

Inquiry is therefore a form of silent action that can afford to explore and wander, *precisely because it does not lead to consequences*. Judgment is the completion of inquiry, as it is the act by which the agent takes the resolution to *apply* their symbolic plan to the situation experienced. It is this decision that is the model of the traditional logical judgment, and not the other way around. So it is not because it is imperfect that deliberation proceeds by hesitation and error, and iterative reworking of the problem. It is *entirely* a movement of reconstruction, within a given network of meanings, of the problematic situation until the solution – the decision to act – appears "evident". Thus, when deliberation is completed, its *a posteriori* formulation in the procedure or in the demonstration essentially erases its own traces – all memory of its wanderings, which retrospectively appear as blunt and useless errors. The procedure is thus *not* an act of experience, like deliberation, but an *ideal* act proposed to the external examiners of the judgment – either because they want to ensure its legitimacy, or because they want to be able to reproduce it or draw inspiration from it in the future (section 6.2). Mathematics are thus, es-

2 On this point, see Madelrieux 2016: 97.

3 I will cite this work by the abbreviation *LTI*, followed by the page number in his edition of Dewey's complete works by Ann Jo Boydston. I follow the practice of designating the volumes of this edition by the abbreviations denoting its three series, EW, MW, LW, followed by the volume number, then the page. The *Logic* corresponds to volume LW 12.

sentially, procedural means that enable the systematic control and formulation of ideal reconstructions of judgement. In this perspective, the constructive theory of types appears as a development and exact formulation of Dewey's logical intuitions (section 6.3).

6.1 Experience and meaning

This section shows the fundamental unity of inquiry in all its forms, from the simplest to the most elaborate: it is nothing more than the discursive form of an elementary phenomenon of the living realm, that by which the agent attempts to restore equilibrium with its environment. This leads to set out some fundamental notions of Dewey's theory of knowledge: the unity of lived experience, which logically precedes its parts (6.1.1), action as the resolution of a tension in experience (6.1.2), inference and meaning, which enable reflective action (6.1.3) and which require language to do so (6.1.4). Finally, an outline is given of the two forms of propositions that follow from the nature of language as a system of meanings, namely existential and universal propositions (6.1.5).

6.1.1 The unity of experience

The notion of problem-solving is so fundamental for Dewey that he grounds it in the elementary activity of living organisms striving towards the re-establishment of their equilibrium with their environment: hunger, danger, sexual desire, and so on. At this stage there is, then, strictly speaking, no *problem* since there is no consciousness or reflection, but a *problematic situation* that presents itself to the individual as a need, a discomfort, a desire. Intentionality, Anscombe's *trying to get...* is already there, in this elementary relationship of the living being with its environment (see above 5.2.1).

Dewey describes the problematic situation as a break or imbalance in the *unity of experience*. This notion is fundamental and has several dimensions. First, it denotes the harmony between an individual and its environment, the fact that the activity (or passivity) of the former is not disturbed or prevented by the latter – such as easily finding food where one expects it to be, or sleeping undisturbed. In this state of continuity the individual's conduct is entirely *habit*, or activity adapted to its environment. Habit is rooted in the individual's organic life, whose organs behave in a regulated way and have a margin of adaptation to the situation (breathing, salivating, sleeping...). These organic activities are integrated into more global units, coordinated reflexes, sensorimotor activities and learned behaviours. The first continuity of experience is therefore spatio-temporal. It makes no sense to define the activity of eating as a sum of mastications, or as the joint functioning of the mouth, stomach, liver, etc. These distinctions are made *a posteriori* in the rational description of the experience of eating, but this experience itself, when left undisturbed, is one and continuous through time and through the different organs that contribute to it. Moreover, in these habitual situations the individual is not aware of any separation between itself and its environment, and its experience is neither objective nor subjective: these categories of discursivity have no application here. Thus the fundamental experience is not that of fire, but that of "it burns", which jointly includes the heat, acridity, luminous and mobile intensity of the flames, phenomena of which it makes no sense to ask whether they are subjective or objective. This continuity means that it is also irrelevant to oppose active experiences of the organism (such as walking) and passive experiences of observing changes in the environment (such as burning). One should say "it walks" just as much as "it burns" to describe the fundamental, infra-discursive experiences of the individual, and Dewey conceives of them all under the same category of *activity*, without considering that there is a fundamental difference between those attributed to the organism and those attributed to the natural environment: all occur only in the interaction between one and the other. It is an abstraction to think that "walking" is not as much an experience *of* the environment as experiencing the heat of fire, and conversely, that *standing* in front of fire is not as much an activity as walking.

This point is essential for understanding Dewey's logic and it is therefore worth dwelling on.[4] There is nothing like a *purely objective experience of the world* where the individual would receive *data* from reality in total passivity, without adding anything to it of its own. Dewey thus operates a systematic deconstruction of terms that evoke such an idea, such as *observation, perception, sensation*. It is easy to recognise the operational and intentional dimension of *observation*, which is in the same register as an experimental protocol. Observing a fact requires attention directed and motivated by inquiry, a protocol, even a summary one, and sometimes instruments (see for example LTI: 116, 136). As for *perception*, if it means the general awareness of the external world, then it is also active, in that awareness is directed attention, which comes into focus only intermittently.[5] If one is talking about elementary sensible perceptions, such as Russell's *"red – here,"* Dewey argues that psychology studies them as if they were observations (LTI: 153), ignoring the context required for them to happen in the first place. Noticing red in the visual field is

4 See on this Madelrieux 2016: 123.

5 LW 1, 233. There are many occurrences of this equivalence, for example in LW 1, 226–235.

a rather exceptional event, which presupposes an unusual context or a particular interest. The “here” in “*red – here*” points to active motor elements in perception (LW 4: 172). And if one wants to refer not even to perception, but to pure sensation as a peripheral bodily stimulus, as opposed to central attention, then one has to explain how to distinguish it from sensations of hunger, fatigue, emotions, and above all how it could reach the nervous centres of the organism *pure* of any of these factors. The organism's state cannot be understood as a sum of passive data converging towards a single decision-making centre which would send back instructions for action to the organs of the body, but is itself entirely in a state of activity which filters and composes thousands of internal and external excitations according to its overall disposition, itself influenced by its past situation and by its anticipations of the future (LW 1: 251).

6.1.2 Action

Action is the animal's response to the problematic situation, its attempt to resolve it and to re-establish balance with its environment. Behaviourists analyse action as conditioning, and habit is indeed reinforced as a generic way of acting when it is effective in a given situation. But conditioning is rather a consequence of action than the other way round, because there is no such thing as a “given” situation, and its very perception is already shaped by a certain horizon of expectation: which objects, or colours, or smells are noticed, how they are associated with each other, all depends on the tension in which the organism finds itself. The *same* sight of prey will excite the animal or not, depending on whether it is hungry. It should be rather said that it is the generic intention to adapt to its environment, to respond to its different needs, which leads the animal, by successive attempts, to recognise different relevant situations *and* their appropriate responses, to distinguish them into increasingly particular situations. The identity of current experience with past experience is first postulated (MW 6: 381), and it is only the failure of a generic action in an insufficiently analysed situation that calls for a more precise action, related to new traits recognised there. The lion chases zebras, but more specifically those that are isolated, because it *perceives* that they are more vulnerable. The action is given from the outset as a generality that is gradually refined.

The problematic situation and the action that responds to it thus form together a “whole” that is the experience of a tension. Complex experiences should not be seen as compositions of atomic experiences that would be the sensitive stimuli, because these are controlled by the global action that responds to a global situation. Any experience is also a network of potential experiences, which action can develop in one direction or another.

Dewey illustrates this thesis by taking the example of the pursuit of a prey:

> The so-called stimulus, being the total state of the organism, moves of itself, because of the tensions contained, into those activities of pursuit which are called the response. [...] The stimulus in its relationship to special activities persists throughout the entire pursuit, although it changes its actual content at each stage of the chase. As the animal runs, specific sensory excitations, those of contact and those that are olfactory and visual, alter with every change of position; with every change in the character of the ground; with changing objects (like bushes and rocks) that progressively intervene; and they also change in intensity with every change in distance from the hunted object. (LTI: 36–37)

One can also see from this example, which shows a form of continuity in the predator's tension, that habit, as an experience of continuity of the organism with its environment, must not be opposed to action, as an experience of a directed attempt to re-establish this continuity after a rupture, in a binary way. Action must always take place on the basis of the organism's most elementary habits, and on the other hand, with the possible exception of deep sleep, any state of contentment must at all times face alerts, excitations, micro-ruptures. There is always separation as well as continuity in experience.

Furthermore, any successful action has an effect on habit. It is learning, in that it refines the animal's relationship with its environment: the animal's coordination improves in response to the environmental conditions it recognises, but these conditions themselves are also modified, in that it learns to recognise them better, to seek them out, to provoke them (for example by active olfaction, or a new body posture). These adjustments between the animal and its environment in turn create new problematic situations and so on – this process is nothing else than the animal's development – its continual adaptation (LTI: 41–42).

The unity of experience means the unity of its duration, of the individual and their environment, of activity and passivity, of interior and exterior; it is even the unity of unity and rupture in experience, as just seen, and the integration of action into habit. All these categories belong to discursivity, which is merely a modality added to experience. They are therefore irrelevant for describing it. It is in the observation of the animal that this original unity of experience can best be conceived:

> To grasp the sources of esthetic experience it is, therefore, necessary to have recourse to animal life below the human scale. The activities of the fox, the dog, and the thrush may at least stand as reminders and symbols of that unity of experience which we so fractionize when work is labor, and thought withdraws us from the world. The live animal is fully present, all there, in all of its actions: in its wary glances, its sharp sniffings, its abrupt cocking of ears. All senses are equally on the qui vive. As you watch, you see motion merging into sense and sense into motion – constituting that animal grace so hard for man to rival. What the live creature retains from the past and what it expects from the future operate as directions in the present. [...] The past absorbed into the present carries on; it presses forward. (LW 10: 24).

6.1.3 Inference and meaning

The text just quoted suggests that, in the unity of experience, the animal is in expectation, or in anticipation of the experience to come. However, apart from this text, Dewey associates few of these terms with animal life. For an expectation, strictly speaking, is the representation of an absent thing. Thus one would not say that the lion *expects* its claws to tear its prey apart. An expectation must therefore be marked in some way, even if it is simply *remarked.* This happens for example when, on seeing smoke, one expects to encounter fire, if one goes in such and such a direction. Dewey speaks here of a *natural sign.*

There is *inference* when this expectation is formulated as *to take place in a general way*. The basis of this generality in expectation necessarily lies in an operation: "If I do this, then this will happen", because, as seen above, action is essentially generic. Habit reacts generically to a situation that it recognises generically. Thus the child infers that fire burns from the expectation they form that, every time they bring their hand close to the flame, they will experience a burn. Dewey gives another illuminating example following this one:

> The Egyptians looked forward to the occurrence of eclipses at specified dates. In so far as past occurrences had been analyzed sufficiently to furnish the ground for the expectation, the latter partook of the nature of inference. In as far, however, as merely temporal occurrences were the ground of the prediction, the latter was not inference in its definitive logical sense. It became such inference when certain constant modes of natural operation were ascertained to be the reason why certain conjunctions of circumstantial conditions could be used to ground a prediction. (LTI: 250)

Dewey distinguishes here between two levels of inference. In the first, we rely solely on a certain temporal rhythmicity of eclipses; there is inference because we know how to *count* time, but this is not a "natural operation". The latter expression is not very common in Dewey, but there is confirmation in another text, chronologically close to this one, that it means natural changes in "physical" things, as opposed to biological operations which are ultimately grounded in them (LW 11: 108). In the second form of inference, which is the one arrived at by the Egyptian astronomers, the "constant modes of natural operation" thus refer to the movements of the stars, a certain configuration of which (the moon passing in front of the sun) makes it possible to predict an eclipse, thanks to calculations (LW 8: 199).

There seems, however, to be a trick in Dewey's reasoning, for the operation of the sun seems to be comparable only metaphorically to that of counting, or to that of the child advancing their hand towards the fire. But it is precisely here that the preliminary remarks on the nature of experience become important. The movement of the sun must not be dissociated from the *act of its observation*, which certainly cannot be limited to simple perception, but requires a protocol for measuring and recording the result so that it may be compared with past acts. The inference of the eclipse then represents a complete operational device of observation and calculation, which is Egyptian astronomy.

Inference is therefore on the threshold of logic, that is, of language. Carrying out an inference, asserting the general character of an expectation, presupposes in fact that we can *mark* experiences at will.

6.1.4 Language

This symbolic competence, language, is a fundamental fact of Dewey's anthropology, and it must be related to two other characteristics of human beings: their life in society and their technical competence. Dewey understands language in a broad sense, including in it not only speech and writing, but also drawing, rituals, monuments, anything by which people communicate with each other (LTI: 27–28). It is therefore correlated with the emergence of a cultural community, *i.e.* the formation of social habits, traditions and even institutions. All of this indeed requires that people be able to *share* their intentions and therefore put in question their frame of reference in order to bring about the emergence of common points of reference that are sketches of an objective world and which symbols help to fix in the flow of subjective experience.

This objective world is born in particular in technical experience. Dewey says that language is "the tool of tools", in that it is required for active control of experience to take place, and in particular for operations of observation and inference to be repeated and verified.

> Anything whatever used as a tool exhibits distinction and identification. Fire existentially burns; while fire which is employed in order to cook and keep warm, especially after other things, like rubbing sticks together, are used as means to generate it, is an existence having meaning and potential essence. The presence of inflammation and terror or discomfort is no longer the whole story; an occurrence is now an object [...]
> As to be a tool, or to be used as means for consequences, is to have and to endow with meaning, language, being the tool of tools, is the cherishing mother of all significance. For other instrumentalities and agencies, the things usually thought of as appliances, agencies and furnishings, can originate and develop only in social groups made possible by language. (LW 1: 145)

This text shows the close link made by Dewey between human sociability, technique and language. The first systems of meanings are technical and cultural, and their common matrix is language.

The eminent role of language stems from the fact that symbols – words, drawings, monuments – have extraordinary properties. Unlike natural signs, they do not refer to anything *per se*, and their meaning has to be instituted. Moreover, they can generally be invoked at will (spoken, written, thought), which makes them easy to manipulate – which is why monuments are relatively less often conceived as elements of language. But above all, the invocation of a symbol in concrete experience does not as such create the expectation of an experience, like smoke creates that of fire. It merely *suggests* it, according to the expression proposed by Dewey, or *evokes* it; its function is to *direct* attention towards certain aspects of the situation or towards certain possible actions (LTI: 114). In order to create an expectation proper, the symbol must be referred to present experience, as when one shouts "Fire on the horizon!" pointing one's finger in one direction. There is an assertion here, marked by the presence of a deictic. It is asserted that whoever looks in that direction will see smoke, or that whoever is near that smoke will experience the heat of fire, and so on. But the expression "fire on the horizon" as such is a simple *meaning*, points only to these operative rules. Meaning never affirms anything about present experience, but indicates some rules that should be relevant in its consideration.

The deactivation of the immediate relationship of symbols to experience is the very source of the power of meanings, for they can thereby be invoked and manipulated at will through symbols. Thanks to them, the inquirer can suggest or evoke inferences between experiences by means of relations between meanings, without committing themselves to the concrete expectations which they ultimately evoke. They *can* thus put them at some distance in order to verify them, identify conditions of invalidity, corroborate them with other inferences, etc. (LTI: 250)

It should be noted that there is a dual power of language here. At the first level, the sequences of meanings, which are the reasonings, make it possible to "rehearse" mentally an action before acting, in the theatrical sense of the term (LTI: 63). When the consequences represented are not those one hopes for, one can thus modify one's conduct in advance. At a second level, language itself, as a system of meanings, is itself open to critical examination: the inferences it suggests can be criticised and modified: this is the object of that critical, second-order inquiry that is the basis of the scientific attitude.

Meanings can simply evoke experiences without generating their expectation, because fundamentally they designate *the operations to be carried out* in order to have such and such an experience, be it a physical action, an observation, or even simple attention to this or that aspect of the situation. They have the status of *rules*: *"If such and such an operation is performed, then such and such an experience will follow"*. To verify a meaning in experience, that such and such *is the case* (e.g. that the temperature of the room is 25°), is simply to perform a certain operation in order to experience a certain experience (consulting the thermometer and checking that the height of the mercury is at gradation 25). Remember that the ultimate source of these rules lies in the fact that every quality, even the most elementary sensitive quality, is a certain response of the environment to a certain generic action, which the organism has learned to discriminate among other possible responses.

Thus common nouns designate species which are conjunctions of possible operations:

> The propositions "This has vitreous lustre; cannot be scratched by a knife; scratches glass; is not fused by a blowpipe; breaks with conchoidal fractures" are propositions which, taken separately, set forth special modes of change. Applied concurrently and cumulatively to this they yield the set of conjoined traits describing the kind quartz. (LTI: 291)

Meanings can only exist in relation to each other. The properties of quartz denote operations that move a material from one state to another, these states themselves being so many potential experiences. For example, "conchoidal fractures" in turn signify a certain shape of the pieces that can be compared to that of a shell, or from which a certain curvature and polish can be measured. So it is always signs of operations, and the changes they bring about in experience, that are related to each other.

It should be noted here that Dewey reserves the term *inference* for the action of evoking and expecting a certain experience from another. It is therefore situated on the level of experience, and not of meanings. On this second plane, Dewey speaks of *relations*, which are nothing else than the *rules* controlling inference. Indeed, one remembers (see 5.2.4) that in mathematics rules can be (and usually are) read as relations.

6.1.5 Existential and universal propositions

Relations serve to control inferences, to ground them, and possibly to discover new ones. Dewey distinguishes two forms for this role.

The first function of a relation, and in particular of a proposition, is to guide the inquirer in their analysis of the experience, by making explicit the inferences to be drawn from it. The proposition *"This is quartz"* establishes a relation between a material object located in experience and the set of expectations that it evokes, and which make it possible either to verify the proposition – if one wants to be sure that it is indeed quartz, or to indicate the possibilities open to action or reflection – what one can do with quartz.

This type of statement must contain a reference to experience. This may be explicit, as when a deictic or proper noun is used, or a circumstantial complement. However, these terms, such as *this*, *here*, *John*, *Athens*, or *In the third century BC*, by themselves do not refer to anything; they give indications that allow the recipient of the proposition to situate it in their own experience, before they can consider its truth. As Dewey says, *this*, encountered in a proposition, first designates a problem to be solved: to identify what it refers to. It is useful here to note that Dewey strongly criticises the artificial examples of propositions found in logic textbooks, such as *"Socrates is a man"*, because a proposition only acquires its full meaning in the course of the inquiry that motivates it. So, for example, *"There is a man here"* is a proposition that has a specific meaning in the context of a completely dark room. It is then an *inference* from some sense of a presence, that invites its verification. In another context, it can be uttered by a walker in the forest, who is asked by guards if they have seen anyone pass by, and who points to a shed. The statement is an observation for the walker, but it contains an inference, that the person is still there. The guards must also make other inferences: that the walker is telling the truth and that the person is the one they are looking for. It is the whole situation that is modified by the proposition, which Dewey makes explicit by saying that these kinds of propositions make it possible to identify *what is in question*, the subject of the inquiry.

There are also implicitly referential propositions. Saying *"Athenians are proud"* presupposes that one can locate this community in time and space (are we talking about the Athenians of today or of the classical age?). But even a proposition as innocuous as *"swans are white"* mobilizes a concept of natural history, and refers to beings existing in a certain time and place in the universe. These propositions indicate inferences that can be made if, in the inquiry, one is dealing with Athenians or swans. They populate the background of our experience with constraints and possibilities for action.

There is a second form of relations between meanings, which does not have to do with the inferences one can make in experience, but with the connections there are between the signified operations alone. For example, to say that *"who goes up will have to go down in order to come back"* is a relation inherent in the *meanings* of going up and coming back – at least when they are understood in the topology of the usual space. It is therefore not an inference in the strict sense, but a simple explication of the constraints inherent in the joint operation of these two meanings. Or saying that *"the distance of a return trip is identical to that of an outbound trip"* points to the equivalence of the measurement of a distance in both directions; it relates to what it means to measure, not to a connection of things in experience. Moreover, it could be that empirically this assertion is false, if for example there are forbidden directions on the route: but such a remark refers to some information not contained in the initial proposition, which modifies the standard meaning of *trip*. Another example, to leave the field of travel, is the proposition that if some horses have harnesses, then some harnesses are on horses. This statement has nothing to do with the existence of horses or harnesses, but with the equivalence of finding a harnessed horse and a harness on a horse. These examples show that Dewey is not here in the Kantian perspective of analytic propositions – indeed, he does not use the term – for this is not a mere propositional reformulation of the contents of a definition of the terms involved, but an operational analysis, which would fall rather on the side of synthetic propositions *a priori*.

Before going any further, it is useful to recapitulate the three relationships that Dewey's semiotics distinguishes: 1) the *connection* between objects of experience, *which alone is the object of inference*; 2) the *reference* that a meaning has to experience, by means of one or more verification operations that it indicates; 3) the *relation* that meanings have with each other (LTI: 60–61). These are of two kinds: either they indicate that such a relation can itself be verified in experience, and are then called existential propositions, or that there is an internal relation between the operations by which they are verified, and are then called universal propositions.

6.2 Inquiry and judgment

The availability of a network of meanings such as language enables reflection to recognise a problematic situation as such, and to institute it as a problem (LTI: 111). This is an integral stage in its resolution via meanings. This moment makes it possible indeed to *determine* the terms of the problem which are considered firm, *i.e.* non-problematic. They are facts that can be observed and that serve as a basis for its resolution. Dewey calls this moment observation, but this operation must not be restricted to perceptual activities: it can also involve accepted hypotheses or preconceptions through which one understands the initial situation.

Any problem evokes one or more suggestions for its resolution, because any meaning by which it is described immediately points to possibilities for action. A suggestion that is examined becomes an idea, and this examination is reasoning, which must represent it symbolically in order to manipulate it. This idea can immediately be considered adequate, and inquiry is then complete. However, a reflective attitude requires one to inspect this idea in all the relationships it contains with the initial conceptions, and with the system of conceptions in which it appears. These relationships indicate ways of testing its adequacy to the situation, they indicate possible operations:

> Both observed facts and entertained ideas are operational. Ideas are operational in that they instigate and direct further operations of observation; they are proposals and plans for acting. (LTI: 116)

This examination provokes new observations, which make it possible to re-express the problem and arrive at a new understanding of the initial situation, where some new reasoning is required – and this iterative movement is repeated until the solution appears as evident in the final re-expression of the problem and reasoning.[6] This moment is the judgment that closes the inquiry. The situation appears as "unified", according to Dewey's expression, *i.e.* there is no longer any disagreement between the inquirer and their environment.

> The new order of facts suggests a modified idea (or hypothesis) which occasions new observations whose result again determines a new order of facts, and so on until the existing order is both unified and complete. In the course of this serial process, the ideas that represent possible solutions are tested or "proved". (LTI: 117)

The inquiry thus comprises five moments which are not so much successive as interdependent:

1. the recognition of a problematic situation,
2. the position of the problem,
3. the suggestion of the solution,
4. its elaboration and verification in reasoning, and
5. the judgement or decision, which closes the inquiry.

In what follows, the focus will be on the two properly discursive moments of inquiry, first the position of the problem which mobilises *material and procedural means* (6.2.1), and then reasoning which properly constructs the solution, or rather *reconstructs it* on the basis of the suggestion found (6.2.2). Inquiry will then be studied from the point of view of its outcome, judgment, which is essentially a decision that modifies experience (6.2.3).

6.2.1 Material and procedural means

Any inquiry is based on a system of meanings, whose *material and procedural means* it mobilises. It is useful, in order to introduce these important terms in the *Logic* of 1938, to start again from an analogy that Dewey draws between the form of practical judgment and the subject / predicate structure of the proposition. In the case of the practical judgment, the subject is the problem of the inquiry, which is to modify the present situation in order to obtain the desired one. The predicate of the judgement *is the course of action* itself, *i.e.* the solution to which the inquiry leads. This solution

6 See on this the analysis of Madelrieux 2016: 119.

relates as much to this construction as to the initial identification of the problem in the experiment. Before establishing a plan of action, one must first recognise *what is at stake* and *interpret* the situation so that it can be described in the system of meanings available, which will provide the materials for this construction.

One can take here an example which will become canonical for cognitive science in the 1960s.[7] It is the very simple case of a banana hanging from the ceiling, which can be unhooked using a stick if the chimpanzee climbs onto a chair. The inquiry must begin by recognising the relevant elements in the situation and what can be done with them: a chair that can be moved and climbed on, a stick that allows to reach distant things, a banana that satisfies hunger. This is the part of the inquiry that *identifies* the subject of the judgement. Its predicative part consists of putting together a plan of action that links these means together. This part works within the problem space that has been described, *i.e.* it merely links together the meanings given. Dewey calls material means the propositions that make it possible to attach meanings to elements of experience – that the chair has four legs, a seat and a backrest, for example, that a stick is an oblong piece of wood, etc. As for procedural means, they tell what one can do with these objects: sit on the chair, climb on it, grab the stick, and so on.

In the case of a trial, the subject of the judgment (the problem to be resolved) is the situation created by a crime, which requires reparation. The crime is described by all the facts that have been established. The predicate is the legal reasoning that uses different articles of law and case law to legally qualify these facts as a crime. The final sentence is the attribution of this reasoning to this situation, so as to resolve it – so that reparation is obtained. Of course, this is not a simple proposition, but a complex argument that interweaves facts and reasons. Here, the logical function of the copula lies in this referential attribution of a (complex) meaning to a (complex) situation. Material means here are all the testimonies on the crime and the relevant laws to be applied. Procedural means are the rules that allow attorneys and judges to conduct the trial, exchange arguments and reach a final decision.

If one takes a scientific example, such as the problem of calculating the speed of a galaxy, the subject of the judgement consists of the experimental facts established and the theoretical hypotheses accepted. These facts are themselves the result of complex observation procedures, in particular the measurement of interference lines in the electro-magnetic spectrum, which requires technical instruments. Assumptions can be made, but they have to be accepted by the scientific community,

7 This example, which is not used by Dewey, was popularised by Newell and Simon in their studies of the GPS program (see above 2.1.2 and appendix A). They were inspired by experiments carried out by Gestalt psychologist Wolfgang Köhler with chimpanzees in the 1910s.

such as that the size of the galaxy studied is similar to those of the same type. It is clear here that the situation is not made up of "raw" facts, but may itself result from a very vast complex network of previous judgements, which is here the very corpus of science. As for the predicate of the judgement, it is the result as well as its complete organised proof (observations, principles, hypotheses, calculations). The material means here are the instruments of observation, the mathematical theories used, the physical principles; the procedural means are the operations (physical and symbolic) by which all these means are mobilised (for example, with the approximations mentioned) to arrange the proof.

The duality of material and procedural means is not limited to their discursive form, but permeates all the elementary practices of life (LTI: 81). In productive activities, material means thus represent tools and materials, and procedural means the techniques that put them to use in complex endeavours (LTI: 383). This particular case highlights two important points. First, these two classes of means are always adapted to each other: we have the techniques that our tools enable, and our tools are designed according to our techniques. This adaptation is dynamic, in that the researcher has the possibility of modifying them according to their needs. For Dewey, the inquirer is never locked into a rigid problem space to find its solutions, but is always able to reconfigure it in the course of the research (LTI: 392). Secondly, the essence of procedural means is to establish a *serial order*, which characterises both technical procedures and rational discourse in general (LTI: 117, 384). This order is imposed by the very fact that all inquiry is *directed* towards solving the problem at hand, which Dewey even calls the "principle of direction" (LTI: 312–313). He compares it to that which orders the degrees of a scale, and contrasts it with the material order which organises facts and data, and which Dewey describes as the order of a whole rather than a series – one would say today a *structure*.

These examples show that one of the roles of a system of meanings – everyday language for elementary practical cases, the law, scientific theories – is to provide the inquirer with the appropriate concepts to represent problems and construct their resolution procedures. It saves the inquirer from constructing their own concepts, by drawing them from a network of already established and controlled relationships.

The duality of *procedural means* and *material means* runs through the *Logic* of 1938 in its entirety. It is also expressed by other pairs of terms: existential and ideational, factual and conceptual, description and definition, and so on.[8] This duality thus plays a fundamental role in the conceptual framework of the book, and in the interpretation of classical logic. At the level of propositional structure, it ex-

8 See in particular LTI: 272, 275, 283.

plains the duality of subject and predicate. At the level of the typologies of propositions, it explains the duality between existential and universal propositions. Finally, the classical syllogism itself denotes this fundamental structure of thought, as the major premise represents a solution available (*i.e.* already constructed) in the system for a given problem (for example: *"When one is ill, one must see a doctor"*), and the minor premise, the identification of the situation with this problem (for example: *"I am ill"*), and the conclusion the decision to apply the proposed solution to the situation (*"I must see a doctor"*) (LTI: 167). Of course, this way of presenting problem-solving is a caricature, and Dewey denounces it as such. Things rarely happen like this, except in cases one faces a paradigmatic problem foreseen by the theory, such as, for example, the construction of an equilateral triangle in the *Elements*. In the general case, however, the situation is only progressively identified, with several meanings of the system being gradually recognised one by one. The construction of the major is also complex, because these properties of the problem have to be linked together by recognising a possible sequence of operations concerning them, and the whole of this twofold task is iterative.

Procedural means tell how to arrange material means to solve a problem, for example heating metal *in order to* pass it through a mould that will give it the required shape once it has cooled. They represent the usual ways of doing things, the known methods and heuristics. In contrast, the material means set out the basic rules governing the objects of the problem, which are used by the procedural means. They tell what inferences can be legitimately made about them, for example that metal becomes liquid when heated, and that a liquid takes on the shape of its container.

However, material means should not be seen as "inert" materials on which procedural means would act. As Dewey keeps repeating, facts are operational in that they *suggest* possible operations. There is in the idea of metal that it can be heated, and the hammer, according to the time-honoured proverb, invites one to look at any problem as a nail. The procedure cited above *is not deduced* from the two material propositions quoted afterwards, but is constructed from them. If material means were inert, this could not happen. On the contrary, they already signify which elementary actions can and must be taken into account to determine a plan. As this terse formula of Dewey's puts it very well: "Powder is what will explode under certain conditions" (LTI: 132).

Nor are these means "material" in that they would denote an ultimate and inanalysable term of thought, as the English empiricists for example were able to speak of sense data as "material" of thought. Dewey's critique of this position has already been mentioned. Dewey calls them such because they make it possible to *recognise* the "subject-matter" of the problem, *i.e.* to frame it using a system of meanings.

6.2.2 The reconstruction of inquiry in reasoning

The position of a problem, according to Dewey, suggests solutions, or rather leads of solutions, for such is the function of material means: to provide tools for action. As soon as I lay down a system of differential equations, I know that there are certain transformations I can do with them, by recognising, for example, forms similar to equations already solved. The solution is then found, or I end up with another system of equations, perhaps simplified. This is a new problem, which may in turn suggest other operations to carry out, and so on.

To take another example, if I need to plan the supply of my retail warehouses for the coming month, the immediate suggestion is to continue with the previous month's plan. However, when I check the capacity of my factories, I realise that two of them will be undergoing maintenance. So I have to look elsewhere for additional production, which means checking the availability of factories further away in my logistics network.

In these iterative approaches, reasoning is each time the performance of symbolic operations suggested by the path under consideration: in the mathematical example, it is a matter of performing algebraic transformations on my system of equations to see what happens; in the logistics example, checking the availability of my factories in the forecast reports. In inquiry, the role of the reasoning stage is to activate the material and procedural means of the problem in order to examine the solutions they suggest. Its role is not so much to find the solution as to verify it formally (without recourse to experience). This idea is clearly expressed in this passage from *How we think*:

> Reasoning has the same effect upon a suggested solution as more intimate and extensive observation has upon the original problem. Acceptance of the suggestion in its first form is prevented by looking into it more thoroughly. Conjectures that seem plausible at first sight are often found unfit or even absurd when their full consequences are traced out. Even when reasoning out the bearings of a supposition does not lead to rejection, it develops the idea into a form in which it is more apposite to the problem. [...] The development of an idea through reasoning helps at least to supply the intervening or intermediate terms that link together into a consistent whole apparently discrepant extremes. (MW 6: 240)

It is clear here that reasoning has first of all the role of verifying the suggested solution, which must be distinguished from the experimental validation of the judgement, the latter involving empirical observation. This verification develops the consequences of an idea, which may fail with regard to the data of the problem, or on the contrary indicate the precise relationships that enable it to be transformed into a solution. Reasoning therefore *constructs* the solution. Working iteratively with the suggestions for a solution (step 3 of the inquiry) and the reformulation of the problem, possibly thanks to new empirical observations (step 2), it may be led to reconstruct the entire problem until a satisfactory solution appears in it. This solution (and this is the second role of reasoning) does not only exhibit the point of arrival that it is possible or suitable to reach, but also the sequence of means for doing so, or put metaphorically, the path to get there. This is the *procedure*, a term that should be taken here in its broadest sense, which can just as easily be a mathematical proof procedure, as a procedure of actions to be taken to solve an everyday problem.

Moreover, the problem itself is completely reconstructed in the inquiry. At the end it no longer resembles what it was at the beginning. Then it was only partially understood, there were ambiguities, incidental elements appeared to be salient while crucial elements were hidden, there were suggested paths of resolution that turned out to be dead ends, and so on. At the end, the inquiry produces a new description of the problem, a final *specification* from which a procedure can be clearly and directly detailed. But this procedure says nothing about the successive iterations, the exploration, the doubts, the abortive attempts, which characterised the inquiry (LTI: 124, 161). It simply describes the relationships that link the problem, as it appears in its final form, with its solution.

The arguments developed by the judge and which form the basis of their decision are not the same as the trial, or even its minutes, because the trial was made up of debates, questions, retellings and arguments, from which the judge took only what was useful in preparing their judgement. Similarly, the action plan is not the same as the deliberation that produced it, nor does the scientific paper describe all the steps that led to the discovery.

The procedure is therefore the *ideal reconstruction*, in the field of meanings, of the inquiry itself. It indicates the approach to be followed that makes it possible to arrive at the same result as the inquiry while avoiding the difficulty and chance of the research. This simple fact explains why empirical studies rarely detect deductions, or even syllogisms in cognitive processes. A syllogism like: "*When one is ill, one must see the doctor. Now I am ill. Therefore, etc.*" does not at all convey the thought process of anyone wondering whether they should go and see a doctor or not, because that would mean that there had been no inquiry as such, but a kind of automatic habit, like a response to a stimulus (LTI: 167). The syllogism could instead be formulated in the advice of someone we would ask for an opinion, or in the explanation we would give them of our resolution. This interpretation is in line with the thesis expressed by Sperber and Mercier (2017), that reason has an essentially pedagogical and justificatory function. It is however too hasty to conclude, with these authors, that rationality has no place in the process of inquiry, and that the latter would be entirely controlled by cognitive habits, and therefore biased. The rational procedure, or the demonstration that rigorously links meanings together, is more the endpoint than the means, the stake in a battle against our inferential habits that inquiry aims precisely to control.

This process of reconstruction does not happen in all inquiries. Most of the time, the inquirer is content with the solution found, which they apply immediately to the problematic situation. On the other hand, this reconstruction is essential to certain forms of inquiry, for example those of law or science, which are committed to producing judgements in due form, *i.e.* which contain their own justification. The reconstruction of inquiry is therefore like a second inquiry, in a reflective posture with respect to the first. It aims to determine what were the decisive elements in the formation of the final judgment, what were the intermediate, significant judgments, etc., and what were the reasons for the final judgment.

In this second inquiry, attention is also generally paid to the *possible generalisation* of the problem, if it is parametric. In this case the reconstruction movement takes place at an even higher level of reflection, since it is not just a question of observing what the decisive stages of the reasoning were, but of finding out how and why these stages were themselves identified, since they could be different in other cases. In other words, the inquirer is interested here in *heuristics, i.e.* the movement by which the problem was posed, reconfigured, etc.

These distinctions lead back to the three classes of imperatives that Simon had distinguished in the resolution of any problem: imperatives (A), which indicate the objective to be achieved, imperatives (B), which require the plan of action for doing so to be made explicit, and finally imperatives (C), which ask how to constitute, in a generic way, such a plan of action (see above 2.1.3). These three classes are precisely articulated and explained in Dewey's theory of inquiry: (A) designates an imperative that directs a simple inquiry, (B) one that directs its proceduralization, and (C) one that aims to describe the method of the inquiry itself, for a more or less general class of problems similar to the one initially solved.

Finally, I define (and this definition is mine) *ratiocination* as an inquiry that proceeds through the scrupulous execution of the rules of a procedure already known. I have deliberately chosen this old-fashioned term, occasionally used by Dewey, leaving aside the pejorative meaning it has today, because it evokes well the form of reflection aimed at here, that which is accomplished step by step and in scrupulous compliance with prescribed rules of inference. It encompasses a broader class of inquiries than that denoted by the term *calculation*, for at least two reasons – setting aside from the outset its initial association with the categories of number and quantity. Firstly, most procedures do not have the Procrustean character of calculation rules,[9]. Most ratiocinations require the inquirer to interpret the rules or to

9 On Procrustean rules, see above 1.2.

complete them in unforeseen cases. Secondly, calculation proceeds by transforming input data into output data without any possible or relevant interruption before the result is reached. Now, just as an inquiry is carried out in constant interaction with the situation, a ratiocination can generally also be open to regulated interactions. Thus, when a foreman organises the work in a workshop in accordance with the engineer's instructions, or when a sales agent, in the dialogue with a customer, carefully applies a *script* provided by their management, one can say that these are cases of ratiocination, but not of calculation.

In what follows, the execution of effective procedures will be qualified as *pure ratiocinations* to distinguish them from the broader class of ratiocinations.

Ratiocination is a stage in learning, *i.e.* in the resorption of problematic situations into habitual experiences, which no longer pose a problem. Thus, learning some calculation methods involves their repeated application, and the laborious discipline this imposes, until they are integrated into our cognitive skills and their application becomes "natural", effortless, and almost unconscious. The procedure is a step in *internalising* the appropriate response to a hitherto unseen situation. It replaces the complex and costly inquiry and makes possible the emergence of a habitual practice which, repeated a certain number of times, can progressively become an unconscious automatism.

6.2.3 Judgement as an act in experience

Dewey defines judgement as the act that closes the inquiry. In the chapter of *Logic* devoted to this notion (called, significantly: *The construction of judgment*), Dewey takes the example of a trial. The final judgment *has* consequences: releasing a defendant or sending their to prison, demanding damages, and so on. Dewey also uses the terms *decision*, *settlement* (in the sense of settling a case (LTI: 124)), or *resolution* to convey the enactive force that he intends to give to judgment. On the other hand, the judgement is *justified* by the inquiry, it can avail itself of the path taken to explain its decision – even if it could appear summary and "unreasonable" by today's standards, such as the invocation of taboos in traditional societies to justify a way of acting (LTI: 68).

There are therefore two aspects to the thesis that judgement is an act. Firstly, it is an action that takes place in experience, transforming an indeterminate situation into a determinate one. Secondly, its justification is inseparable from the procedure that recapitulates inquiry.

Practical judgement, *i.e.* the resolution to act in accordance with a plan, is an example of the first point. The meanings here represent possible actions and the modifications that each would bring about in experience. The inquiry consists of linking them together into a plan that will hopefully lead to the desired situation. Judgment completes it with the resolution to act.

A judgement must therefore be radically distinguished from a proposition. A judgement *may* be expressed by a proposition, but it cannot be reduced to one. A proposition is a relation between meanings, and therefore has no intrinsic reference to experience, as discussed above. A judgement, on the other hand, is motivated by a problem that arises in experience, and which it resolves. It is not a matter of judging someone *in abstracto*, as may be done in fictional case studies, but of judging such-and-such a person who has caused such-and-such harm. It is not a question of knowing what you would do if this or that situation ever happened, but of deciding what course of action to develop in the present situation. The judgment therefore has a direct experiential importance that the proposition has not:

> Judgment may be identified as the settled outcome of inquiry [...] In this sense [it] is distinguished from propositions. The content of the latter is intermediate and representative and is carried by symbols; while judgment, as finally made, has direct existential import. [...] I shall use assertion to designate the logical status [of subject-matter which has been prepared to be final]. (LTI: 123)

In the course of an inquiry, one can certainly *consider* propositions which seem relevant to the present problem. For example, the proposition that the accused was in such and such a place at the time of the crime may be put forward; but in order to become a usable piece of evidence in the trial it must be established, that is to say become the subject of a partial judgment which will eventually support the argument of the final judgment.

Dewey's thesis is that all authentic judgements are such *acts* that come to transform experience, including scientific judgements. Take again the example of calculating the speed of a galaxy. This calculation may be done as an exercise, or to calibrate new equipment. In this case, the judgement is not about the result, but about the satisfaction given by the student or by the machine. Perhaps this calculation was made simply out of *curiosity*. But this must then be itself qualified in order to assess the judgement that resolves it: was it an anomaly that the astronomer thought they had detected? in this case, the calculation is part of a wider inquiry, which concerns the consistency or falsification of a set of scientific hypotheses. Was the aim to use this result to calculate other physical quantities, such as the speed of expansion of a particular region of the universe? In this case as in the previous one, the judgement serves to construct a more complex judgement, of admittedly entirely "theoretical" interest but which modifies the astronomer's knowledge, by confirming or invalidating their theories, and which is intended to be published, and which changes the state of science more generally.[10]

Judgement is also an action, not only in that it modifies experience, but in that it is itself a process, distinct however from inquiry. Dewey makes this visible in the surprising analogy that he proposes between judgement and an industrial organisation:

> [The copula] [...] might be regarded as what sets the subject-and-predicate contents at work *executing* their functions in relation to one another. In complex undertakings a plan for division of functions is usually laid out on paper. But this plan is not the actual division of labour. The latter consists in the actual distribution of the active factors of what is doing in their cooperation with one another. [...] The plan may be set forth and explained in propositions; its propositional exposition may be a means of criticism and of re-arrangement of the plan of distribution. But the actual division can only be enacted. [...] But it no more is a functioning division of labor than a blueprint is a house in process of building or a map is a journey. (LTI: 138)

What Dewey is trying to make clear is that the intermediate stages of reasoning that lead to the final result are *conserved* in the judgment, not merely as documentary, but as active entities. The origin of this analogy is found a few paragraphs earlier in the quoted text, where Dewey criticizes Bradley's thesis that there would be an inherent contradiction in the idea of judgement, since it consists in distinguishing features in experience before reuniting them (as one distinguishes the redness in the apple before saying that the apple is red):

> Since [subject and predicate] are functional and operative, there is no more conflict than there is in the fact that in the course of any complex productive activity, industrial or social, divisions of labour are instituted which are nevertheless functionally related to one another. For they are instituted as means of cooperation with a view to a common unified result. [...] Judgement, like inquiry, is temporal. It is temporal, not in the external sense that the act of judging takes time, but in the sense that its object undergoes reconstitution in reaching the final state of determined resolution and unification which is the objective that governs judgement. (LTI: 136–137)

If the judgement is never confused with the crude result of the inquiry, such as the resolution to do such and such a thing, or the sentence of the condemned man, or a number representing a speed, it is because it also includes its justification, it is because it is the entire construction that makes it possible to ensure this result.[11] As Dewey says:

10 On Dewey's doctrine of science, see below, section 7.2.

11 See on this subject Madelrieux 2016: 103.

> Through a series of intermediate meanings, a meaning is finally reached which is more clearly relevant to the problem in question than the idea originally suggested. It indicates the operations that can be performed to test its applicability [...] In other words, the idea or meaning, when developed in discourse, directs the activities which, when performed, provide its material evidence. (LTI: 115)

Thus, the resolution to act made in the practical judgment differs from a mere impulse *because* it is placed under the aegis of a considered plan of action, the court judgment differs from an arbitrary sanction by the argument that the judge constructs at the end of the trial, and the scientific result differs from a mere estimate by the replicable method it exhibits.

> A scientific work in physics or astronomy produces a report of calculations and deductions derived from past observations and experiments. But it is more than a report; it is also an indication, a summons to make further observations and experiments. No scientific report would obtain consideration if it did not describe the apparatus by means of which the experiments were conducted and the results obtained; not because the apparatus is revered, but because this *procedure* indicates to other inquirers how to set to work to obtain results which will agree or not, in their own experiment, with those previously attained, and thus confirm, modify and rectify the latter. The scientific result recorded in a report is in effect the designation of a method to be followed and a prediction of what will be found when specified observations are performed. (LW 1: 39, my italics)

The plan, which is the reasoning itself, or the final ordered discourse, is therefore this sequence of meanings *to which the inquiry leads* and which the judgment attributes to the situation. Any reasoning, as a chain of propositions, has a similar status: it is entirely hypothetical, and all its material premises (which refer to experience) must each be the object of an assertion, in order for a conclusive judgement to be truly justified.

These remarks make clear that pure ratiocination by itself does not contain an ounce of judgement. It merely operates the instructions of the effective procedures until it achieves its given output. It is only then that judgement may occur, deciding or not to attribute this result to the problem which motivated the call of the procedure. For example, a businessperson may compute the expected profit of their business plan given certain assumptions about sales or costs, and this computation may follow very elaborate rules modelling the operations of the business, tax, required capital, and so on; but dissatisfied with the outcome, they may judge that they must aim at higher targets and revise their business plan again. Judgement is bracketed out of such procedures, which is precisely the goal of ratiocination.

6.3 Mathematics and procedural means

This section is devoted to Dewey's conception of mathematics, which is characterized by the central assertion that mathematics are *procedural systems* (6.3.1). Dewey conceives of proofs as forms of procedures (6.3.2), which leads to wonder about Dewey's relationship with intuitionism, first historically and then logically (6.3.3): it seems that Martin-Löf's theory of types is one of the possible formalisms in which any inquiry may be procedurally recapitulated.

6.3.1 Mathematics as procedural systems

Dewey gives a central role to the term *procedure* in the conceptual framework of his theory of inquiry. Not counting the phrase *procedural means*, it appears frequently in the work (85 times, compared with 137 and 260 occurrences of *process* and *method*, respectively) at a time when it is in the process of acquiring its modern meaning (see above 1.2). Dewey understands *procedure* in its usual sense at the time, as progression in the course of an action, and indeed sometimes interchanges it with *process*.[12] As a *manner of proceeding* it is then closely associated with problem-solving, as when he refers to doctors and lawyers whose good procedure "essentially consists in formulating the right questions" (LTI: 172): here the term has the very general role of a precept, or good practice in the conduct of an inquiry. However, he associates it particularly with the term *method*, with which it often appears in conjunction, and this in the context of scientific inquiry (LTI: 26, 65, 390, 404, etc). In this context, Dewey does speak of procedure in the sense of establishing a sequential order in the progression of reasoning or action, and evokes the image of "rungs of a ladder", which does refer to the distinct stages of a procedure as it is currently understood.

Mathematics have a procedural status for Dewey, which shows their affinity to techniques (LTI: 384), as can be seen from the following statement:

> The conceptual nature of the material data of mathematics means that they are determined exclusively and wholly in reference to the possibility of operations of transformation, *the latter constituting procedural means.* (LTI: 403, my italics)

Mathematics are an entirely procedural system in that their material means have no marks by which they can be referred, directly or indirectly, to concrete experience – they denote pure possible operations, such as counting. Unlike a theory of

12 For example LTI: 15, 436.

the natural sciences, they do not have concepts such as, for example, those of *animal species*, *cell*, *mass*, *charge*, or *energy*, all of which must be recognised or measured in experience (albeit sometimes through complex and indirect observational or measurement procedures). On the contrary the material means of mathematics signify only possible operations on elements that are defined exclusively with regard to them.

> The condition of transformability which mathematical contents must satisfy demands that there be "data" which are determined exclusively and exhaustively by reference to the operations and rules of operations executed or to be executed with and upon them. (LTI: 402)

Transformation is a term that Dewey introduces specifically in his development on mathematics to characterise the progress of ordered discourse (reasoning) that substitutes meanings for one another by controlled relations. Mathematics is thus about inferences from one possible operation to another: *If I can do this and get this result, then I can do that and get that other one*, as for example: *If I can add 2 and 3 and get 5, then I can add 2 and 4 and get 6*. The numbers 2, 3, 5, etc. here are defined exclusively with respect to the elementary operations instituted on them, those of Peano's axiomatics, for example.

This characteristic explains the properties by which mathematics are usually described. On the one hand, they cannot be invalidated by experience, because they are free of any *reference* to it. One may make an error while counting horses in a meadow, by confusing one with another; but one cannot err in judging that the five horses (if correctly counted) and three more coming (if also correctly counted) will not be eight, because this relation is derived directly from a chain of relations about the act of counting, not about the horses.

This also explains the universal applicability of mathematical theories, such as graph theory, in the sense that their results may be asserted in experience as soon as their elementary operations may be recognized there. They say nothing more than: *if such operations are effective there, then so are also such other operations*.

On the other hand, for this very reason, mathematics, as a science, *i.e.* as an approach to the systematization of relations between pure operations, present themselves as "formal" symbolic systems, which may prompt the belief that they manipulate symbols without meaning, *i.e.* that their meaning is fully conveyed by the rules of their syntax. But one should rather say, in Dewey's vocabulary, that mathematics have no *reference i.e.* no relation to the matter of experience, for they are indeed endowed with meaning, they are even pure meaning, since they are entirely operational:

> In the sense which "meaning" bears in any conception having even indirect existential reference, the terms have no meaning – a fact which accounts, probably, for the view that math-

> ematical subject-matter is simply a string of arbitrary marks. But in the wider logical sense, they have a meaning constituted exclusively and wholly by their relations to one another as determined by satisfaction of the condition of transformability. (LTI: 395)

This property explains the character of mathematical objects of being defined in relation to other mathematical objects, and this at an indeterminate level of depth. This is possible because it is only ever a question of operations being defined in relation to other operating procedures, and so on. Take the example of a differential equation, which establishes a certain relationship between functions and their derivatives. Derivation refers to an operation that can be performed on real functions, under certain conditions. In turn, a real function refers to a more or less complex operation that can be performed on real numbers. The real numbers are certain sequences of integers, and the integers are finally defined by the elementary operation of counting. *All this* is present in the relation defined by the differential equation and is available for solving the problem it can represent. These signs of possible operations are like the industrial plan that Dewey takes as a model of judgment(see above 6.2.3); they are available for the judgment that will "put them to work" in the procedure solving the problem.

An important point is therefore that any mathematical theory, however complex and infinite its objects, is constructed from a finite number of operational rules about them. π is not countable, but it can be manipulated in algebraic equations and, when necessary, approximated with the desired precision by an increasingly long, but always finite, procedure. One reasons about these operations, not about their objects, and this is why one can "calculate" with them, *i.e.* apply procedures to them, even though they are most often themselves infinite.

6.3.2 Logical calculus and proof

Mathematics are regarded as the standard of scientific rigour because they deal only with possible operations and possible transformations of operations, any reference to possible objects in the experience of these operations being suspended. Inference may thus be totally controlled, since it does not point to a specific connection between things in experience. They relate to reasoning itself, of which they are the ideal expression:

> Ordered discourse is itself a series of transformations conducted according to rules of rigorous (or necessary) and fruitful substitution of meanings. Such transformation is possible only as a system of interrelated abstract characters is instituted. (LTI: 392)

This explains the key role of proof in mathematics. Dewey defines it as *the cumulative and procedural production of propositions* (LTI: 310), which transforms an initial formulation of the problem into a terminal formulation that is immediately applicable to it – an action, an already available object, etc. (LTI: 313). This definition, together with the insistence on the role of substitutability in the progress of the procedure may bring here to mind the Hilbertian conception of proof:

> A formula will be said to be provable if it is an axiom, or results from an axiom by substitution, or is the final formula of a proof. (Hilbert [1922] 1996: 1137)

However Hilbert speaks rather of substitutions of formulae than of transformations of meanings, and this difference is of importance. Dewey refuses to separate syntax and semantics, and to conceive of logic as a simple system of regulated rewritings of formulae whose meaning would be suspended.

Meanings are purified and transformed, but never become a pure system of signs. Indeed they are only valid because they point to a possible action, and that is what a meaning is in the first place. If logic is symbolic and propositional, it is not because it is basically just a grammar, but because the essential property of discourse as a whole is to deal with possible actions and not with real things. It examines what may happen if one did something, without actually doing it (LTI: 392).

In a symbolic system, transformations of discourse are carried out using logical (AND, OR, etc.) and mathematical operators, specially designed to make them rigorous (without indeterminacy) and productive. The meaning of these operators is given by universal propositions:

> Universal propositions about logical forms are propositional functions and as such are in themselves neither true nor false. They state modes of procedure in inquiry which are postulated as applicable and as required in any controlled inquiry. (LTI: 159)

If it is remarked that, for Dewey, a universal proposition is nothing else than the expression of an operative rule (LTI: 379), then one gets exactly Martin-Löf's perspective, namely that the significance of logical constants lies entirely in their rules of introduction and use in reasoning. They are therefore functions that allow the inquiry to progress, *i.e.* to reformulate its expectations until they lead to known solutions. The algebraic interpretation of logical connectors allows Dewey to approach this intuition when he interprets the validity of the conjunction in relation to an act of proof:

> The validity of the proposition "James, John, Robert and Henry were present on such and such an occasion" is impeached *when it is shown* that any one of the four was absent. (LTI: 337, my italics)

Here lies the very spirit of the intuitionist interpretation of the logical constants of negation and conjunction. Asserting the negation of a judgement consists in showing the impossibility of a proof, and asserting a conjunction requires exhibiting the proofs of each of its terms. Another example shows that Dewey had quite clearly perceived what corresponds to the introductory rules in the constructive theory of types:

> Take, for example, such a postulate as the following: "If a and b are elements of the field K, then ab ($a \times b$) are elements of K." The postulated elements are a, b. The postulated operations are represented by "and" and by "×" or ab. The primitive proposition does not first postulate certain elements, and then by means of another primitive proposition postulate a certain operation in two separate postulates. The elements and the operations are laid down in a single postulate in logical dependence each upon the other. a is defined to be such that if the operation designated by and is applicable, then the operation symbolized by × is necessarily applicable. The elements are instituted in relation to the operations by which they are related and the operations and their rules are determined in reference to the elements. (LTI: 403)

This latter idea, that the terms and operations of an algebra are co-defined exclusively by each other is well known to mathematicians. Dewey attempts here to forestall the interpretation that this postulate would describe abstract relational structures, in other words, to use today's consecrated terms, that its meaning would reside in the models that satisfy it. On the contrary, he insists that these are stipulative definitions (cf. the terms *postulated, instituted*) which authorise the use of an operation if its elements satisfy certain conditions.

6.3.3 Dewey and intuitionism

One can recognise in this account several features of intuitionistic logic noted above (5.3.2). First, for Dewey, there is no fundamental distinction between theorem and problem, and the proof of a judgement is therefore a procedure for the same reasons that Martin-Löf evokes: it is a construction that exhibits its object, the proposition. Secondly, the judgement, as an act of knowledge, must be understood as the couple constituted by the object *and* its construction procedure (or again: by the proposition *and* its proof). Finally, there is no strict distinction between objects and properties, (individuals and predicates in first-order logic), since all are meanings pointing to lived experiences and possible operations in experience. It may also be added that truth is defined as *warranted assertability*, a famous expression of Dewey which intuitionists have often taken up to designate their own conception of truth (Posy 1998: 303).

These affinities could be extended on other themes, for example concerning the rejection of formalism, *i.e.* of the separation of syntax and semantics,[13] the rejection of the principle of excluded middle, or the definition of numbers and infinity.[14] What, then, is the sense of these possible connections between Dewey and the constructivist school in logic? There is no indication that Dewey had read Brouwer or Heyting, or even heard of their work. However, he cites Poincaré in connection with the principle of induction, which he accepts, like Poincaré, to regard as primary, but which he refuses to justify by an appeal to intuition (LTI: 405). This criticism may be considered benign, and presupposes a fundamental agreement on the conception of numbers and mathematical reasoning that Poincaré proposes. Dewey was certainly not hostile to the French mathematician, and it may even be argued that his reading oriented his conception of mathematics since he refers to his works several times in other texts, and places him among the scholars whose reflections on "the logical significance of the procedural modes" of his science have most advanced contemporary logic.[15]

Beyond this influence of Poincaré, the meaning of these similarities comes foremost from the fact that, adopting the same fundamental perspective on the nature of thought as activity, Dewey and Martin-Löf arrive at the same results, or at any rate at theses that signal each other.

They thus both affirm the unity of theoretical knowledge and practical knowledge, once the latter is properly understood as *problem-solving knowledge*. Judgements are the complex acts that carry out this resolution, and they have logical primacy over their result, *i.e.* the proposition that this act constructs. Above all, like Dewey, Martin-Löf does not reduce the scope of logic to acts of knowledge alone, but considers it to be immanent to rational activity in general, because it does nothing more than describe the consequences of what one purports to do (see above 5.3.3).

However, Martin-Löf does not develop a philosophy of action, and his indications for extending the theory of types towards the natural sciences are too fragmentary. For his part Dewey, who remains technically dependent on Boole's algebraic logic, and the simple propositional calculus, has no access to the first-order predicate calculus and quantification. He cannot therefore arrive at the functional operators (abstraction and application) of the λ-calculus, nor therefore at a general and contemporary view of computability.

From all these connections, can it be inferred that type theory, or an equivalent constructivist logic, has the capacity to express, in formal terms, the content of the

---

13 See on this subject Sundholm (1997: 191–192).

14 See, respectively, LTI: 344–345, 362, 409.

15 MW 3: 68. See also MW 4: 262.

theory of inquiry that Dewey develops? The question, put this way, does not seem to make much sense, since Dewey assigns a very precise place to symbolic forms in inquiry. Inquiry may develop systems of symbols to represent possible operations, mental or concrete, as well as their potential sequences. But to look for a symbolic system that would represent these operations of inquiry in the process of developing a system would be like trying, like Baron Münchhausen stuck in a mire, to pull himself and his horse out of it by his own hair. At any rate, this is the impression given by Burke (2002)'s curious attempt to find correspondences between Dewey's ideas and various contemporary formal logics, for example when he proposes to represent the *theories* available to an agent by a formal structure of sets, which would represent, among other things, a network of pre-theoretical meanings, a collection of ontologies and reasoning structures, etc. (Burke 2002: 137).

Rather, the perspective should be reversed: *it is the inquiry that, once it has reached its conclusion, re-expresses itself*, in the available symbolic systems, *in a procedure that recapitulates it*. The procedure is not always made explicit, and most of our inquiries are forgotten as soon as they have been satisfactorily completed. But it is sometimes found useful to record some of them, in order to check that the sequence of inferences made is correct, or to share it with others who will check it, or to teach it, or to remember it so that the inquiry won't need to be repeated in a similar situation, and so on. This is the function of proofs, but also of computational procedures, for example, or of scripting complex collective actions.

Such an approach is not associated with any specific symbolic system. It may take the form of geometry, arithmetic, propositional calculus, graph theory, and so on. The specific foundational ambition of constructive type theory, which distinguishes it from set theory, and brings it closer to λ-calculus or Post's systems, lies in the claim of its ability to express *any procedure*, which implies also that it be able to represent any symbolic system that describes its field of meanings, and in which it is expressed as a composition of activities and specifications. The extent of this ambition may be gauged from the difficulties already mentioned that type theory encounters in representing Euclidean axiomatics (see section B.1).

The idea that it *should* be possible to represent any effective procedure in a single symbolic system has become intuitive since the 1930s and the work of Church and Turing. It is on this account that type theory, or other equivalent systems, could claim a special relationship with the theory of inquiry as expounded by Dewey. This framework should not be interpreted as the one to be *used* in practice by the mathematician to search for proofs and solutions, even though it may be useful for their to do so in certain cases; nor even as the one in which they *must* express them once they have found them in order to make sure that they are correct, for there may be other, more convenient systems of verification available in their field; but as the one in which they *may* always express them, whatever the subject. Type theory claims to be an ideal and universal form of the notion of procedure, a framework in which any procedure may be the object of an exact and exhaustive explication, the one that the mathematician can produce if they are asked to make entirely explicit the judgment of knowledge that they claim *to be able* to make.

It is therefore only by this long diversions that the notion of procedure, as Dewey draws it out, can be effectively connected to the modern notion of an effective procedure or algorithm.

There is in Dewey's conception an essential epistemological progress, in that it avoids the moment of deliberation being wholly discounted, as it was the case with Martin-Löf's and Sundholm's's account of judgement (see above 6, introduction) Dewey's theory of inquiry makes it possible to distinguish clearly between inquiry and procedure, between the complex, tortuous process of the former and the neat, mechanical execution of the latter. This progress is decisive because *this gap* is the reason for the cumulative dynamic of knowledge which, in each judgment or act of knowledge, is capable of recapitulating itself into a compact procedure, possibly capable of becoming a habit, or of being signified as a simple operation that can be manipulated in turn: of becoming an object.

Thus, the paradox noted previously (6), that the ideal act of proof is precisely that which a mathematician *never* deigns to achieve in its tiresome exhaustiveness, has a very simple answer in Dewey's perspective. Inquiry aims to draw out this ideal act, but should certainly not be confused with it. The procedure represents the inquiry once it has been completed, and once all the mistakes it may have made along the way and the blind alleys it has explored have been eliminated, for its function is simply to help future inquiries form by "marking out" the right path, to enable its verification, or even to reduce it to pure ratiocination. This functional separation is very clear in the specific case where the role of mathematical inquiry is to determine the optimal methods of calculation for a given problem, so that they can be performed by *computers* – human persons, before the advent of computers, such as the famous "hairdressers" that Prony had hired to calculate logarithm tables under the direction of mathematicians (see above 3.3.1). Above all, these hard-working calculators are not asked to conduct the inquiry again, but instead to follow to the letter the instructions given by mathematicians. This is a real division of labour, according to Prony's own admission, who says he got his idea after reading Adam Smith.

7 Systems and problem-solving

A clear view of inquiry as a unique activity, which extends to all human practices, however intellectual or material, is now available. Distinctions between descriptions and prescriptions make only sense within a specific frame of inquiry, the first ones to explicate constraints and goals to achieve in a situation, and the second ones the possible transformations of this situation (chapter 5). The fundamental nature of procedures has also been clarified. They are reconstructions of solved inquiries aiming at the facilitation of future ones. Dewey's interpretation of mathematics, as occupied with the clarification and construction of pure procedural means is convincing (chapter 6).

It is now possible to return to the guiding question of this reflection: how do effective procedures relate to procedures in general, and systematic problem-solving to problem-solving in general? What is it about systematicity that makes it play such a key role in the transformation of traditional practices and techniques into gigantic industrial and digital complexes? Why are scientific theories so powerful in practice, while they seem to have the utmost disregard for practical applications?

This question is not easy to answer at first sight. The pursuit of systematicity is usually seen as a moment in scientific activity when it resolutely turns its back on practical concerns in order to focus on its most theoretical goal of explicating its own principles and inner organisation. Pragmatist philosophers often downplay its importance as a housekeeping task. The philosopher of mathematics H. Wang ([1961] 1998: 133) loathes theoretical axiomatisation as a belated activity of compiling and organising knowledge guided by "aesthetic needs" and a "librarian" conception of science. Such a view is of course reductive. According to Nicholas Rescher (1979: 27), the systematisation of knowledge is a central task of inquiry because it provides a "quality control" over acquired knowledge, indeed the only one possible if one argues, as Rescher does, that internal coherence is the criterion of truth. This does not mean that systematicity must take the vertical form of Euclidean axiomatisation, where elements are included in the system only if they derive from some basic, first-order ones, as theorems derive from axioms in a theory. Rather, Rescher favours what he calls a network approach, where the strength of horizontal connections between elements collectively builds the system. The important result here is that such systems, which Rescher (1979: ch. 3 and 4) calls Hegelian, have an organic quality that makes them evolutive and quite immune to the loss of any particular element. Systematicity is not a yes-or-no quality of collections, but a gradient, capable of more or less.

More recently, Hoyningen-Huene (2013) has also argued that science is essentially about systematicity. His perspective is empirical: what distinguishes scientific

https://doi.org/10.1515/9783111494500-007

explanations and predictions from everyday ones, for example, is their orderly use of already established theories or empirical generalisations, or the carefulness of their argumentation and reference to documented sources. For Hoyningen-Huene, this systematic, orderly, methodical attitude is paramount, even when it is not based on constituted systems of knowledge such as theories, and even when it does not aim to build one (as is the case in the historical and human sciences). In particular, it promotes a critical attitude towards everyday knowledge, openness to criticism, and specific standards for the generation of new knowledge (e.g. in data collection). It is also concerned with completeness and interconnectedness (it abhors both gaps in knowledge and isolated pieces of knowledge). Given the great diversity of scientific disciplines (natural sciences, mathematics, history and natural history, human sciences, etc.) and their evolving standards over time, Hoyningen-Huene (2013: 30–34) believes that a deeper characterisation of systematicity at this generic level is not possible, at least not without adopting a normative perspective.

The purpose of this chapter is not to examine and criticise these views of the role of systematicity in science. Rather, it is to examine the use of systems in problem-solving (whether scientific or not), which is mentioned only in passing by Rescher (1979: 14, 32) and by Hoyningen-Huene (2013: 121–122). But this leads to a qualification of the view that systematicity has the sole function of being a criterion of good knowledge (aesthetic, efficient, coherent, etc.) – a quality control function, as Rescher puts it. It does, of course, have this role, but its success in modern science also derives, as I will argue, from its instrumental usefulness in promoting a new way of solving problems.

This conception of systems as the proper object of reason, in its project of unifying the rules of judgement, detached from the daily work of understanding, is often associated with Kant (section 7.1). Dewey's views are helpful in grasping the close connection between the *global inquiry* of reason, as I will call it, and the *local inquiry* of problem-solving. While consistent with Rescher's and Hoyningen-Huene's theses about science (both of which are indeed in the pragmatist tradition), it is also broader in that it claims that a core function of knowledge systems is to provide problem-solving with controlled meanings and inferences in order to increase its effectiveness (section 7.2).

The history of the first systems of knowledge in the early modern period is telling of the difficulty to design systems fruitful to problem-solving (section 7.3). Of particular interest is one such source of the idea of *system*, that of the *reduction in art*, which designated the reasoned description of an art or practice. The problem or such descriptions were their inability to practically help to solve new problems beyond those already listed. Section 7.4 exposes how geometry provided eventually the model of systems geared towards problem-solving, with the rise of modern military engineering in the 17th century. The special power of geometry is investigated by comparison to simpler deduction systems. Four generic principles for universal systems of inference (that can express any effective procedure) are explicated.

7.1 Systems as objects of global inquiry

The thesis defended in this section is that it is not the traditional opposition between truth and usefulness that distinguishes science and *techné*, but rather the gap in their systematic ambition (7.1.1). But even this distinction is not sufficient, for the concept of a theoretical system of nature is only a regulative idea of reason (7.1.2). What is rather critical is the division of the work of inquiry between the construction of systems (which I call *global* inquiry) and the *local* resolution of problems. These two orders of inquiry are transversal to any discipline of knowledge, whether "theoretical" or "practical" (7.1.3).

It is however not clear how to conceive such a complementarity. While it is easy to understand the subservience of local inquiries to the build-up of a global one, it is not clear what use global inquiry (the construction of systems) might have for local inquiry, and what would thus be this *systematic* way of solving problems.

7.1.1 The limited horizon of technical knowledge

According to Aquinas, it will be recalled (5.1.1), the finality of knowledge is the principal criterion by which theoretical and practical knowledge are distinguished. Since such a criterion is external and accidental to the very content of that knowledge, the fundamental unity of knowledge is thus demonstrated: "It is accidental to an object grasped by the intelligence whether it is ordered to action or not" (*Summa Theologica* I q. 79, a. 11).

The strength of this position can be seen all the better if one affirms, with Kant, that concepts are packages of rules (see 5.1.2). Take the example of a *prey-predator* model of population dynamics, which makes it possible to determine the equilibrium of two animal populations, one being predatory on the other. This model can be used by an ecologist who wants to account for observed fluctuations in animal populations, as was done in the study of lynxes and hares in Canada (Finerty 1979). But it can also be used by an ecologist who *wants to intervene* in this balance, as part of a rewilding project, see for example (Pereira and Navarro 2015: 76). *They will use the same program* to determine the appropriate parameters for their intervention. In other words, the same knowledge (the differential equations and the program that models them) appears as theoretical or practical depending on the problem that is to be solved, depending on the intention that presides over the mobilisation of this knowledge, whether it is to explain or to act.

However this is not exactly Aristotle's assertion, who readily concedes that the craftsman goes through the same series of causes as the scientist (see above 4.1.1). His interest lies in the *dynamics* of the constitution of knowledge and its horizon. Technicians stop their inquiry as soon as their interests are satisfied. This marks the limits of their knowledge, which may therefore fail to capture essential characteristics of the thing at hand. The scientist, aiming for its true nature and devoid of any other interest, is situated in an entirely different dynamic.

Kant himself seems to recognise this essential distinction of interest when, in the chapter on the *Transcendental Method* in the *Critique of Pure Reason*, he contrasts *technical* systems of knowledge whose unity is determined by empirical circumstances, with the aim of an *architectonic* unity of knowledge, which would be founded on an idea *a priori*:

> Under the government of reason our cognitions cannot at all constitute a rhapsody but must constitute a system, in which alone they can support and advance its essential ends. I understand by a system, however, the unity of the manifold cognitions under one idea. This is the rational concept of the form of a whole, insofar as through this the domain of the manifold as well as the position of the parts with respect to *a priori*. [...]
> The schema [of the system] that is not outlined in accordance with an idea, *i.e.*, from the chief end of reason, but empirically, in accordance with aims occurring contingently (whose number one cannot know in advance), yields *technical* unity, but that which arises only in consequence of an idea (where reason provides the ends *a priori* and does not await them empirically) grounds *architectonic* unity. What we call science, whose schema contains the outline (*monogramma*) and the division of the whole into members in conformity with the idea, *i.e.*, *a priori*, cannot arise technically, from the similarity of the manifold or the contingent use of cognition *in concreto* for all sorts of arbitrary external ends, but arises architectonically. (*Critique of Pure Reason*, A 832–834 / B 860–862, AA 03: 538–539, transl. Guyer and Wood 1998)

Technical systems are systems which lead idea or core motivation is not given by reason, but by some empirical goal. This goal may be driven by the desire to know, e.g. classify all mammals on earth, but as long as these animals' distinctive principle has not been itself derived from the laws of biology, this taxonomy serves a limited end of reason. It is not difficult to find here the old Aristotelian opposition between the limited generality of the knowledge of technical means and the true universality of the knowledge of true causes. The opposition between useful and disinterested knowledge is therefore transposed to a higher plane. Though the distinction between utility and truth does not allow theoretical laws to be clearly separated from technical rules, it can nevertheless be observed at the level of the very dynamics of the systems of knowledge and of their resulting form.

Such a view has come back forcefully in contemporary reflections on technology. They seek to clarify the intuition that technology embraces a domain of knowledge which dynamic is autonomous from sciences – opposing thus frontally the *applicationist* schema stemming from 19th century positivism (according to which technical knowledge would be a mere derivation from science). In his study of discussions on this issue, the philosopher of technology Wybo Houkes (2009: 312) gives this recurring argument the name "TU intuition" (for Truth *vs* Usefulness).

For example, philosopher Roel Wieringa proposes the following distinction:

> Practical problems call for changes in the world, so that it better suits the goals of a given stakeholder. Knowledge problems, on the other hand, do not call for changing the world but for changing our knowledge of the world. (Wieringa 2009: 1)

This distinct dynamic of the constitution of technical knowledge makes it always dependent on external circumstances, even when they aim to constitute themselves as a system. This idea was already encountered in the discussion on Aristotle, namely that a technical system of knowledge, if it were conceived, would achieve only a mediocre degree of universality (see above 4.4.2). Contemporary philosophers of technology clarify this idea by noting several epistemic characteristics by which technical knowledge is differentiated from theoretical knowledge. 1) The most important is that of *locality*, in other words that the concepts, representations and symbolisms mobilised by technical deliberation are always mobilised from a given context, *i.e.* a problem to be solved, which dictates their relevance, whereas theoretical inquiry aims to establish adequate representations independently of the given context. This is exactly what is expressed in this remark by the philosopher of technology and engineering Sjoerd Zwart:

> Descriptive knowledge is valued insofar as it depends as little as possible on context. Newton's mechanics produces concrete results only if it is provided with precise initial conditions. Prescriptive knowledge, on the other hand, is valued if it is adapted to an exactly circumscribed type of context. (Zwart 2020: 9)

The "precise initial conditions" are reminiscent of Kant's gunman, who has to decide which concrete parameters to take into account, such as air friction, so that the ballistic model allows them to predict the exact trajectory of the shot. This requirement runs counter to that of theoretical exposition, which tends to focus on simple problems where laws can be applied directly.

2) It is *superficial*, in the sense that it is indifferent to the justifications for the knowledge it mobilises in problem-solving, and is content for it to be *warranted*. This superficiality is expressed, among other things, by the *black-box* method, where

one is interested in the useful properties of the object, but not in their underlying mechanisms, as philosopher Mario Bunge notes:

> Theories of technology are definitely poorer than those of pure science; they are invariably less deep, and this because the practical man, for whom they are intended, is chiefly interested in net effects that occur and are controllable on the human scale; he wants to know how things within his reach can be made to work for him, rather than how things of any kind really are. Thus, the electronics expert need not worry about the difficulties that plague the quantum electron theories [...]. Consequently, whenever possible the applied researcher will attempt to schematize his system as a black box; he will deal preferably with external variables (input and output), will regard all others as at best handy intervening variabes with no ontological import [...]. This is why his oversimplifications and mistakes are not more often harmful-because his hypotheses are superficial. (Bunge 1966: 333)

3) The third specificity of practical knowledge follows from the previous one. The same problem may have a multiplicity of possible methods of resolution, depending on the diversity of stakeholders' criteria evoked by Wieringa. Even more generally, the same technical object may be modelled in several different ways, depending on the problem to be solved. This is what Hendricks, Jakobsen and S. A. Pedersen (2000: 302) call the "epistemic pluralism" of technical knowledge, giving the example of civil engineering models that model concrete as a plastic material or as an elastic material.[1]

For Hendricks, this pluralism implies a dynamic of progress of technical knowledge different from that of fundamental science, as it was described by Kuhn. Whereas the latter progresses from "paradigm to paradigm" through periods of profound questioning, the former progresses in a more evolutionary manner, because it is "poly-paradigmatic" (Hendricks, Jakobsen and S. A. Pedersen 2000: 284). These three epistemic specificities of technical systems – locality, superficiality and epistemic pluralism – derive from their fundamental contextual dimension, which theoretical research, on the contrary, aims to *eliminate*. It is these specific features that allow to distinguish, empirically, between theoretical and practical knowledge systems.

7.1.2 Scientists, artisans of reason

Against this argument, which places contemporary philosophers of technology in a distant continuity with the Aristotelian tradition, Aquinas and Kant may retort that

1 *Elastic* and *plastic* refer here to properties exactly defined in solid-state physics. For more details, see the preface to Kostic's thesis (2009: i).

the myopia of technology is itself relative to the horizons of its stakeholders. Technical systems may, in their best interests, borrow their methods from the sciences and set themselves the goal of higher-order problems (problems to solve problems). In this way, many theoretical judgements and concepts, and even entire theories, are developed with a more or less distant practical end in view.

This is obvious in all applied theories, such as metallurgy or hydraulics, but there are also many systems of knowledge, traditionally described as scientific, which are deeply rooted in the resolution of useful problems. In biology and chemistry, for example, there is a continuous transition from theory to practice. In mathematics, graph theory has its origins in problems of determining the "shortest path", but it has subsequently developed as a theoretical inquiry in its own right.

As already noted, this observation of the complementary nature of theory and practice was made very early on, in the Arab world (see above 5.1.1). When Aquinas asserts that it is possible to have speculative knowledge of technical knowledge (thus the architect who can consider the different ways of building the house, not in order to do so, but only in order to know them), he is merely noting the reality of the reductions in art for which Vitruvius provides the model from Latin antiquity. Art is not necessarily short-sighted, then, and can engage in speculative research when it sees its interest in doing so, even if it is more remote. It is through this argument of the "useful detour" that Bacon thus makes the apology of science:

> I propose a natural history which is not so much [...] to make the results of existing experiments better, as to illuminate the discovery of causes [...] For although works and the active part of the sciences are mainly what I am after, I nevertheless wait for the ripened crop, and do not try to mow the moss or green corn. For it is enough for me to know that axioms rightly discovered bring floods of works in their wake; and reveal them not in dribs and drabs but in whole clusters. (Bacon [1620] 2004: 39)

In this text, the image of *theory* as an object of knowledge made up of two phases, one of construction, and the other of use, is superimposed on the economic image of investment and profit, accumulation and rent. This schema is diametrically opposed to that promoted by Aristotle and medieval theologians, who compare knowledge to elevation, liberation or purification. If, then, the differential in systematicity between science and technology appears relative and accidental, it is because the scale on which it is measured is itself entirely borrowed from the reign of technology.

How is then one to distinguish the "disinterested" desire for knowledge from its long-term usefulness? At the institutional level, fundamental research today is funded by public and private institutions professing the belief that sometimes the most abstract and "gratuitous" problems of mathematics and science appear to bear an extraordinary usefulness only a few years ahead. This ambiguity does not only appear at the institutional level. On an individual level too, researchers' motivations are often mixed, and even difficult to separate: they may be sincerely interested in finding out the causes of a phenomenon, and at the same time have chosen this research subject among others because it has practical implications, or even because it is well-funded, because it will advance their career, and so on.

The contemporary debate on technoscience is structured by these questions concerning the possibility of distinguishing science and technology and – a question that determines the previous one – that of disentangling a motivation proper to science (the "pure" will to know) from its socio-political determinations.[2]

All these questions and debates do not have much sense from a Kantian perspective, for which the ambiguity concerning the *sincerity of our desire to know* cannot be resolved empirically, either by psychology or sociology or even by examining the productions of science (where one would attempt, for example, to distinguish "useful" theorems from those that would be "gratuitous" or "beautiful").

To assert, as Wieringa does (text quoted in 7.1.1), that changes in knowledge are not also changes in the world, seems to betray a particularly ethereal conception of science. It seems to count for nothing the publications of scientists, their impact on laboratory practices, the reorientation of research policies, but also their dissemination to the public and their subsequent influence on culture, without even mentioning their technological implications. Could it seriously be argued that all this is merely *accidental* consequences of scientific inquiry, whose essential motivation would be solely the "change in the cognitive state" of the people who access this knowledge? Without even subscribing to Dewey's philosophy, one can see what is problematic about the binary separation of the "mental" effects and the "concrete" effects of the application of a rule.

As for the epistemic peculiarities of technical knowledge mentioned above, they may also be observed in the sciences. As regards "superficiality", a chemist will use the laws of atomic physics with the same indifference to their causes as an engineer uses the laws of mechanics. Epistemic pluralism also exists in the sciences, as can be seen in the multiplicity of models of a theory, or even in the multiplicity of its *versions* (Vorms 2009), as for mechanics its Newtonian or Hamiltonian formulation – not to mention the survival, entirely pragmatic, of classical mechanics alongside the theory of relativity.

Kant seems to have been quite aware of the historically situated character of the sciences; thus, immediately after defining what he means by *technical system* (see

2 For a useful clarification, see Parent 2004.

above 7.1.1), he refers to scientists as *Vernunftkünstler, artisans of reason*, implying that their theories are partly determined by "accidental goals".

> The mathematician, the naturalist, the logician are only artisans of reason, however eminent the former may be in rational cognitions and however much progress the latter may have made in philosophical cognition. There is still a teacher in the ideal, who controls all of these and uses them as tools to advance the essential ends ofhuman reason. Him alone we must call the philosopher. (*Critique of Pure Reason*, A 839 / B 867, AA 03: 542–543, transl. Guyer and Wood 1998 modified)

Kant's position is thus that the problems scientists give themselves are partly determined by their own tradition and the socio-cultural conditions of their practice. In this respect there is no solution of continuity between science and technique: both proceed from the same faculty of knowledge and both are entirely committed to the resolution of problems that are given to them by someone else. Science can *also* be guided by reason, which seeks its autonomy by aiming at the idea of an architectonics of knowledge. This idea, however, is not empirical. It is an object of study for the philosopher, who is here an ideal and problematic figure. To go further requires a metaphysical commitment concerning truth and the relation of the thinking subject to reality, as Houkes sees very well:

> The difference in aims [between truth and utility] seems to presuppose a realist conception of science, according to which scientific theories should be interpreted as descriptions of (the structure of) reality, and science as a continuing enterprise to construct more accurate theories. […] The broad spectrum of realist conceptions can be contrasted with another vision of science: instrumentalism. Instrumentalists seek to decouple scientific inquiry from truth, and instead insist on its connection to utility (Houkes 2009: 318).[3]

These two visions, for Kant, are equally possible, whether one considers the requirement of reason to *postulate* the intelligibility of reality, and interprets the advances of science as manifestations of this possibility, or whether one simply seeks to describe the empirical reality of science, insofar as it fits into the social, political, economic or psychological network of human intentions.

7.1.3 Global and local inquiry

The results of the previous discussion show that the concept of a theoretical system of nature, which would be established independently of any human intention,

3 See also Niiniluoto 1993: 3 and Hendricks, Jakobsen and S. A. Pedersen 2000: 288.

falls under the status of a regulative idea of reason. Any concrete scientific theory is "technical", as Kant says, because it develops *from* a certain state of research, its questions, its conceptual and technological tools and its socio-cultural conditions. Conversely, the long-term planning of human utilities is capable of developing technical systems to the point of making them (quasi-)autonomous sciences, and it may even, in its own well-understood interests, encourage basic research. This means that the concrete, observable differences between technical and scientific knowledge-based enterprises are merely differences of degree on a continuous scale. Their only essential difference is lodged in the eye of the inquirer, depending on whether they aim to know things as they are, or as they would like them to be.

But here another *articulation* within knowledge emerges, which passes close to the old Aristotelian distinction between theory and practice, without fitting exactly into it. It is in many respects very elementary: for while on the one hand there is the resolution of particular problems, which may be theoretical as well as technical-practical, there is on the other hand that higher-order inquiry, which aims to establish a certain unity of knowledge over a more or less wide perimeter, which criticises the knowledge already acquired, fixes the guiding problems of research and constantly reformulates them in the light of their results. This research may serve technical purposes, explicitly subordinate to problem-solving, or theoretical purposes, the evidence for which is more ambiguous. Whatever the case, this inquiry has an undeniable reality, which must be distinguished from problem-solving, which has only a local scope. I shall call such an inquiry *global*, although this term is somewhat inelegant. The other terms that might suit it: *architectonic, systematic, theoretical* cannot be used as their very meaning is the object of this inquiry. The advantage of *global* is that it stands in direct opposition to the *local* character of problem-solving. Global inquiry does not solve particular problems – which is the job of the understanding. It corresponds to what Kant calls *reason*.

Before their rework by reason, the empirical rules about experience formed by the understanding are indeed the result and the material of local inquiries. They emerge from the subject's confrontation with experience, in which they must situate practically and conceptually. This does not mean that these rules are singular, since it is precisely the task of reflective judgement to create the conditions for a coherent apprehension through the diverse flow of experience. However, they are always conditioned by their context and the problems for which they have been forged. More generally, they have the properties of practical knowledge already noted above (7.1.1), such as the superficiality of causal analysis and epistemic pluralism. In contrast, the global inquiry of reason aims to decontextualise these rules in order to elevate them to an unconditional value, at least ideally.

Global inquiry is not separate from local inquiry, which is problem-solving. Indeed, it can only progress by identifying particular problems that must be solved one after the other. As mathematician Reuben Hersh (1997: 6) says:

> Mathematics is a vast network of interconnected problems and solutions.

According to Hersh, mathematical theories are not primarily axiomatic systems, but fields of problems themselves. They evolve through the joint efforts of mathematicians. Mathematics progresses in two ways, on the one hand by solving problems which sometimes open up new avenues of research, but also by inventing new theories which are nothing else than new fields of research. Axiomatization and formalization are late activities in mathematical theorizing. They do not come first, but last.

Therefore, in scientific endeavours, local problems are ordered to global inquiry just as local derivatives are the singular moments of a curve that envelops them. Any research mobilises experiments, reformulations of hypotheses, the writing of reports, readings, presentations at conferences, etc. – all particular and "practical" problems that are nevertheless ordered to a higher-order theoretical inquiry. Anyone who would ask the researcher *when* he is researching outside all these activities would be asking an irrelevant question, like Ryle's tourist who, after seeing the various colleges and faculties scattered across Oxford, would express a wish to see now the *University* (Ryle [1949] 2002: 16).

In particular explanation, which is so important in the sciences, is not an activity distinct from problem-solving, but one of its species. First of all, there are also technical explanations, which can sometimes be difficult to distinguish from scientific explanations. But there are also mythological, political and practical explanations. Each of these refers to different systems of meanings, and an inquirer can even cobble together an *ad hoc* theory when needed. Explanation consists in *locating* a phenomenon in a system of meanings using the concepts and rules of inference it provides.

7.2 Systems of meanings

To better understand what global inquiry is, it is useful to return to Dewey's *Logic* and to his views on the nature of scientific inquiry. This section begins with an early text which suggests, somewhat bluntly, that the activity of *theorising* is not a *sui generis* and late invention of highbrow intellectuals, but that it is already at work in the most minute activities of life (7.2.1). In later writings of Dewey, the idea of *theory* is replaced by the more generic expression of *system of meanings*, which designates the background context of the inquiry. The scientific approach consists in establishing controlled systems of meanings (7.2.2). From this derive its demand for systematicity (7.2.3) and its remarkably cumulative dynamics (7.2.4).

It should be noted here that this exposé of Dewey's doctrine concerning the role of scientific theories in problem-solving does not rule out that they have other functions and other drivers of their dynamics; it is only asserted that they have *at least* those set out in this section.

7.2.1 The theory of the situation

An early text expresses Dewey's general idea in the pithy formula that *there is theory as soon as there is action*:

> Where […] there is anything which deserves the name of conduct, there is an idea, a "theory", at least as large as the action. Because the theory is narrow in scope it is not lacking; and it is narrow only so far as the corresponding act is abstract and partial. The average man can walk without much theory, because walking is not an act of great content […] For any *act* (as distinct from mere impulse) there must be "theory", and the wider the act, the greater its import, the more exigent the demand for theory. (EW 3: 96)

This thesis is asserted in the context of a discussion about the existence and nature of moral theories. Dewey argues that they cannot be a sum of abstract principles, but that each situation calls for an analysis of what should be done, taking into account all the particularities, constraints and objectives involved. Moral rules are tools to guide moral insight but cannot determine action, because they are, like all rules, abstract or hypothetical. It is only in situated moral reflection that an integrated and concrete idea of what to do can emerge. No doubt such an analysis will be incomplete, but one always aims for such an idea, and this is what must be first called theory, the *theory of the situation*, the idea that guides action. When one learns to walk or count, an intellectual effort is made to understand how to direct one's body and mind. In this way, one constructs a theory of what works and what doesn't, even if it is not articulated and is directly incorporated into the habit. These theories do not appear to be moral because they are aimed at obtaining a general skill, although they in fact are, because these skills are good for the person.

The influence of Hegel is still prevalent in this analysis, for example in the idea that moral action must incorporate all practical constraints and other determinations of the situation – as opposed to the abstraction of the Kantian moral law, or that concreteness is an extra, not a defect, of rationality. From this perspective, then, Dewey blithely overlooks the traditional meaning of *theory* as a discursive system of representations, and the fact that many actions dispense with such systems entirely.

However, this is not the essential point, and if *theory* is taken in the broader sense given by Dewey, this thesis follows directly, at several levels, from the epistemological considerations outlined in chapter 6. The essential idea is that all action, and even all behaviour, unfolds against a background of meanings, inferences, or habits already acquired, which determine both its aim and its means. Any rational resolve to act requires a network of meanings from which the inquiry draws its material and procedural resources. In the case of rapid, intuitive reflection, one relies on the suggestions that arise in the immediate apprehension of the problem. In the case of habitual or routine conduct, there is, so to speak, a kind of pre-framing of experience – e.g. that, if I walk, the ground will stand firm in front of me. And taking this metaphorical register to its logical conclusion, one might even go so far as to formulate that bacteria following an ionised gradient in an aqueous medium "have the theory" that such a gradient leads to food. Of course, the main idea through all these examples is not that there is in this case reasoning, representation, or even sentience, but that action, essentially generic, projects an equally generic reading of the environment in which it takes place – that it adapts to the features only that it knows how to recognise.

The point of these extrapolations is that, once turned around, they show that scientific theories are the highest form of a background of meanings or habits that has always been there, and which is necessary for all action and inquiry, rational or not, discursive or not, because they draw their means and materials from it. Scientific theories aim to reform the natural inferences inscribed in language, but language itself is built on the practices and interactions of a community, and these in turn are collective adaptations to the pressures and opportunities of a given organic environment.

The shortcoming of the 1891 paper is that, while it emphasises the necessary complementarity between theory and practice, it fails to account for the dynamic of progress set in motion by the reflective competence of the individual, capable of bringing their own practices to language, and therefore of observing them critically, and even of observing their own language in order to check its relevance and reliability. If everything is theory, how can the specificity and superiority of scientific theories be explained?

This is maybe why Dewey, in contrast to his very broad use of the term *theory* in 1891, restricts it to the extreme in the 1938 *Logic*, pairing it almost exclusively with the adjective *logical.* The expression *scientific theory* only appears in one place (LTI: 70–71), in the context of a historical discussion. There are also pejorative mentions of political or psychological theories at the end of the book, and more generally, a non-technical use of the term, where it is taken as an expansion of the notions of idea, hypothesis, scientific law. This omission is certainly deliberate. Dewey meticulously specifies what he means by scientific inquiry or method, but he combats the identification of science with bodies of general propositions (laws, theorems) that are associated today with the term *theory*. Such a "dogmatic restriction of science" (LTI: 434) in effect establishes an artificial split within the various scientific approaches, rejecting outside "science" not only techniques, but also the experimental activities of the sciences themselves.

This distrust stems from Dewey's perception of what persists in *theory* of the old θεωρία, its contemplative dimension which establishes an untenable face-to-face relationship between subject and object, and which leads to what he famously calls the *spectator theory of knowledge*:

> The theory of knowing is modelled after what was supposed to take place in the act of vision. The object refracts light to the eye and is seen; it makes a difference to the eye and to the person having an optical apparatus, but none to the thing seen. The real object is the object so fixed in its regal aloofness that it is a king to any beholding mind that may gaze upon it. A spectator theory of knowledge is the inevitable outcome. There have been theories which hold that mental activity intervenes, but they have retained the old premise. They have therefore concluded that it is impossible to know reality. Since mind intervenes, we know, according to them, only some modified semblance of the real object, some "appearance". [...]
> All of these notions about certainty and the fixed, about the nature of the real world, about the nature of the mind and its organs of knowing, are completely bound up with one another, and their consequences ramify into practically all important ideas entertained upon any philosophic question. They all flow – such is my basic thesis – from the separation (set up in the interest of the quest for absolute certainty) between theory and practice, knowledge and action. (LW 4: 19–20)

7.2.2 Systems of meanings

However, the idea of a background of meanings necessary to the development of inquiry is still there, and Dewey expresses it by the phrase *system of meanings*. It features strongly in *Experience and Nature* and in the *Logic*, and is found regularly throughout his other writings. In *Experience and Nature*, it is the intellect itself that is qualified as such (LW 1: 230–231). In the *Logic* of 1938, it is associated with the language that unites a given human community (LTI: 56). Beyond these primitive matrices of knowledge, specialised systems of meanings are instituted by human communities – for example, subgroup-specific codes or rituals, specialised languages or scientific theories, which aim to reform meanings so as to better control inferences. The establishment of a "self-illuminating coherent system of meanings" is the aim of science (MW 2: 307).

What transpires through all these uses of the expression is that it is opposed to inquiry and action as, in Aristotle, potency to act. Whether language, mind, codes, science – all these denominations cover no concrete phenomenal object in particular, but designate what is available to the individual through habit, collective practices, and material devices such as books or libraries. However, contrary perhaps to the Aristotelian idea, potentiality here is not passive, but a dynamic that orients and elicits action, that shapes experience.[4] The passage where Dewey calls the intellect a *system of meanings* is perhaps the most illuminating in this respect:

> The relation between mind and consciousness may be indicated by a familiar happening. When we read a book, we are immediately conscious of meanings that present themselves, and vanish. [...] But we are capable of getting ideas from what is read because of an organized system of meanings of which we are not at any one time completely aware. Our mathematical or political "mind" is the system of such meanings as possess and determine our particular apprehensions or ideas. There is however, a continuum or spectrum between this containing system and the meanings which, being focal and urgent, are the ideas of the moment. There is a contextual field between the latter and those meanings which determine the habitual direction of our conscious thoughts and supply the organs for their formation. One great mistake in the orthodox psychological tradition is its exclusive preoccupation with sharp focalization to the neglect of the vague shading off from the foci into a field of increasing dimness. (LW 1: 231)

In this text appears quite clearly the idea of an active power, situated at the edge of consciousness, surrounding and pressing this clearing. The description is typical of Dewey's style of thinking, which avoids straightforward categorical oppositions, and always seeks to bring the dynamics of experience and thought to the fore.

Science is part of the very broad process of reflection to *take control* of this background of meanings by making it explicit and subjecting it to critical examination. The three examples of inquiry given in 6.2.1 are useful for illustrating the critical gradation that leads to science. The inquiry that seeks to establish a plan of action may be called critical in a figurative sense, in that it suspends thoughtless action in favour of reflection. But apart from that, it is purely constructive. It assembles into a solution inferential materials that it finds available in the situation in an uncritical way; it acts within the system of common sense. In the case of a trial, on the other hand, it is a question of verifying and validating a single complex inference, which starts from the material of available facts and leads to the guilt or innocence of the defendant. The trial is based on a system of meanings, the law, which is carefully established and controlled. Finally, in the scientific case, the aim is to determine the facts or laws that are intended to be inserted into such a controlled system of meanings. This is theoretical inquiry proper. There is a fourth critical level of inquiry: logic, as meta-theory, is an inquiry into inquiry, and thus aims to draw out

4 One may refer here to the analysis by Hildebrand (2008: 24–25) of Dewey's concept of habit.

the whole of these critical movements, as well as their universal patterns – their *guiding principles*, an expression that Dewey borrows from Peirce.

The gradation just sketched is intended to be suggestive, and should not be understood as a rigid classification. Some common sense judgments take on the seriousness and meticulousness of the court. Judicial decisions may be given a value of jurisprudence and thus added to the system of law. It is sometimes difficult to distinguish between applied scientific problems and those intended to strengthen the system of science. And sometimes reflections on the method of science – logical reflections in Dewey's sense – enables it progress. On the other hand, complex inquiries can become habitual and enter common sense, if their use becomes frequent enough: "The enterprises of one age become the sport and pastime of another." (LTI: 70)

These examples show that reflection is given the opportunity to take a critical look not only at its own productions, but also at the systems of meanings that underlie them. There is no question of developing here all the epistemological implications of this fact, as well as the different modalities of critical reflection already outlined. The discussion will simply attempt to characterise the nature of science as a *criticism of systems* and its cumulative dynamic.

7.2.3 Decontextualization and systematicity

The first point may be addressed by the formula that science criticises systems of meanings *in that they are not systematic enough.* Dewey returns throughout his *Logic* to this requirement. As such, natural language is so lacking in systematicity that Dewey sometimes substitutes for it the suggestive expression of a *constellation of meanings*:

> The customs, the ethos and spirit of a group is the decisive factor in determining the system of meanings in use. The system is one in a practical and institutional sense rather than in an intellectual sense. Meanings that are formed on this basis are sure to contain much that is irrelevant and to exclude much that is required for intelligent control of activity. The meanings are coarse, and many of them are inconsistent with each other from a logical point of view. One meaning is appropriate to action under certain institutional group conditions; another, in some other situation, and there is no attempt to relate the different situations to one another in a coherent scheme. In an intellectual sense, there are many languages, though in a social sense there is but one. This multiplicity of language-meaning constellations is also a mark of our existing culture. A word means one thing in relation to a religious institution, still another thing in business, a third thing in law, and so on. This fact is the real Babel of communication. (LTI: 56)

Language, then, is not really a system, but a simple "constellation of meanings" that expresses, in a disparate and ambiguous way, all sorts of practices and operations from different social groups, because its function is precisely to enable their communication, to make the businessman, the priest and the craftsman converse about things whose potentialities and consequences they each consider different. As a result, language increases a defect that all inferences have, as they are precisely adapted to a definite context, and as they say nothing about their own validity under new conditions and circumstances; practice has a certain tolerance for exceptions (MW 1: 23).

The role of scientific theories is to ensure that the inferences on which actions rely are valid, *i.e.* successful. This requires them to explore exhaustively their conditions and consequences, and thus to radically reformulate them:

> Each meaning that enters [scientific] language is determined expressly in its relation to the other members of the language system. In any ordered reasoning or discourse, this criterion takes precedence over that instituted by its connection with cultural habits [...] The ideal of a scientific language is the construction of a system in which meanings are related to each other by inference and discourse and in which symbols are such as to indicate this relationship. (LTI: 56)

It appears here that systematicity means the *control* of the relationship of each meaning with the other members of the system, and that this relationship is either established by inference (an act in experience) or by discourse (a relation of meanings). The duality between material and procedural means allows to characterise what it means to *systematise* by a double work of control. On the one hand, it is a matter of *ordering* empirical inferences in relation to one another, and on the other of making explicit the internal necessity of this order, *i.e.* of reducing it to a logical one. All this may not appear very original, unless one remembers that meanings deal with *operations* possible in experience and their consequences (which point to other operations). The first part of the systematization of a theory consists, therefore, in determining which operations are primitive or primary in the establishment of facts, and the second, in showing how (*i.e.* by what rules) all the others can be *generated* or examined on their basis.

Of course, in the course of this inquiry, it may become apparent that the operations initially candidate to be the bedrock of the system suffer from exceptions, or that their relationship to other operations is more complex than anticipated. They are then corrected or rejected accordingly. This explains on the one hand the increasing abstraction of science, which seems to define theoretical terms with no apparent relation to immediate experience, such as *"mass"* or *"force"*, and on the other hand its mathematisation, since it now only expresses relationships between these operations, and no longer immediately realisable inferences. Dewey gives several examples of this work throughout *Logic*. Two of them will be now discussed.

With regard to the establishment of primitive facts, Dewey takes the example of the evolution of the conception of metals (LTI: 274). The malleability of a metal was significant in the ancient conception. It allowed the craftsmen to recognise a material as a metal but also to discriminate between different metals by their greater or lesser torsion at an equivalent pressure. This property came directly from practical experience of metals as materials that could be shaped for a variety of purposes and uses. In the modern conception, this type of property is no longer essential. Metals are described using chemical formulae that allow to directly deduce the results of a large number of possible operations on these bodies, including operations related to malleability, but also their reactions with chemical compounds, such as oxygen or sulphur, or electrical conductivity. As a result, a greater number of ways of recognising metals are made available, making it possible to discover new ones, to distinguish between their different isotopes, or to describe them much more precisely than by their former immediate qualities. On the other hand, it has also greatly increased the possibilities open to us for using metals and analysing the phenomena that affect them.

Chemical formulae have meaning only within a given theory, chemistry, which explains all their consequences. It tells us, for example, how to calculate the conductivity and density of a metal from its formula, or its reactivity with oxygen, and therefore establishes the necessary links between these operations. The chemical formula is thus a meaning that lies at the intersection of all the operations that can be carried out on the elements that it describes, enabling them to be related in a regulated, and no longer merely empirical, way. The final stage of systematisation consists in having a *disjunctive* definition of metals that allows them all to be identified *a priori*. This definition is given by the periodic table of the elements, where the group of metals is distinguished by the small delta of energy separating the valence band and the conduction band; it is also realised, on the occasion of this systematic definition, that there are semi-metallic elements, *i.e.* which can behave like metals under certain conditions. This example perfectly illustrates Dewey's position. It is also consonant with the Kantian definition of the system as "architectonic unity of knowledge founded on an idea" (see above 7.1.1), which regulatory and critical function of natural concepts is clearly apparent here. It is not just a question of reorganising knowledge according to aesthetic aims, but of reconstructing it.

Another example Dewey takes (LTI: 304) to illustrate this logical reduction of connections between different possible operations, is the proposition *everything scarce is expensive*. This proposition can be interpreted in two ways. Either it is empirical: it asserts that, in the vaguely delimited whole of experience – for example in today's Western economies, this principle has been verified and will continue to be verified. Or it means that there is a necessary connection between the difficulty of finding a thing in trade and the price that must be paid for it, because, for example, of the competition between buyers to obtain it. In the second case, a universal proposition is established, which applies even if there is no longer any scarcity of goods. There is no need to verify such a proposition in experiment, and if the latter invalidates it, it is because one was talking about something else (LTI: 301). In a certain sense this link may be qualified as analytical, but that does not mean that dearness can be deduced from the definition of scarcity by syllogism. Dewey moreover does not use this Kantian adjective. Instead, he insists on the *analysis* of what is involved in the operations at stake – sourcing, negotiating, setting a price, etc. which makes it possible to establish an interrelated set of definitions in a system of meanings where the law in question can then appear as a theorem. Although Dewey does not develop this example further, it is interesting to examine how Walras ([1874] 1988) deals with the question: he constructs a mathematical model of market equilibrium, where buying and selling take place according to an auction mechanism that makes it possible to show the explicit relationship between price setting and the tension of supply and demand. It is clear here that, in the process of scientific inquiry, definition is an arduous process. It may require, not only making distinctions of meaning between concepts that were initially thought to be identical, or considering as incidental what was thought to be essential, but sometimes reconfiguring a system of meanings altogether and reducing it to a purely mathematical level in order to ensure the robustness of the inferences it seeks to warrant.

Walras' model is something of a fiction, because prices are never fixed by the model of auctions he describes. It has the same status as the principle of inertia, which describes the movement of an isolated body, which will never be found in nature. These are not abstractions or idealisations, because the point is not to describe situations that would be close to these ideal cases, but operational principles guiding inferences that could be made between different operations of measurement (LTI: 302–304), as in geometry.

Another lesson to be learned from this example is to show clearly in what sense the theoretical attitude "decontextualises" knowledge. By focusing solely on the validity of an inference, theoretical inquiry must make explicit the different contexts in which it may arise, *i.e.* the problems that mobilise it – with their purposes, their presuppositions, their constraints. It thus re-expresses the inference in a form that suits them all and depends on none, even if it means specifying it or even diffracting it into new, more specialised networks of inferences. The assertion *"everything scarce is expensive"* is uttered by businessmen or consumers who are worried about their interests, for example about bread in a year of drought, a diamond deal, or a trip to the Moon. These situations have very different price formation dynamics and the strategies of their protagonists may be incomparable.

However, the value of the general price equilibrium model is that it is suitable for all these situations, ignoring the actual behaviour of the players but showing their underlying logic.

Organising problems in this way, showing the identity of their structure through the apparent diversity of their conditions, or establishing their dependence on one another so as to factorise the most elementary problems, is core to the systematic attitude.

In the natural sciences, the work of constructing systematic theories from a field of problems, and conversely the mobilisation of theories to solve particular problems, relies on *modelling*, which can be thought of as a hinge that allows problems and theories to be articulated in a flexible way.

Models have been described very generically by Morgan and Morrison (1999: 10) as *instruments of inquiry*. More precisely, Varenne defines a model as

> a type of formal construct possessing unity and homogeneity. This unity and homogeneity are chosen in such a way that they enable this model to answer certain questions or fulfil certain precise functions of understanding, action, communication, management or decision in relation to an observable, conceivable or imaginable object, system, phenomenon or situation. (Varenne 2013: 299)

In both definitions, the "questions" or the "inquiry" play a central role: a model only makes sense in relation to a class of problems that it intends to solve, however general it may be. As the scientist Barry Jones also says:

> A model is the way you set up a problem for inquiry.[5]

Modelling is therefore nothing else than constructing a problem space, since it involves describing experience through a set of meanings governed by internal operating rules, from which new relationships (and therefore ultimately inferences) may be established. Modelling is used for both practical and theoretical problems, but in the second case, that of scientific explanation, the model is the real end of the inquiry. It is the demonstration of a specialised network of inferences that can be used to solve a wide range of problems.

Modelling is therefore part of theoretical inquiry, which is also a constructive activity, in that it aims to establish and verify a system of meanings. This may explain why, for many scientists, there is only a difference of degree between model and theory, which concerns their generality, precision and confirmation (Bailer-Jones 2002: 291–293). A theory could be mobilised for a very broad class of problems, whereas a model would generally have a fairly narrow scope.

5 Interview reported by Bailer-Jones (2002: 282).

The common background of theories and models is that they are all systems of meanings. However, it does not seem to us that they are distinguished by a different degree of generality, since a theory is not explicitly oriented towards problem-solving, or "adapted" to a local problematic, however extensive it may be. As Varenne expresses following his definition:

> Constructed around a unitary formal structuring, the model is distinguished from the mathematised theory by its character of locality, by its prior adaptation to the questions initially posed and by its powerlessness to lead directly to theorems (statements demonstrated to be true and valid in all generality within a fixed theoretical framework). (Varenne 2013: 299)

A theory, as already said, is presented as neutral of any intentionality and independent of any given class of problems, however broad, and it is this essential feature that distinguishes it from problem spaces and models. It is this neutrality that allows it to claim that it aims at "true theorems", as Varenne notes, for it seems to be interested simply at things "as they are", independently of any other interest and any context. It is not possible in the context of this study enter into the vast debate about the relationship between the notions of theory and model, of which Morgan and Morrison (1999: 18–20) give an overview: models in some cases allow theories to be constructed, implications to be explored, tested or interpreted. In turn, theories, in the semantic interpretation to which Van Fraassen and Kuhn subscribe, are collections of models. Whatever these relations, instrumental or constitutive, it seems effective to operate such a division of intellectual labour between models, which are contextual, instrumental, and, all in all, contingent systems of meanings, and theories which are systems of reference whose role is to integrate the knowledge contained in the former in a coherent and perennial manner.

7.2.4 The dynamics of science

From this perspective, it can be understood why theoretical activity has been characterised as disinterested and separate from the world of practice. Entirely occupied with the critical task of making explicit the implicit and natural theories of a community in order to control their inferences, it appears detached from action, just as meaning is detached from experience. But it is actually just another form of practice:

> Controlled inference is science, and science is, therefore, a highly specialised industry. It is such a specialised practice that it does not seem to be a practice at all. This high specialisation is partly the reason for the current antithesis of theory and practice. (MW 8: 78)

This core insight provides a convincing solution to Anscombe's struggle about the status of *pure observation* and scientific knowledge (see above 5.2.2). Science is certainly a practice but, as its core intention is to make inference robust *in all relevant contexts*, it organizes and strives around an ethos of neutrality and non-intentionality.

A central characteristic of the scientific approach is therefore that it constructs networks of representations that no longer have any obvious link with everyday experience. The facts that it conjures up through its controlled observations and experimental protocols no longer have much to do with the facts and actions that are familiar to common sense. This is to be expected, for the very purpose of controlling inference involves decentring from one's qualitative, natural and social environment, towards an objective, neutral, impoverished world that divests itself of the language of emotions, sensations, social conventions, etc. in order to exhibit only the conditions for the validity of the inference (MW 8: 81).

The fallacy of scientism, according to Dewey, is to believe that scientific systems of meanings represent the real world, and that the situation in which each person moves is merely a fiction, an appearance, to be got rid of. But there is in fact no privilege here, simply another mode of access to the world, another "form of life", Wittgenstein would say. The world of science itself only has value insofar as its meanings are experienced and put into practice – experimentation being a form of practice.

This is not to say, however, that scientific endeavour is subject to the utilitarian ends of the current practices of human persons – a reproach often levelled at Dewey.[6] On the contrary, just as any problem solved leads *also* to a new way of perceiving the environment, and recognising new problems in it, scientific activity very quickly emancipates its initial questions and motivations to generate its own body of questions:

> This mode of practice known as theoretical activity emancipates experience – and enables constant progress. No matter how much specialized skills improve, we remain limited by the constancy and fixity of our ends. Significant progress, progress that is more than technical, depends on the ability to foresee new and different results and to organise the conditions for achieving them. Science is the instrument for increasing our technical capacity to achieve results that are already known and appreciated. More importantly, it is the method for emancipating ourselves from our slavery to customary ends, to ends established in the past. (MW 8: 81)

In other words, the object of science, which is the control of inference, does not only make it safer to go from *A* to *B*, but also allows to discover inferences from *A* to

6 See on this subject the clarifications of Madelrieux (2016: 134–135).

C, D, E,.... terms that are invented in the process and open up new possibilities for action and experience. Science certainly answers our questions and problems, but also reformulates them and invents new ones: it builds theories in that it expands and reconceptualises the naïve theories of common sense.

This dynamic is best seen in the twofold relationship between science and common sense. On the one hand, it is a systematic reconstruction of the networks of meanings of common sense, not a new foundation on foreign soil. Dewey frequently takes up the classical idea, inherited from Aristotle, that science developed from *technè*. A theory is therefore never really constructed, but always reconstructed from a previous structure. This operation is first destructive (Dewey also says: *disintegrative* (LTI: 81)) of previous aggregates of meanings before it is reconstructive. The above example of the definition of metals is illuminating: it was because the operations performed with metals had expanded, that it had begun to be noticed, for example, that they reacted differently with certain chemical compounds, that it became necessary to revise, step after step, their characterizations until the hierarchy of their properties was reorganized and they were linked together in a new system, which better corresponded to the new problems of science.

On the other hand, this reconstruction is not limited to a simple rearrangement of structures on a plot of land given once and for all. It is also an opening up of perspectives, an enlargement, a decentring, so that the appropriate architectural metaphor, in this respect, might be that of the successive enlargements and embellishments of the Château de Versailles under Louis XIV, covering and annihilating the modest hunting manor that existed at the origin of the project. It is the conception of this double dynamic of reconstruction and enlargement of systems of meanings that allows Dewey to suggest an interpretation of scientific revolutions that echoes, by anticipation, that of Kuhn:

> The history of real scientific progress is marked by the adoption and invention of material devices and related techniques [...] The new data thus instituted do much more than provide facts to confirm and refine old conceptions. They institute *a new order of problems* whose solution requires *a new conceptual frame of reference*. In particular, it is through the use of new instruments and techniques that changes and relationships in these changes have been revealed in what was previously considered fixed; a process that has continued at an accelerated pace since the 17th century. (LTI: 389, my italics)

It is now possible to come back to the question on the unity of problem-solving, whether systematic or not. Problem-solving is nothing else than rational inquiry itself, taken in its local dimension, *i.e.* occupied with particular objects in particular situations – that was already grasped in its essential unity at the end of our reading of Aristotle (see above 4.3.4). This is not to say that problem-solving is *practical*, in the Aristotelian sense, or concrete ("material"), or occupied with individual and singular objects. There are problems in "theoretical" disciplines too, as it is evident in mathematics. Practical problems can also be posed in an abstract way, such as shorter-path problems. Finally practical, concrete problems can relate to general situations, such as whether there should always be alternative routes to the main routes in a logistics network. And there are practical problems that reduce to the simple exercise of judgement, as the example of the *check-list* has shown (see above 5.2.1).

There are thus two orders of inquiry:

1. Its *local* and elementary order is problem-solving; it is called *deliberation* in domains having to do with action. The generic term I will henceforth use synonymously with *problem-solving* is *inquiry*, in reference to the Aristotelian ζήτησις which is the search for the middle-term or cause (see above 4.3.1). I will specify that this is a *local* inquiry where confusion is possible. In a Kantian register, this order corresponds to the activity of the understanding, which establishes and mobilises rules in its confrontation with experience.
2. The *global* and higher order of inquiry is the construction of systems from knowledge already acquired, *i.e.* the organisation and coherence of rules in relation to one another. This higher-order inquiry is immediately broken down into lower-order inquiries, and ultimately into specific problems such as assertions to be demonstrated, predictions to be verified, explanations to be given, but also reports to be written, experiments to be conducted and communications to be made. Global inquiry is the activity of Kantian reason.

These two orders are transversal to any form of discipline and field of knowledge. There can be a practice of chemistry and astronomy, as discussed earlier (2.3.2), just as there can be a theory of the piano. Here practice and theory must be understood in the sense that emerges from Al-Farabi, as the two complementary parts of any discipline of knowledge, one occupied with solving the problems that arise in its field, the other with organising and interpreting the knowledge acquired. The distinction of knowledge according to its ultimate goals is not essential here. It could certainly be said that the practice of astronomy is designed to increase astronomical knowledge, whereas the theory of the piano, on the other hand, is only of interest if it serves its practice in a more or less direct way. However such assertions could be contested. The case of chemistry, for example, is much trickier to decide.

For the purposes of this research, it is more interesting to distinguish between areas of knowledge that lend themselves to fruitful global inquiry, *i.e.* that converges to first principles endowed with significant power as regards problem-solving. Remember that Aristotle had reserved this property for knowledge of natural or eternal essences. For him, knowledge directed towards the realisation of material human intentions was imbued with such contingency that any global inquiry would be lost in the meanderings of ever more numerous particular factors (see above 4.4.2); hence his idea that technical knowledge was capable only of a poor and local universality, that is, of "rhapsodic" systems as Kant puts it, who seems to be of the same opinion as Aristotle on this subject.

But how can global inquiries be fruitful to solve problems in technical domains and human affairs, where exceptions are so numerous and where changes of meanings and intentions so frequent? This question is dependent on an even more basic one: *What use are systems for problem-solving*? To speak of *systematic* problem-solving seems to assume that a prior global inquiry has made available a system of knowledge from which local inquiries can benefit. But the nature of this benefit is not yet clear. The ideas of *application* and *deduction* by which this relationship is often described seem to be fuzzy and misleading.

The "librarian" conception of theories and systems is that they only give more easy access to some already existing knowledge. Dewey has a more elaborate interpretation of their value, in that they make inferences more secure. In this search of the control of inference, science also gives access to new possibilities of inferences and action, which narrow practical concerns and local inquiries would never have found out by themselves.

But such an interpretation is still insufficient. Science helps problem-solving by making it more *efficient* and by expanding its scope, but that is not the end of the matter. It also transforms problem-solving into a modality which I have provisionally called systematic, and which essentially consists in the design of effective procedures (see above 1.2). The reason for this labelling still needs to be justified.

7.3 Systems and theories

This section sketches a genealogy of the tangled relationships between global and local inquiries, *i.e.* between system building and problem-solving, by tracing the evolution of the term *system* in the early modern period. It is useful to say at the outset that this discussion does not aim to make a historical point, but rather to question history about the difficulty of this articulation. After a brief review of its mainstream meaning in the period, that of *hypothesis* (7.3.1), the discussion focuses on its early association with *method* in Ramist circles, in the context of the humanist-inspired trend towards the rationalisation of the arts (7.3.2). Of particular interest for this study is the explicit tension between the internal and unique method that guides reason in all its inquiries, and the diverse and external systems that instantiate it in particular disciplines (7.3.3). This tension is present in Leibniz's thought, which he interprets in the classical direction of the *mathesis universalis* (7.3.4). It is finally, and ironically, when the notion of *system* is discarded by late 18th century philosophers to the benefit of *theory*, that its true usefulness is also articulated: it does not lie in sclerotic and dogmatic systems, but in the *systematic spirit* championed by d'Alembert in the *Preliminary Discourse* of the *Encyclopédie* (7.3.5).

7.3.1 The notion of system in the early modern period

The term σύστημα is common in ancient Greek, and generally refers to any kind of organised wholes, be they human institutions (the army, a state), natural things or representations. The Stoics played a particularly important role in giving the term its modern meaning. On the one hand, *technè* is defined as a system of representations:

> Τέχνη ἐστίν [...] σύστημα ἐκ καταλήψεων συγγεγυμνασμένων πρός τι τέλος εὔχρηστον τῷ βίῳ.
> Art [...] is a set (*system*) of notions put into practice for some purpose useful to life. (Lucian, *The Parasite*, ch. 4, SVF 1: 273)

On the other hand, the expression *world system* also originates in a Stoic context:

> The world is the system of heaven, earth and the natures therein.
> Κόσμος μὲν οὖν ἐστι σύστημα ἐξ οὐρανοῦ καὶ γῆς καὶ τῶν ἐν τούτοις περιεχομένων φύσεων.[7]

Victor Goldschmidt sums up very well the systematic character of Stoic philosophy, within which the coherence of the preceding quotations can be understood:

> More than any other ancient school, Stoicism insisted on the systematic character of its philosophy. [...] [The system] takes two forms: as a methodical progression, on the one hand, and on the other, and more profoundly, as an organic totality. Under the first aspect, system means the *order of reasons*. The various theses are not set out in just any way, and even it is impossible to develop any particular thesis separately, detached from the whole. [...] The order of reasons conforms to the order of things. But this very fact warns us that this order, founded in being, covers a *deeper system* which, for its part, is not a progression that is accomplished, but a completed totality. [...] The order of reasons is brought back to the total *Logos*, and methodical progress is concentrated and absorbed in the perfect wisdom that must be present all at once, and whose instantaneous completion corresponds to that of the universe. It is only while waiting to achieve this [perfection] that the apprentice-philosopher exercises himself in what might be called methodological time, and traverses in different directions reality which, though always the same, he does not yet possess in the form of totality. (Goldschmidt [1953] 1989: 61–63)

---

7 Pseudo-Aristotle, *De Mundo*, 391b9. The *De Mundo* is generally regarded as a treatise of Stoic origin. Cf. also a corresponding passage in Diogenes Laertius, who attributes the expression to Posidonius, *SVF 2: 526.*

Since σύστημα was translated into Latin by various terms meaning *assembly*, such as *compages, collectio*, the word *systema* was not generally used in the Middle Ages, except in its musical sense, denoting a certain form of harmony.[8] It did not reappear until the end of the Renaissance, when it became acceptable to introduce Greek words into Latin practice. It was then reactivated through two parallel and, as far as I know, independent routes.

The best known route runs through the discussion of the *world system* proposed by Copernicus. It was not Copernicus himself who used the expression, but his disciple Rheticus in his 1540 work setting out some of his master's theses. This book did not meet with great success at the time of its publishing, but it was regularly reprinted and read by astronomers, so that the term became central to the "world systems" controversy that developed in the 1580s and 1590s: people then spoke of the *Ptolemaic system*, the *Copernicus system*, and so on.[9]

This expression allowed to recover the two stoic meanings identified by Goldschmidt. On the one hand, *world system* can refer to things themselves insofar as they are organised, *i.e.* the solar system, in today's sense. By extension of this meaning, *system* will soon refer to other organised *objective* wholes, such as the *digestive system* (see below).

On the other hand, *world system* can refer to the *conception* held by different astronomers of the organisation and order governing the movement of the earth, sun, moon and planets. As such, an *epistemological modus vivendi* was very quickly established, designed to affirm the possible coexistence of several cosmological conceptions – the supporters of the Church, because they could not deny that modern mathematics vindicated, in terms of simplicity and elegance, the views of the early moderns, and the latter, because they wanted to protect themselves from accusations of heresy. So *system* very quickly became, in its philosophical usage, synonymous with *hypothesis*.

It is this definition that French dictionaries retain as the principal one at the end of the 17th century, as for example Furetière, who sketches a broad panorama of its uses (this entry is quoted here partially):

> System, s. masc. Astronomical term. A supposition, or hypothesis made by astronomers of a certain order, or arrangement of the parts of the universe, on the basis of which they explain all the phenomena, or appearances, which are found in the course of the stars or in their changes. There is no difference between system and hypothesis, except that the hypothesis is a more particular system, and the system a more general hypothesis. The system of Ptolemy, Copernicus, Tyco Brahe and Fracastor. Father Deschales claims that as many as twenty systems,

8 Hager and Strub 1995: 827; Lerner 2005: 410.

9 See the history of the expression *systema mundi* presented by Lerner (2005).

> or hypotheses, can be invented to explain with equal precision all the appearances of the stars – by considering as immobile some of the nine terms we have, namely the seven planets, the Earth, and the firmament.
> This word comes from the Greek and means composition.
> In physics, the system of the senses, movement, food, etc. is also called the way in which the organs are assumed and conceived to be arranged. So it means constitution, situation.
> There are doctors who follow the system of flavours; others who follow the system of the four qualities; others the system of acids and alkalis. *This doctor makes a new system of fevers*, that is to say, he assumes certain principles, according to which he explains all the nature and symptoms of fevers. Theologians have also developed various systems of Grace. [...]
> This word is also used figuratively. The system of Court affairs [...] (Furetière 1701)

In addition to its sense of *hypothesis* the "objective" sense of *system* appears in the example of physiological structures, such as the digestive system, from which is also derived its political or institutional sense ("the system of Court affairs"), which again appears as "figurative". It is interesting to note the care Furetière takes to distinguish physiological systems from the systems "of physicians", which are not organic structures but medical hypotheses – what would be today called theories, such as the "system of fevers".

These two main meanings at the end of the 17th century in France had strikingly different fortunes. The use of *system* as hypothesis has completely disappeared. As "objective structure", it has become ubiquitous in the natural sciences, technology, and the human sciences, where it refers in particular to institutional and social phenomena.

7.3.2 La Ramée – method and system

In addition to the "systema mundi" debate, there is a second route by which σύστημα was reintroduced into the Renaissance world. It is more convoluted; but it crucially anchors its meaning in proximity to *methodus*, another word of Greek origin that also reappears in the late Middle Ages.

As already mentioned, for Aristotle, μέθοδος means first *inquiry* and *approach to inquiry*.[10] It has been shown in which sense it may connote procedural reflection, notably in the *Topics* and *Second Analytics*, but it remains cloistered to logic. In particular, μέθοδος (or its variants ἔφοδος, ὁδός) is not part of the standard vocabulary of Greek mathematicians. Only Ptolemy and Hero, at the beginning of our era, use it to designate heuristics but also certain numerical algorithms. In Roman times,

10 See section 4.3, and Ritter et al. 1980: 1306.

the term *methodus* enjoys a certain fortune in medicine, since it refers to a specific school, the methodical school, opposed to the dogmatists and empiricists. The term thus remains common in this discipline, and in the Middle Ages it is generally associated with art much more than with theory.[11] In this context, referring to its etymology (ὁδός = road, way), the Scholastic philosophers associate it with the notion of efficiency, defining it as a "shortcut" among the possible routes for doing something.

Many books, as early as the beginning of the sixteenth century, bore the title *Method*, and aimed at the reasoned description of an art or craft (Gilbert 1960: 69).[12] The *reductions in art* are undertakings for the reasoned and methodical description of the practical know-how in a given discipline. Hélène Vérin has written a history of this great intellectual movement[13] which runs through the early modern period. The idea as well as the expression *to reduce in art* (*ad artem redigere, réduire en art*) originates in Latin Antiquity. Cicero argues for this need for rationalisation in the context of law, but it is Vitruvius's treatise on architecture that gives posterity the model of *reductions in art*. This genre expands enormously with the invention of printing (Vérin 2008: 13, 16). All the arts are concerned, liberal as well as mechanical: agriculture, mining, war, fortifications, shipbuilding, rhetoric, dance, fencing, etc. but also, by extension, ethical disciplines such as ethics and politics. The book of Vérin (2008) describing reductions in arts has as its sub-title: "A project of rationalisation of practices".

> Reducing in art means gathering, putting in order and disseminating in writing; gathering scattered, confused or even profuse data, ordering them, clarifying them with the help of rules and precepts, in a brief and methodical presentation, and disseminating the result of this work in writing in a language accessible to all, so that everyone has the means to act in the best possible way, their choices thus facilitated, guided, regulated. Through this work of reduction, the author contributes to the dissemination of knowledge and thus to its possible improvement, thanks to the successive – and cumulative – work of generations; the end goal always mentioned is the public good. The point of view is that of action: on ideas, on people, on things. (Vérin and Dubourg Glatigny 2008: 19)

Reductions in art correspond thus to what Aquinas identified as *speculation taking practice as its object* (see above 5.1.1), at a time when the relationship between theory and practice was rapidly changing:

---

11 See Hamou 2014: 2; Ritter et al. 1980: 1305–1308.

12 See appendix (233) for a list of such works, which Gilbert says is suggestive only. See also Ong 1958: 228.

13 The two reference works on this question are Vérin 1993, 2008.

> The *De Re Metallica*[14] took root in humanist thought, between university, administration and medicine, and even more concretely, in the network of Erasmus' circle. The major concern of these intellectuals of the 16th century lay, prior to any distinction between science and technology, in the necessity they felt to evolve theoretical thought, to draw it toward practical concerns, and to equip it with the means to engage with reality. (Garçon 2008: 269)

The function of these works is multiple. As well as safeguarding and disseminating practical knowledge, they aim to educate administrators, lawyers and the sponsors of ever larger and more expensive projects, who must therefore be able to understand the standards of the trades with which they are dealing.[15] Overall, the aim is to increase control over the action and the assurance of its success, its *effectiveness*. Dubourg Glatigny and Vérin (2008: 50) talk of the emergence of "a shared rationality, centred on effectiveness and compliance with rules". By setting out the current state of the arts, these works hope to spark invention, particularly from scholars, mainly geometers, capable of examining the collected facts and reasons and of correcting them.[16]

From a methodical point of view, authors of reductions in art are inspired by ancient sources where they exist, but also and above all by contemporary practice and inventions. They visit workshops, interview craftsmen, and even practise the art themselves, although this is not generally the case. Reduction in art requires foremost the specific skill of dialectic (Vérin and Dubourg Glatigny 2008: 21) which is capable of seeing the universal in the particular, of making the worker say what goes without saying for them, of describing delicate operations that are transmitted by gesture and not by word. Reductions in art therefore involve a major effort to build up and standardise technical vocabularies (Dubourg Glatigny and Vérin 2008: 59). They are also an important milestone concerning the visual and diagrammatic transmission of knowledge, representing decisions by decision trees, by quantitative tables, and by engravings representing machines, workshops and the workers' gestures; in the case of performative arts such as dance and fencing, an entire language, both visual and textual, is invented (Dubourg Glatigny and Vérin 2008: 60).

It is against this rich background that Pierre de La Ramée reintroduces the concept of *method* into logic.[17] Invention, judgement and method are indeed the three parts of his new logic, presented in his *Dialectique*, rewritten several times throughout his life, from 1543 until his death in 1572.

---

14 One of the emblematic works of the 16th century, by Georgius Agricola (1494 – 1555), which undertook to explain the art of mining.

15 Vérin and Dubourg Glatigny 2008: 25, 35; Dubourg Glatigny and Vérin 2008: 56.

16 Vérin and Dubourg Glatigny 2008: 36; Dubourg Glatigny and Vérin 2008: 56.

17 I follow here the suggestion of Vérin and Dubourg Glatigny (2008: 15).

This tripartition of dialectic follows the logical order of the constitution of knowledge. Arguments are found by invention with the help of *common loci*; simple judgement (the name given by La Ramée to syllogism) allows to verify their validity, by inferring them from arguments already known, and finally the method organises all the arguments into a homogeneous, coherent and ordered whole based on axioms. For La Ramée, method deals solely with the order in which an already constituted science should be presented, particularly with a view to pedagogical purposes. In a famous image, he shows it at work in the reconstitution of the right order of the propositions of Grammar, which one would have written on small strips of paper, and mixed in an urn. Method does not help to formulate these propositions (invention), nor to prove their validity (judgement); it only makes sure they can be easily *accessed* by reason – it shows *the right way* to find them infallibly.

La Ramée's concerns are not directed towards the inquiry of new knowledge. His focus is on learning and on transmission. It implies that the scholar's main mission is not to seek, but to "argue well". La Ramée sometimes asserts that rhetoric has pre-eminence among the arts of the *Trivium*. However, "disputing" should be understood in the Platonic tradition, which defines thought as a dialogue with oneself, *i.e.* as discursivity in general, and "thus the art of knowing, *i.e.* dialectic or logic, is one and the same doctrine for perceiving all things."[18]

With *method*, it is indeed the idea of *system* that is already there. La Ramée knows the Stoic definition of τέχνη (see above 7.3.1), and translates it as such:

> *Ars est comprehensio praeceptorum in rebus aeternis, propriorum et ordine dispositorum, ad utilem vitae finem spectantuim.*
> Art is the understanding of precepts in things eternal, proper and orderly, aiming at the useful end of life.[19]

La Ramée uses *comprehensio* to translate the Greek term in Lucian's quotation, but after this quotation he refers to Book II of his *Dialectic*, which deals with method: the link is therefore explicit. Moreover, Melanchthon, his contemporary and correspondent, translates Lucian's definition as early as 1534 using the Latinised form *systema*. As the latter was widely read in Protestant countries, it is assumed that this is one of the main sources of diffusion of the term in this sense, well later though, around the years 1570–1580 (Ritschl 1906: 10–24). It is then used, in Protestant circles, to designate works of Christian didactics, setting out in an orderly fashion the truths of religion, which are conceived, alongside the image of the Church as the body of

---

18 *Dialectic* 1996: 121. This passage is quoted and commented on by Bruyère 1984: 223.
19 Cited by Schmidt-Biggemann (1983: 44, n. 162). The original text of the *Praelectiones* of 1572 is difficult to access.

Christ, as the *corpus integrum* of the truths contained in the New Testament, none of which could be removed without the whole edifice collapsing.

Given the ramist background that makes the term very close to *method*, it is not surprising that once adopted, *systema* quickly secularises and generalises to mean the ordered exposition of a field of knowledge. The three "sytematists" best known in this respect are Bartholomeus Keckermann (1572 – 1609), Clemens Timpler (1563 – 1624) and Johann Heinrich Alsted (1588 – 1638), who were the first to publish works bearing the title *System*. This title promises the reader an orderly and didactic exposition of the discipline in question. Keckermann, for example, is the author of *Systems of Logic, Rhetoric, Politics*, and so on. But the great work of this moment in the history of the compilation of knowledge is Alsted's *Encyclopaedia*, published in seven books in 1630, on 2500 pages. This work does not only aim to bring together the essential knowledge of all sciences, so as to render all other books useless, but it also presents itself, in a distant Lullist inspiration, as the very complement of philosophy, an "encyclopaedic course" that should educate the faculties of the mind – invention, judgement, memory (Schmidt-Biggemann 1983: 131–139).

At the most basic and pragmatic level, the aim of these systems is to help readers *find* the relevant information in the knowledge available to them. They are at least partly motivated by the awareness of the abundance of books made possible by the printing press and by the excitement and anxiety it caused. Printing would also enable the invention of modern typographical tools that helped navigate through this abundance of knowledge and somewhat cope with it: the table of contents, the thematic index, of places, of people, etc. (Blair 2010: ch. 3)

Among these visual devices, tables and trees play a special role, as they aim to map information according to their internal logic. Tables, which index atoms of knowledge through two or more dimensions, rely on a combinatorial scheme which can be traced back to Raymond Lull, in the 13th century (Uckelman 2010: 429). His idea of a primitive alphabet of knowledge, which combinatorial manipulation would generate truth, is even older, linking up with theosophical and particularly Kabbalistic thought (Zweig 1997). Trees can lay claim to the Platonic *diaeresis*. La Ramée made such extensive use of this scheme to organize knowledge that it became proverbial. It made patent the defects of this method, which Aristotle had already pointed out (see above 4.4.1): La Ramée would constantly change his principles of division throughout these works, so that an impression of arbitrariness takes hold, which his opponents laughed at (Ong 1958: 200).

7.3.3 Systems and method

What relation can these ideas have to local inquiry, and especially to systematic problem-solving? Method, in the Ramist sense, does not by any means seem to imply the modern sense of *procedure*; and the vast compilations called systems do not seem to be of any use in solving any problem.

These conclusions should, however, be qualified. For the early moderns – and for common sense too – a solved inquiry results in a piece of knowledge which the system is designed to record in order to relieve memory. Therefore, what systems propose *is* in a sense procedural: they promise to guide the *search* for any piece of knowledge, which is the only *meta-procedure* that people need when knowledge is available "somewhere". Samuel Johnson's maxim is representative of the early moderns' new approach to knowledge, which was one of *accessibility*:

> Knowledge is of two kinds. We know a subject ourselves or we know how to find it. (Boswell 1887: 2: 365)[20]

Moreover, in their Ramist initial and demanding version, systems promise to index knowledge, not according to some arbitrary order, e.g. alphabetical, but according to reason. In other words, the inquirer does not need to have any prior knowledge before starting the search, such as which word correctly describes the object of their inquiry if the order is alphabetical (a common difficulty with technical dictionaries for beginners). They only have to let their mind follow the logical order of reasons proposed by the system to find the answer to their questions.

The spirit in which methods and systems are written does not aim at the specialisation of knowledge. On the contrary, it aims to show that all knowledge is accessible to anyone who knows how to use their reason well. In particular, reductions in art serve and inspire the humanist cause of a unified and non-disciplinary ideal of knowledge. Above all, the aim is to develop the ability to think for oneself, and the liberal arts are seen as the ideal places for the exercise of the mind, leading to its full development. For La Ramée, particular methods are merely instances of the one and only Method, which in principle can deal with any subject of knowledge; they merely guide and support it in these undertakings.

Indeed,[21] the "artificial" dialectic which La Ramée presents and teaches is merely the image of a natural dialectic which every man possesses: the former is there only to develop and guide the latter when it is perverted. Dialectic is, rather than a faculty, a *virtus*, a term in which one must perceive the original *vis* of the

---

20 Cited and commented by Blair (2010: 263).

21 This discussion relies on Bruyère 1986.

active mind, naturally disposed to knowledge, *i.e.* a *force* that must be exercised in order for it to become habit and effective knowledge. It follows that the method is also the image of a *natural* method, one that requires the mind to gather its knowledge into a coherent whole, ordering it according to what it finds clearest, which happens also to be what is most general and primary, and from which everything else proceeds – a presupposition with Platonic overtones. The method is therefore unique and universal, because it is the very principle that makes possible the multiple methods of the reductions in art:

> This method is singular and unique in doctrines well instituted because through its singular and unique [way], it is proceeded by things antecedent of the whole, [which are] absolutely clearer and remarkable, to clarify and illustrate things consequent and unknown. (*Dialectic* 1996: 121)
> This is why this method is to be admired not only for its clarity, but also for its conciseness [...] Through this way, it will be permitted to know more quickly and more concisely. (*Dialectica* (1554), quoted and commented by (Bruyère 1984: 137))

This relationship between the natural and artificial methods is also present, though in different terms, among German systematists. For example, for Keckermann, *Logic* must be understood in a twofold way. First, it is a natural disposition which habit, acquired by precept and exercise, must develop. Secondly, it is the *explicit system* of these precepts, *i.e.* their *external* didactic exposition. The system thus plays in Keckermann the same role as the artificial method in La Ramée: it is the discursive externalisation, in the precepts of art, of the inner and true know-how of a given practice (Ritschl 1906: 27–29).

This duality reflects the tension between the inner methodical disposition invoked by La Ramée and the external system that is its tangible result, couched in rules ordered in textbooks. For while the former manifests the autonomy of the active mind, the latter might lead one to believe that it is now possible – thanks to its very existence – to do without it: there would be nothing left, in effect, but to apply the rules.

This tension is explicit in the 17th century. At one end, for Descartes the Method can never consist in an inert corpus of already-formed propositions. It must direct one's mind in its search for truth by adopting the order imposed by the clarity and distinction of the ideas themselves. Descartes does follow the inspiration of La Ramée in this sense, but he conceives of the mind as its own Master, who must teach itself how to think. Cartesian method is therefore first and foremost "a practice that aims to sharpen the mind's faculties, to make it more aware, more attentive, so to speak" (Hamou 2014: 6). The *Rules for the Direction of the Mind* are meant only to guide judgement, not to provide any catch-all recipe for solving problems. They do not aim to freeze human behaviour in the repetition of servile gestures, but to liberate from them. Surprisingly for us, the *routine* – a term attested in French since the middle of the 16th century – is precisely what the rule is opposed to, as the *Dictionnaire de l'Académie Française* (2001: 417) writes:

> Routine. n. Usage acquired by long habit, without the aid of study and rules. *He has never studied thoroughly, but he has acquired I don't know what routine of discourse, of speaking. He does this by routine. he knows no music, but he sings by routine[...] An old routine. The routine of the world, of Court.*

Routine therefore refers to the know-how acquired through experience and custom. This knowledge is particular to a given place and time, and cannot reach universality. It is *mechanical* in the medieval sense that this term had, *i.e.* servile. A workman, remarks Nicolas Aubin, author of the first french *navy dictionary* in 1702, is at a loss when told of the practices of his trade current in other places, even when they are close.[22] Routine has no rules, because there is nothing universal about it.

However, at the other end of this tension about rules and method, also stemming from La Ramée, Bacon's *Novum Organum* (2004) sets out to design a "new toolkit" for thought which, contrary to Descartes' belief, is necessary in order to free the inquirer from their own illusions and prejudices. This toolkit includes instruments, but also data tables, diagrams, encyclopaedias, all these graphic inventions made possible in particular by printing. As historian Guido Giglioni (2022) says:

> A method for Bacon is not of an intellectual nature, nor should it be a creation of the intellect; rather, it has to be something that provides the intellect with helps from the outside. In this sense, it is a *machina*, a device that is capable of restoring our clear perception of the true nature of things. [...] Technology is the remedy for ingenium. A limping man walking along a safe road will reach their destination more easily than an athlete running in a wild territory; using compasses and rulers, an ordinary person will trace circles and straight lines with more precision than a well-trained artist by hand alone [...] There can be real change or reformation (*instauratio*) only if the power of innovation (*i.e., ingenium*) and the ability of the mind to elaborate representational models of reality (*i.e., intellectus*) are channelled and harnessed through extra-mental ways of controlling the way in which we gather information and keep control of our passions.

In this respect, the tables and tree diagrams may have a richer role than guiding the search for a piece of knowledge. In reductions in art, they come to classify rules and precepts for decision-making (Vérin and Dubourg Glatigny 2008: 35). A diagram that gives the solution to a practical problem is a kind of *questionnaire* that asks what

22 Cited by Dubourg Glatigny and Vérin (2008: 59).

are the decisive characteristics of the situation, that invites *to think* about it in a new, ordered way, and that leads to the right decision.

> Any reduction in art must define and order its operations. But again, definitions and ordering are interdependent. The path to follow depends on the content of the operations previously selected, more precisely, on the power or potentiality that characterises each of them. This rationalisation of successive choices is often presented as an interlocking of operations that go from the general to the particular, as expressed in tree tables. They offer an ordered system of classes of increasingly specific operations. The tables of Simon Stevin and Jean Errard for fortification, those of Testelin for painting are illustrations of this. They are present in all fields where the moment of conception is decisive. The aim is not to eliminate the exercise of judgement but to guide it, shorten it, facilitate it. (Dubourg Glatigny and Vérin 2008: 62)

However, it may happen that by dint of abbreviating the judgement, judgement is rendered useless. The example of Testelin's tables, cited by the authors, illustrates this possibility. Published in 1680, these six tables aimed to give the rules of painting, in the context of the Royal Academy created by Colbert (Michel 2008: 110). Each presented in tree-like form, they set out the rules of line, expression, proportions, light and dark, subject order and colour. According to Christian Michel, who has studied them in detail, they constrain the painter's choice in varying ways: while those for the proportion of human figures according to age, race, sex, explicitly leave a fairly wide margin for personal creativity, "it is different for certain parts of other tables, particularly those where the search for rules [...] turns into recipes" (Michel 2008: 115).

In that second case, trees and tables tell exactly what to do. The decision tree is a straight line that the rule traces towards the resolution of the problem, a line that the diversity of circumstances does not manage to break, but by which it develops through the succession of its branches. The "rationalisation of successive choices" that the tree diagram operates leads and shortens reflection because it avoids unnecessary detours and dead ends, that is to say what may be called *false problems*, *i.e.* so many fewer problems to solve: places where reflection could have erred or hesitated but which, once all the reasons have been properly ordered, turn out to have only *one way out* and to require, in fact, *no judgement at all.*

7.3.4 Leibniz: the Characteristic against procedures

This tension between two possible interpretations of the meaning of the method and the consequent value of systematic knowledge is clearly visible in Leibniz's thought.

On the one side, the understanding of method as a step-by-step guide for solving problems is clearly expressed by Leibniz. He arguably comes close to the mordern idea of *procedure* when he demands that method guides the mind in the manner of a calculus. Significantly, however, he cannot find a term to express this idea and has to resort to a metaphor:

> The true method must provide us with a *filum Ariadnes*, that is to say, a certain sensible and coarse means, which leads the mind as the lines drawn in geometry, and the form of the operations which are prescribed to apprentices in Arithmetic. Without this, our mind cannot travel a long way without going astray. We see this clearly in Analysis, and if we had characters such as the ones I conceive in metaphysics and morals, and what depends on them, we could make in these matters very assured and very important propositions; we could put the advantages and disadvantages into account when deliberation is involved. (*Letter to Jean Gallois*, Sept. 1677, D.II.1: 381)

Though the idea of applying numerical procedures to speculation dates back at least to Raymond Lull, it is with Leibniz that all these avenues find a first reflective interpretation. For him, the affinity of thought and calculation is not an analogy, nor an identity, but a research program, which he expresses through the objective of expanding the certainty of mathematical procedures to all areas of knowledge:

> I have noticed that the cause which makes it so easy for us to err outside Mathematics, and that Geometers have been so happy in their reasonings, is only because in Geometry and other parts of abstract Mathematics, we can make continuous experiments or proofs, not only on the conclusion, but also at any moment, and at every step we take on the premises by reducing the whole to numbers; [...] The only way to rectify our reasonings is to make them as sensible as those of the Mathematicians are, so that one can find one's error at a glance, and when there are disputes between people, one can only say: let us count, without further ceremony, to see which one is right.[23]

What is expressed here is the certainty that comes, not from intuition, as in Descartes, but from the rigour of the procedure itself. Significantly, when Leibniz wishes to show how the rigour of mathematical thinking could be transposed to other fields, he takes the example of two practical disciplines, law and medicine. Their operating rules, according to Leibniz, aim at the same ideal of rigour and universal validity as mathematical demonstrations, but they are hindered by the multiplicity of their distinctions and exceptions. This state, far from being essential, merely marks the primitive state of their advancement:

> It is true that there are rules [of jurisprudence], which have exceptions, especially in questions where there are many circumstances [...] But to make their use safe, these exceptions must be determined in number and in meaning, as far as possible: and then it may happen that the

23 *Project and Essays to arrive at some certainty to end a good deal of disputes and to advance the art of inventing*, C: 175.

> exception itself has its sub-exceptions, that is, its replications, and that the replication has duplications, etc..., but in the end all these exceptions and sub-exceptions, well determined, joined with the rule, must complete universality. (*New Essays on Human Understanding*, IV, 7, §11, D.VI.6: 419)

What Leibniz expresses clearly in these lines is the idea of a systematic procedure that he deems analytically contained in the very notion of a rule. In an early writing,[24] he had already spoken out against the saying "there is no rule without exception", demonstrating its absurdity. Here he conceives of the rule in an even more general way, as containing its own exceptions, but also those contained within them themselves, *i.e.* as a complete path that infallibly determines the solution sought. Thus he continues:

> But if these sorts of rules, loaded with exceptions and sub-exceptions, were to enter into academic disputes, it would still be necessary to argue with pen in hand, keeping as it were a record of what is said on both sides. And this would still be necessary moreover, by disputing constantly in form by several syllogisms [and prosyllogisms and proprosyllogisms), mingled from time to time with distinctions, where the best memory in the world must be confused. (D.VI.6: 420)

In this second paragraph is clearly expressed the idea of a deterministic course of reasoning, following a rule that can be said to be universal because its exceptions are in *finite* number, which requires a *graphical memory* to reach its conclusion. Even if Leibniz formulates the idea hypothetically, even incredulously as to its practicability, there is something here that closely resembles what would be today called the execution of a program.

In this example from jurisprudence, it is finally grasped, in a clear way, that deliberation may be expressed procedurally, and that it has a deep affinity with mathematical demonstrations. Writing a procedure and writing a demonstration are essentially conceived of as the same thing.

However, from another point of view, Leibniz also remains infinitely distant from the modern idea of an effective procedure. While considering Descartes's method empty,[25] he still asserts its unicity ("the true method"), a thesis dear to La Ramée and to Descartes alike. He only makes it point to his grandiose project of a Universal Characterstic. Whereas indeed we conceive of procedures as being as diverse as the problems they aim to solve, for Leibniz on the contrary, a single universal calculation, a single procedure, should make it possible to solve all questions, as long as they were analysed in their primitive (and directly operative) meanings.

---

24 *Nova Methodus discendae docendaeque Jursiprudentiae* (1667), D.VI.1: 308–310.
25 *Letter to Swenligius*, G.IV: 329.

Procedure and rule are thus destined to be simplified *until they become entirely transparent.* The fact that jurisprudence has so many exceptions, and medicine so many rules of thumb, points towards their perfectibility. Rules become shorter and simpler as they become more general, so that, if the language directly expressing the primitive notions was found, a *unique* procedure of calculation would replace all the others. The language that Leibniz calls for is *universal* not only because it transcends borders and cultures, but because it can be applied as it is to any problem.

A single procedure, of the grammatical type, would thus serve to represent all others, in the same way that particular rules can be derived from more general rules. The language Leibniz dreams of would thus be "very easy to learn."[26] He is not interested in rules and procedures in their own positivity and thickness, as they will be studied later by Babbage and Turing. He can see that they enable the ignorant to solve problems, but that does not interest him. With the Characteristic, ignorance will be made simply impossible:

> It won't be possible for fantasies advanced by one who does not understand them to be written in these characters. An ignoramus will not be able to use them, or striving to do so, he will thereby become learned. (Letter to Jean Gallois, D.II.1: 381)

Rules and procedures must therefore be considered as temporary and imperfect devices that must become eventually obsolete by the Characteristic:

> If the words were made according to an artifice which I see possible, but of which those who have made universal languages have not been aware, we could arrive at this effect by the words themselves, which would be of incredible utility for human life. But in the meantime there is another less beautiful way, but one that is already open, instead of the other that would have to be made from scratch. It is by making use of characters after the example of mathematicians, which are proper to fix our Mind, and by adding to them a proof of numbers. (*Projet et Essais pour arriver à quelque certitude...*, C: 175)

While the possibility of procedural knowledge is thus clearly conceived by Leibniz, it is not the ultimate form of knowledge; it is a support to compensate for the confusion of reflective thought when it cannot immediately run through a long chain of elements. Clear thinking should, however, be preferred as soon as possible:

> [It is a mistake to] want to reduce truths to maxims [...] against the grain and without need, for the human mind considers much all at once and it is a hindrance to it to want to force it to stop at every step it takes and express everything it thinks. It is precisely as if, in making his account with a merchant or with a host, one wanted to oblige him to count everything with his fingers in order to be more sure of it. (*New Essays on Human Understanding*, IV, 7, §11, D.VI.6: 422)

26 *Préface à la science générale* (1677), C: 154.

The sequential character of procedures is no longer relevant in intuitive knowledge, which considers everything at once. There is here a distant echo of the relationship described by Keckermann between the external system and the internal *habitus*, the former being a mere precept for the development of the latter, a relationship which also evokes the one between the artificial method and the natural method of La Ramée.

7.3.5 The criticism of systems and the *systematic spirit*

The conceptual dead-end towards which Leibniz was headed is not specific to the Characteristic Universalis. Actually, the whole concept of theoretical systems, as it evolved in the early 18th century proved unfit for scientific research and finally gave way to the modern pragmatic and malleable concept of *scientific theory*. Behind this notion however, the main useful features of systems were retained; it appears very clearly in d'Alembert's famous formula that one should seek to have a *systematic spirit* but avoid the *spirit of systems*, by which he meant dogmatism. The systematic spirit articulates what I have called above *global inquiry*.

Furetière's definition of *system* (see above 7.3.1) alludes to the "systems of Grace". It refers to the polemic concerning Malebranche's "system of occasional causes" , which was to spread the term in metaphysics. One would therefore talk, from the 1680s onwards, of the "systems" of Descartes or Spinoza, and Leibniz himself entitled his philosophy the "system of pre-established harmony". Philosophers realise at this time that the hypothetical nature of Descartes's "machinic" explanations has two layers: it is not only a question of whether mechanistic explanations of phenomena by "size, figure and motion" are correct (see above 8.1.3), but also of *what would be the very meaning* of saying they are such, since the mind's constructions are incommensurable with the affections of bodies.

This multiplicity of astronomical and philosophical systems, which the exchange of arguments did not seem able to separate, was to encourage the spread of a sceptical posture hostile to the idea of system. Malebranche himself mentions it in his treatise *De la Recherche de la Vérité* in 1675, so much so that it can be said that in early modern philosophy, systems were met with criticism as soon as they appeared.

Two approaches attempted to escape scepticism. The oldest is that of English experimental philosophy, which took shape in the 1660s around the Royal Society, aiming at the regulation of the use of hypotheses in science. The second is eclecticism, which developed around the beginning of the 18th century in Germany. It concerns more properly the philosophical attitude. Finally, in the 1750s, the French encyclopaedists, achieved an uneasy synthesis of these two movements.

1) The English empiricist school inherits from Bacon his distrust of over-hasty systematic constructions and praises the patience required by the experimental method in collecting facts, verifying them and relating them. In particular, the English empiricists prefer to *system* the use of the term *theory* to designate their doctrines. This is a minor terminological peculiarity, as this usage remains rare and is never (to our knowledge) explicitly promoted. It would, however, be decisive in its adoption, a century later, to designate scientific doctrines, and it is therefore useful to take a closer look at it.

Until the end of the 16th century the adjective *theorica* has a natural preeminence over the noun *theory*, which appears only rarely. Its earliest occurrences to designate scientific doctrines, moreover, use the term *theorica* derived from its adjectival form. *Theoria* appears in a specialised use in the 15th century to designate treatises on astronomy (O. Pedersen 1961: 161) and sometimes mechanical clocks such as the astrarium (Shank 2007: 19–21). Glocenius's learned dictionary, published in 1613, mentions only the adjective *theoretical*, which refers to its older, speculative sense.

There is no detailed history of the spread of the term *theory* in the 17th century. It began to be used regularly there in various contexts which derived from its astronomical usage but which were not necessarily scientific. Its course is therefore very broadly equivalent to that of *system*, with the advantage that it is less associated with astronomical quarrels, and that it remains rooted in the mathematical and instrumental conception of this science, simply descriptive of the movements of the planets, and silent on their causes. As such, it also retains its old "contemplative" connotations.

Boyle uses *theory* interchangeably with *hypothesis*. Unlike *system* however, which according to Furetière is the "hypothèse plus générale", a good theory is for Boyle a local and fallible hypothesis, which agrees with past observations, but which must also support the test of future experiments:

> One of the Conditions of a good Hypothesis is, that It fairly comport not onely with all other Truths, but with all other Phaenomena of Nature, as well as those 'tis fram'd to explicate. For this being granted, (which cannot be deny'd,) He that establishes a Theory, which he expects shall be acquiesc'd in by all succeeding Times, and make Him famous in them, must not onely have a care, that none of the Phaenomena of Nature, that are already taken notice of, do contradict his Hypothesis at the present, but that no Phaenomena that may be hereafter discover'd, shall do it for the future. (Boyle 1674: 207)

Newton, meanwhile, used the term as early as 1672 in his *Theory of Light and Colors*, a title perhaps suggested by Oldenburg. However, in his correspondence, particularly with Hooke, he uses it in a sense quite distinct from Boyle, this time contrasting *theory* with *hypothesis*. Hooke had criticised Newton's doctrine by declaring his corpuscular hypothesis of light ungrounded. To which the latter replied that Hooke was making a logical confusion here, as he (Newton) only put forward this hypothesis at the end of his treatise, as a conjecture that could reasonably *be inferred from his theory*. However this theory, based solely on observations and on their mathematical description, did not depend on it in any way. It is therefore because one *already* has a theory that one earns the right to put forward hypotheses. In our opinion, this position is based on the closing remark of his long letter that the doctrine of colours should be regarded as mixed mathematics, whose physical principles are reduced to wholly certain observations of fact.[27] The distinction between *theory* and *hypothesis* is thus clear in Newton: the former is the province of the mathematical sciences, and the latter of natural philosophy. In 1672, Newton imported the term from astronomy to optics and, a few years later, it was in the same sense that he used it in the *Principia*, albeit very discreetly, in connection with mechanics. Generally speaking, the use of the term *theory* by English scientists, although positive, is very rare, and it will have no immediate posterity.

2) In philosophy, another way of escaping the systems dispute was eclecticism, a movement that spanned the 18th century. It first emerged in Germany with the figures of Johann Christoph Sturm (1635 – 1703) and Christian Thomasius (1655 – 1728). Eclecticism promotes the freedom to think for oneself. It enjoins the educated mind not to allow itself to be drawn into one or another philosophical "sect", but to *choose* from the various systems the different propositions which seem to best accord with the truth. Christian Wolff criticised this idea as early as 1729, in his opuscule *De differentia intellectus systematici et non systematici*, arguing that a disordered eclecticism could only lead to syncretism (Wolff does not use the word, but Diderot will) and that the true eclectic mind must also be systematic. It is indeed this systematic mindset that Wolff foremost defends, and systems only by ricochet.

It is the methodical sense of *system* that resurfaces here, the one popularised in Germany by the works of Keckermann, Timpler and Alsted, and which applies not only to metaphysics and natural philosophy, but to every discipline. In his treatise, Wolff gives examples from mathematics, astronomy, law, etc. to illustrate the idea of *system*. And so it is indeed in a ramist spirit of "thinking for oneself" that Wolff can assert that the true eclectic philosopher will be profoundly systematic:

> Those who possess a systematic intelligence are free from the prejudice of authority and are able to proceed as eclectics. Indeed, they admit only that which can be proved to be contained in a system by its principles. They therefore judge for internal reasons, but not for external ones […] The systematic mind quickly skims over the writings of other authors. He skips over

27 *Letter to Hooke*, 11 June 1672, in *The correspondence of Isaac Newton* 1959: 1: 187.

> what is already contained in his system and focuses his attention only on propositions with which he is not yet familiar. If he understands or explains them in his own terms, he tries to deduce them from the concept of the subject by principles derived from his system. If they are found to be true by these principles, he adds them to his system [...] From this it follows how mistaken are those who oppose eclectic philosophy to demonstrative philosophy. (Wolff [1729] 2019: 76)

The French encyclopaedists are heirs to these two critical traditions, which they have difficulty reconciling. On the one hand, the eclectics in the manner of Wolff saw systematicity as the very condition of philosophy. On the other hand, the credit they give to Newton makes them condemn the construction of systems as contrary to the scientific spirit.

Their posture towards the idea of systematicity is therefore ambivalent, which can be seen in their terminological hesitations. In the *Encyclopédie*, Diderot thus presents the eclectic thinker as the true philosopher who avoids both the sectarianism of systems, the scepticism that rejects them all in their entirety, and the syncretism that mixes them without method.[28] On the other hand, in the articles of the *Encyclopédie*, the expression *systematic mind* is most often pejorative, as for example in the article *Philosophy*, which contrasts it with the "genuine philosophical mind", whose description is similar to that of the eclectic:

> The systematic mindset does no less harm to the progress of truth: by systematic mind, I do not mean he who binds truths together, in order to form demonstrations, which is nothing else than the true philosophical spirit, but I designate he who builds plans, and forms systems of the universe, to which he then wishes to adjust, willingly or by force, the phenomena. (Article "Système", E 15: 515a)

In this assertion, the term *system* is so negatively connoted that Diderot attempts to sever its deep links with the notions of *order* of knowledge and *method*, proposing to qualify what Wolff called *systematic* as *philosophical* quite simply.

Such a proposal had little chance of succeeding, and it is rather that of d'Alembert that posterity will remember when, in a famous passage of the *Preliminary Discourse* (1751), he uses *systematic* in the opposite sense to that of Diderot, which he instead calls *esprit de système*:

> It is not at all by vague and arbitrary hypotheses that we can hope to know nature; it is by thoughtful study of phenomena, by the comparisons we make among them, by the art of reducing, as much as that may be possible, a large number of phenomena to a single one that can be regarded as their principle. Indeed, the more one reduces the number of principles

28 Article "Eclectisme", E 5: 271a. See Carboncini 1987 on Wolff's discrete influence on the *Encyclopédie*.

> of a science, the more one gives them scope, and since the object of a science is necessarily fixed, the principles applied to that object will be so much the more fertile as they are fewer in number. This reduction which, moreover, makes them easier to understand, constitutes the true "systematic spirit" (*esprit systématique*). One must be very careful not to mistake this for the "spirit of system" (*esprit de système*), with which it does not always agree. (E 1: vi)

Incidentally, it should be noted that here the opposition between *esprit systématique* and *esprit de système* is not straightforward, since d'Alembert merely remarks that the two "do not always agree", implying that they sometimes do. Further on he calls the *esprit de système the mindset of hypothesis and conjecture* with a positive reference to Boyle's local and fallible hypotheses (E 1: xxxi). Hypotheses do have a positive role to play in science, and it is only necessary to prevent them being confused with "frivolous conjectures" and becoming unshakeable principles, so that the physicist never turns them into "systems". Nonetheless, the formula is striking and captures the mindset of the empiricist school of Boyle and Newton. The systematic spirit consists in this double movement, inspired by Bacon, of first reducing the variety of a given field of experience to the minimum number of principles, in order then to reap the fruits of their universal applicability, which gives so much power to the resolution of problems.

Although this *systematic attitude* is a central virtue of science, the term *system* itself, with such negative connotations, will be eventually discarded so that the ideas it conveys – that of method and that of scientific hypothesis – may regain credibility. And it is indeed also in the *Preliminary Discourse* that one finds, perhaps for one of the very first times, the suggestion to replace the term *system* by that of *theory*, in connection with Newton:

> His Theory of the World (for I do not want to say his System) is now so generally received, etc. (E 1: xxvi)

Perhaps d'Alembert borrows the word *theory* from another author, but it is in any case contemporary with the 1750s. Voltaire, for example, does not yet use it in his *Eléments de la philosophie de Newton* of 1737, though he sounds the charge against systems. Likewise Condillac's *Traité des systèmes* (1749), also so virulent against them, nonetheless points to Newton's physics as a system whose principles are phenomena (Condillac [1749] 2010: 187), for want of a better word. He contrasts this strange expression with the "abstract systems" of the philosophers and with the "systems of suppositions" of the Cartesian scientists.

In any case, d'Alembert's little phrase bears witness to a still uncertain substitution. It became widespread in the last quarter of the 18th century. A century later, it posed no problem for Claude Bernard who still remembered very well the former meaning of *system*, although he falsely attributed it to "scholasticism":

> In the experimental method, the hypothesis is a scientific idea that must be subjected to experiment. Scientific invention lies in the creation of a promising and fruitful hypothesis; it is given by the feeling or by the very genius of the scientist who created it. When the hypothesis is subjected to the experimental method, it becomes a theory; when it is subjected to logic alone, it becomes a system. The system is therefore a hypothesis to which the facts have been logically reduced by reasoning, but without critical experimental verification. The theory is the verified hypothesis, after it has been subjected to the scrutiny of reasoning and experimental criticism. In a word, systems and doctrines in medicine are hypothetical or theoretical ideas transformed into immutable principles. This way of proceeding belongs essentially to scholasticism and differs radically from the experimental method. (Bernard [1865] 2013: 372)

It should be noted that it is the Boylian sense of *empirical theory*, and not the Newtonian sense of *mathematical theory* that Claude Bernard retains here. Their coexistence and their compatibility is a question for the philosophy of science that is still very much alive today.

By replacing *system*, *theory* will take on many of its connotations. In everyday language, a theory is still referred to as a hypothesis that has yet to be verified. The term was to spread outside the scientific field and designate works of *reduction in art*, – thus *Gardening Theories* were written.[29] The term became even more widespread, and theories of happiness or salvation were also published at this time (König and Pulte 1998: 1136).

Theory also retains its ancient meaning, which associates it with the contemplation of principles. A theory is not just an “efficient machinery” for answering questions about nature and solving problems. It also aims to reveal the principles of phenomena, by providing demonstrations that link one to the other – in the essential process of *reduction* of definitions that Aristotle already described in *APo.* II.8 (see above 4.3.3). What is now called *fundamental research* carries this speculative ambition. According to Nicholas Jardine (1976: 305–310), this dual aim of the scientific enterprise was articulated by Galileo through the clever use of geometry: to make progress in understanding the principles of reality *and* to solve problems relating to phenomena. The historian of science Peter Dear advocates a similar thesis: for him, modern science emerges, from Galileo onwards, as an “amalgam of natural philosophy and instrumentality”:

> In modern culture, “science” can be represented in its form of natural philosophy or in its form of instrumentality, but not in both at the same time. When a scientific statement is considered to belong to natural philosophy, it has the status of a description of the natural world. It may be possible to do something with it, but as it stands, it simply has to do with the world as it is. Conversely, when a scientific statement is seen as an expression of instrumentality, it is a report on how to do something, which can also be said to have implications for the way the

29 e.g. *la Théorie des Jardins* by Jean-Marie Morel (1776).

> world is. Scientific statements are inherently ambiguous in terms of the type of statement they claim, natural-philosophical or instrumental. (Dear 2006: 8)

7.4 Geometry and the decoupling of systems and problems

Leibniz's position makes clear the difficulty the early moderns had in properly conceiving the usefulness of systems (whether scientific or technical) for problem-solving.

On the one hand there is the systemless routine, peculiar and irregular, where judgement is enslaved to habit and custom. On the other hand, there is the perfect system as described by Leibniz, which guides judgement so well that the inquirer need only supply the facts and accept the sequence of syllogisms until the final resolution of the problem – an ideal framework, perhaps, but one that does not yet exist and whose very possibility is suspect. Between these two poles, the modern notions of method and system appear problematic: they must certainly guide inquiry, but they cannot do so infallibly in order to provide the solution to any problem: what can be the use of systems in this case?

The systematic spirit of d'Alembert is inspired by Newton's *Principia Mathematica i.e.* by a scientific method grounded in the mathematisation of observed phenomena. It is in mathematics, and in particular in geometry that the systematic spirit has its proper location: Wolff considers Euclid's *Elements* as the pinnacle and model of systematic thinking.

This section shows how geometry provided indeed the right template to think about systematicity, focusing on engineering. The role of geometry in the scientific method is considered in the next chapter.

In 7.4.1, the notion of *decoupling* is introduced to describe the strict separation between the elementary meanings that a system such as geometry provides and the actual resolution of problems. It is this attitude – to apply only pre-agreed and exact means in problem-solving – that gives rise to the modern figure of the *engineer*, itself based on a complex division of work in military fortification projects (7.4.2). The analysis of the underlying properties of Euclidean geometry explain its power to express an indefinite range of problems; they can be related to the very principles of computability, as recently expressed by Dershowitz and Gurevich (2008) (7.4.3).

7.4.1 The decoupling of systems and problems

The *Rules for the Direction of the Mind* give a distinct solution to this tension, because they are concerned with mathematical problems. In this realm, reflection moves within a set framework, which provides both the terms of the problem to be solved and those with which the solution may be composed. Geometry, as a system of knowledge, does nothing more, and indeed this is why the *Rules* must serve as heuristics to guide the mind in its search.

When they aim for exhaustiveness, the tables and dichotomies of reductions in art seek to solve *directly* the problems encountered: thus, for example, the "tables of battle quadrature which make it possible to know instantly, according to the numbers available, the best distribution of soldiers in space." (Dubourg Glatigny and Vérin 2008: 61) When, on the contrary, they merely list relevant considerations about the problems, they fail to pose them in such a way that they can be solved once for all. What reductions in art fail to do, then, is to create a language for *framing* problems that would be rich enough to adequately describe the variety and complexity of the situations to be faced, and yet small enough that their handling could be object of definite and universal rules.

In other words, the shortcoming of these methods is that they do not know how to *decouple* properly the systems of rules they construct from the problems that they help to solve. *Decoupling* means that the system itself is never intended to provide a solution to any problem. This does not, however, imply indifference and abstraction: it must provide the appropriate means of describing *all* problems and of solving them *demonstratively*, a formidable specification if ever there was one.

In the early modern era, geometry is the model of such a system. It is the domain of the certain and exact knowledge *par excellence*, by virtue of the fact that geometers *construct* their objects using the only instruments authorised by Euclid's postulates, the ruler and the compass. Bos (2001) has shown the centrality of this notion of *construction* in the research of geometers until at least 1650, concerning the additional *admissible* instruments for obtaining objects that resisted any construction by the ruler and compass, and the central role played by Descartes in this research.[30] The question facing geometers, according to Bos, was to determine which geometrical objects could be said to be *known* exactly, and the mark of this knowledge was not yet – as nowadays – to possess the equation of the figure, but to be able to construct it. The trisection of the angle, the squaring of the circle, the drawing of certain curves such as the spiral are examples of problems that could not be solved with ruler and compass, but with additional instruments, such as mobile compasses (Bos 2001: 6). *Exactness* is the term used by Bos to designate the goal of this research program the 17th century, *i.e.* to find the minimal construction means relevant for a given class of figures. Exactness, in this sense, has nothing to do with *precision*, as it does not admit more or less. It is a logical property regarding the conformation

30 See also Blåsjö 2022 on this subject.

to pre-agreed operative rules. It is the founding concept for decoupling systems of meanings from problem-solving.

The idea of grounding reductions in art on geometric principles appears quite early in the modern period. It appears, for example, surprisingly, in fencing manuals from the 16th century (Brioist 2008). It was undoubtedly a question, for the masters of swordsmanship (as it was for the masters of many other "mechanical" arts), of gaining recognition for fencing as a liberal art. A key factor may also have been the ambient neo-Platonism of the cultivated spheres of the time, which naturally promoted geometry as the model of knowledge. In any case, the path of mixed mathematics seemed more plausible concerning a bodily discipline than the arts of the *trivium* (logic, grammar and rhetoric), even if they were also called upon. But beyond this socio-cultural context, there was an important intuition in this research: geometry gave the idea of grounding an art on a uniform operative language for describing and solving problems, which also allowed their demonstrative validation. Thus, many fencing manuals from the 16th century are driven by the obsession to reduce gestures to sequences of elementary positions, the guards – an "alphabet" says one of the authors: the fencer's art would thus be reduced to knowing the laws of these sequences, relative to those of the opponent (Brioist 2008: 240–241).

It is, however, the building crafts that can naturally draw the most fruitful inspiration from geometry, and it is here that the decoupling of design and execution can be best seen, with the appearance in the 17th century of the central figure of the modern engineer.

7.4.2 The engineer and the quotation

According to Hélène Vérin, reduction in art of fortification, in the 16th century, consisted in

> proposing, not only useful prior knowledge, but also the means to go back from the concrete cases that are imposed on practitioners when they have to "invent" the fortification of a given place, to known and mastered types of solutions. (Vérin 1993: 137)

In particular, the problem needs to be properly posed by studying the natural strengths and weaknesses of the site in terms of defence relative to the offensive means that the enemy might have at their disposal. In addition, the typical forms of a fortification are described and discussed, as well as the "standard" elements of which it is made up (curtain walls, terrepleins, etc.) (Vérin 1993: 138–141). In terms of means of resolution, simulation is recommended (imagine an attack on the fortress), tables are provided showing the results of *tests*, *i.e.* experiments with specific effects (for example cannon fire on a wall, depending on different parameters), the optimal solution to certain specific problems is given (such as determining the optimal angle of a wall), and finally possible global solutions are illustrated by drawings and plans, which become real supports for reflection (Vérin 1993: 152–166).

It is therefore a matter of *guiding* problem-solving, by helping the inquirer to describe it carefully, in all its relevant dimensions, and to organise the space of solutions into predefined and partially optimised options that they would only have to assemble, according to the recommended heuristic procedures. While geometry does feature in these treatises, its role is both heuristic and instrumental.

The model developed by Galileo in his *Brief Introduction to Military Architecture*, characterised by "extreme abstraction [and] the pre-eminence of geometry" (Vérin 1993: 141) was to mark an important inflection. In the 17th century, the order of exposition of subjects was reversed and treatises increasingly began with expositions of geometry. It was no longer simply a matter of recalling practical notions concerning the art of measuring and determining proportions, but it was the engineer's art itself that was conceived as that of an initially geometrical construction, starting from certain principles, with demonstrable effects, which then had to be patiently accommodated to the accidents of the site, thanks to precise measurement and instruments (Vérin 1993: 167–178).

This development was to pave the way for the transformation of the engineering profession, which comes to occupy, at the end of the 17th century an "place unheard of, where no one had ever dared to put anyone", in the words of the *Encycopédie Méthodique* (Vérin 1993: 181).

The word *engignour* is attested as early as the Middle Ages. It refers to builders of *engines* for sieges, and they were therefore initially carpenters. The engineer became a foundryman when artillery appeared and then an architect when fortifications became crucial in the practice of war. Their title, derived from *ingenium*, connotes above all cunning and deviousness; the notion is insistently opposed to force. For the Arabs, the science of engineering was indeed ancient mechanics, which they interpreted as the *ruses of geometry*. The association with mathematics is therefore ancient (Vérin 1993: ch. 1 and 2).[31]

In the 17th century, the engineer is at the intersection of a "field of diverse forces": the engineer may be military, government clerk, entrepreneur, scientist, technician – and sometimes all of these at once. It is properly up to them to *untie* the knots between these varied utilitarian, economic and regalian interests by imposing a *design* that must satisfy all stakeholders. The emergence of the modern engineer is in particular inseparable, in France, from the emergence of the modern

31 See also Pautet 2018. On the interpretation of mechanics, see above 5.1.1.

state, which seeks to increase its economic and military power through reasoned policies of military, naval and civil construction (Vérin 1993: 184–188). The engineer, as a civil servant of the State, thus occupies a unique and new role: on the one hand, to develop a project that can convince different stakeholders of the optimality of the design in relation to varied and often conflicting criteria – defence potential, cost, practicality, etc. – On the other hand, to put contractors out to tender in a transparent and fair manner, to ensure that the best bid is selected, and to control the way in which the work is carried out. All this converges towards what the great military engineer Bélidor (1698 – 1761) describes as the "engineer's masterpiece", the *devis* or quotation, which describes the project, demonstrates its properties, details the possible stages of its construction, and estimates its cost (Vérin 1993: 227).

The quotation must meet all these objectives and convince all the parties involved of the possibility and optimality of the proposed project. It is, just as much as a plan, a demonstration, even an argument that must also persuade, or even seduce. In Vérin's words, the quotation asserts "a sovereign triumph of the intelligible over the sensible".

Geometry is central to the project. On reading the quotation,

> the geometric structure of the artifice, even if it was thought through at length after many measurements and mature deliberations, by reworked tests, taken up again on paper, until it was judiciously adapted to the premises, nonetheless appears [....] as the irruption of an essentially other space, [...] entirely designed which, superimposing itself on the memory of the place and the judgements made in accordance with it, in a way puts these judgements in abeyance. (Vérin 1993: 233)

The geometric inspiration does not stop there. Upstream of the design, mathematics is also used to describe the functional constraints of the problem to be solved – such as the proportionality between the defensive value of a stronghold and its building cost (Vérin 1993: 368–378). Downstream, they determine the details of the constructions, the materials to be used, according to their properties, as well as the order of construction:

> We have thus moved from dimensions to materials, from the general order to the detail of the particular order [...]; then from the order of all the parts to that of their construction. The latter applies a general method. Throughout, abstraction – geometrical, methodical – regulates under the term of "detail" the diversity of materials and conduits. [...] The establishment of the general level, the profiles of the coverings, their thicknesses, the depth of the foundations were decided after the engineer had probed the grounds and assessed the quality of the soil... (Vérin 1993: 234)

As Vérin notes, such an approach, which excludes the working contractors from any involvement in drawing up the specifications, and forces them to acquiesce to them *en bloc*, is not always conspicuous for its effectiveness. Beyond the historical particularities of the procedures for awarding these public contracts in the 17th and 18th centuries, however, an essential point is conquered here: the establishment of a common, if not standardised, language of construction operations, which enables the engineer to have *a priori* knowledge of the operations and figures that can be carried out, as well as their properties (costs, time, conditions):

> The generalisation of the company contract for the production of artifices required a different social organisation of useful knowledge, which corresponded to a different division of their acquisition and practice, [... and which made it possible] to have, on the one hand, informed workers, capable of reading and executing a quotation, and on the other hand, designers and project managers, responsible for drawing it up. [...] The special feature of contracts is that they are not simply about anticipating, but about agreeing to anticipate. And we only agree by using words, and, in the case of the contract, on a text. (Vérin 1993: 229)

According to Vérin, the contract is an important vector in the emergence of modern technology. Much more than reductions in art, the contract forces "the task to become an operation" (Dubourg Glatigny and Vérin 2008: 58), and forces these operations to standardise and acquire a uniform technical vocabulary. This standardisation does "two things". On the one hand it makes the engineer capable "of having the most difficult work carried out", so that "nothing will escape him" and he "will even go so far as to foresee accidents that might occur".[32] On the other hand, it allows the secret knowledge of the workshops to be revealed and disseminated – thus paving the way for the encyclopaedists.

The quotation, which is the specification of the work to be done in the terms of these standardised operations, is the focal point where a *decoupling* takes place between, on the one hand, the standardised socio-technical system that makes both certain problems and certain solutions possible, and on the other hand the place and time of its mobilisation for the particular elaboration of a problem and its solution. A solution to a problem is not systematic because it would be *given* by the system, like a recipe that would be recorded there, but because it is constructed *in the terms* of the system, which provides the means to verify its soundness.

The modern organisation of a project, as it is put in place around the engineer's quotation, consists in a triple *division of work*. First, there is the organisation of the work itself, made possible by the standardisation of operations mentioned above. While this may not yet be the systematic approach of industry in the division of work (see above section 3.3), it is now made possible in principle. But there is also *a second*, higher-level division of work, which takes place between the engineer and

32 I paraphrase here Vérin (1993: 229) quoting Bélidor.

the contractor, between the design phase and the execution phase of the project. Here there is a clear distinction between the two forms of rules provided by the technical system. On the one hand, there are the standardised rules of execution, which are their common language; they indicate which operations are feasible, at what cost, etc. In particular, they enable the engineer to define the work to be carried out, and to concentrate on the pure design of their work. But there are also, on the other hand, the rules of the art of good design, *i.e.* the precepts of the method that the engineer must follow to ensure the success of their project. The ambiguity that was perceived in the reductions in art regarding the function of the rules – guiding or constraining judgement – is here clearly resolved by the division of work: the design phase concentrates all the liberal dimension of art, the proper domain of the mind that can only be guided, and not constrained, by the rules. The execution phase concentrates its mechanical and servile dimension, which requires standardisation and precise direction. It should be noted that such a distinction is only relative, and that it can arise recursively: you can be a contractor on the one hand and a project owner on the other, because what appears to be a simple operation on the one hand may turn out to be a real problem to be solved on the other.

Finally, a *third* division of work appears, at an even higher level, between the scientist and the engineer. This time it is the inquiry itself that splits into problem-solving on the one hand and systematic inquiry on the other, one that uses the system, the other that elaborates it. It is precisely this division of labour that distinguishes science from practical activities according to Dewey (see above 7.2.4). The engineer is not concerned with the control of inference per se, as geometers are. But they can make great use of such controlled inferences to build strong fortifications.

These divisions of labour at multiple levels, firstly between contractors of different specialities, then between client, engineer and contractor, and finally between scientist and engineer, are articulated and intersect in complex ways within a project. An engineer manages a large number of contractors, may call on several areas of theoretical expertise, and often has to respond to several clients with different objectives. But there may also be several engineers on a project, some organised hierarchically, others specialising in a particular aspect of the project, others intervening at specific times, and so on. All of them, however, act according to a certain *order* of their particular preoccupations, "that of the treatise (teaching the discipline); that of the project (designing the work); that of the quotation (contracting with contractors); that of the execution (managing the work)" says Hélène Vérin (1993: 245), to which could be added at least that of use (the "user manual") and that of expenditure (financing). Modern technology is the interlocking and organisation of *various orderings*, in indefinite number, based on the three elementary schemes that have been described.

But why can geometry represent all the problems in its field, without first indexing them? And why is it possible to construct systematic solutions, but not systematically, *i.e.* by applying a universal recipe for analysing problems? What differentiates it, for example, from syllogistic systems where a systematic procedure like *pons asinorum* or Simon's GPS are possible?[33] And why is it more effective? Are there, moreover, other forms of systematicity even more effective? These are the questions that must now be addressed.

7.4.3 Systematic problem-solving

When one tries to organise a set of rules describing a situation, the methods that come to mind are that of division – a generic case is split into specific cases and then into increasingly particular cases – or that of combination: a finite set of descriptive characteristics of the situation is agreed upon from the outset, and all possible situations are then obtained by combination. As already discussed, these two methods were known to the Greeks themselves, and were used to organise reductions in art until the 17th century.

The problem with these methods is that they do not take account of the dynamics inherent in action or judgement. Nothing new can arise in the situations squared by arborescent or combinatorial methods, since they are finite in number – nothing can be done or judged that has not already been judged in advance. In other words, there can be no *real problems* to solve, only infallible recipes to apply. It is partly for this reason, it may be remembered, that Aristotle remained dubious about any possible systematisation of human action.

This discussion attempts to analyse and generalise what geometry teaches about the possibility of generative systems of rules (allowing an indefinite number of problems to be expressed). In order to do this, the nature of systematic problem-solving must be elucidated in the broadest possible sense.

The systematic mind, according to Kant, is a requirement of reason to examine the sets of rules conceived by the understanding and to organise them. It proceeds therefore from a reflective, second order process. The understanding only knows how to conceive rules, *i.e.* to find out what should be done or what should be judged in a given situation. It also knows how to apply them, *i.e.* to recognise them in the situations it encounters. But that is as far as it goes.

The thesis defended here is that the organisation of rules by reason, beyond the control of inference, allows for *systematic problem-solving.*

33 On the *pons asinorum*, see above 4.3.2 and on the GPS 2.1.2, and appendix A.

The systematic mind requires exactness and exhaustiveness. *Exactness* means that the inquiry progresses strictly through inferences allowed by rules (there is no "improvisation"), and that all the rules used have been defined beforehand – new ones cannot be defined in the course of the inquiry. This condition is not natural. It is when you are thinking about a problem that you generally realise the necessity and validity of a rule; you state it in passing. In geometry, appendix B gives the example of the rule that the sum of two adjacent angles in an interior line is equal to the angle formed by the two exterior lines (see above B.2.2).

The systematic mind is not satisfied with such casualness, because it knows that rules laid down in passing can conceal subtle errors. It demands that all rules be recorded separately and in advance, so that they may be the object of separate control, and that the inquiry be strictly limited to their use – this also implies that they be of finite number. Here is the root of the third division of labour identified around the figure of the engineer, who must *use* the theory *constructed* by the scientist. Euclidean geometry provides the near-perfect model for this division of labour, since all construction and reading operations are exactly defined.

The generativity of geometry (the fact that it makes it possible to describe an indefinite number of situations and problems) lies in the fact that the elementary construction operations it defines (marking points, drawing lines, drawing circles) can be *composed* with each other in order to generate increasingly rich figures: when one has two initial points, one can draw a segment connecting them and construct two circles; but one can then also mark the two intersections between these two circles, and also those that each of them has with the segment extended at its two ends, etc. What is more, any new points can always be marked on or off these figures, so geometry is never limited by a finite initial material.

Two complementary principles are at work here. Firstly, descriptions of situations and actions on situations are defined independently. Situations are defined by an indefinite number (however always finite) of elements of a finite and definite number of types (e.g. points, circles, segments). These elements have properties or relationships which describe the situation (magnitudes in the case of geometry: length, angle, area), and are themselves constrained by rules (e.g. all points on a circle are equidistant with the centre). Actions transform existing elements in a situation or construct new ones (only constructions are allowed in the case of geometry) to generate a new situation; they may be contingent to conditions in the initial situation. The proposition that actions and situations are defined independently means that, in principle, all defined actions can apply to any situation (if the required elements are available and conditions are met) and that any situation resulting from an action can still be described with the same types and relationships; nothing "unheard of" can happen. The actions transform one situation into another that remains available for the operators to apply again. I call this the *principle of se-*

quential composition in the sense of the composition of mathematical functions: an ordered sequence of actions composes a new situation from the sequence of unitary transformations operated by each, and there is no *a priori* bound to the length of this sequence. This principle, however basic, is not met by decision trees nor by combinatorial tables, which transformations can be applied a finite number of times (length of tree and dimensions, respectively). In trees, actions that go from one node to another are usually particular to these two nodes, and cannot be applied to others.

A second principle is that actions are local, *i.e.* their result does not depend on the initial configuration as a whole. An action rule applies only to a finite number of the characteristics of a situation, which it "filters out" so to speak in order to determine the action. This is very clear in the demonstrations of Euclid's theorems:[34] it is by considering only some of the elements of a figure that the inquirer is able to recognise a configuration for which they have a rule. I call this second principle *principle of locality*. Without it, there would be no real generativity, since the action of the operators would need to be defined on each possible configuration. In a tree diagram, it is the whole situation that determines the rule, from its most general to its most particular characteristics. Therefore such schemes are impractical for dealing with cases of polymorphism, where the same relevant difference appears in cases that are very far apart specifically or generically.

These two principles, taken together, explain the generativity of geometry, *i.e.* how a finite number of exact rules can give rise to indefinitely diverse situations. Each construction of a point, line or circle is an action that operates differently from one configuration to another, just as the same movement of a knight in chess, on the same squares, nevertheless has a meaning that varies according to the situation of the other pieces on the board, or just as adding +1 to 5 or to 55 obviously gives different results. This last example shows that these two principles also characterise the two rules that generate natural integers in indefinite numbers; composition is evident in the construction **succ(succ(...(succ(0)))))**; as for locality it can be seen in that **succ()** applies only to its immediate antecedent.

The notion of *exhaustiveness* must now be analysed, which together with *exactness* defines what I have called systematic problem-solving (see above 1.2). This idea is intuitive at first. As soon as a procedure to solve a problem is proposed, its scope must be validated, *i.e.* of the conditions under which it yields a correct solution. Special solutions may exist for particular cases or exceptions; in this case, all these variants are brought together by specifying their cases of application, and a more general scope for the solution is hereby obtained.

34 I develop this idea in appendix B.

Exhaustiveness is relative to a homogeneous class of problems for which a unique solution is expected. It depends therefore on the different meanings that these notions of *homogeneity* and *uniqueness* can take. In a way, geometry provides their model, since Euclidean space is homogeneous and any indicated procedure almost always applies in a uniform manner to all points in space. It may be that, for some constructions, existence conditions are necessary, for example for the intersection of two circles, but these can be included in the statement so as to safeguard the homogeneity of the procedure. On the other hand it may also sometimes be the case that the procedure differs depending on the starting configuration; for example Theorem I.35 of the *Elements* requires three slightly different demonstrations depending on the position of one point relative to two others in the statement. However these "topological accidents" are rare and are not at the heart of Euclidean geometry.

The problem of exhaustiveness therefore arises when there are relatively complex *cases* or *exceptions* to deal with in an otherwise homogeneous situation. If there are a finite number of such cases or exceptions A, B, C, D, E, etc., it is sufficient to enumerate them and specify the appropriate procedure for each. However, given the principle of locality, it is necessary to consider the case where several local exceptions are combined (for example AB, BD, etc.) and require a new rule, specific to their combination. Such complications can in turn be composed among themselves (D, CD, ABE, CDE, etc.) and the question then arises as to how far one should accept describing as "homogeneous" a situation folded multiple times into cases and subcases, and scarred by accidents on all sides with combinations that are themselves sporadic. The example of Deep Blue, the chess program, showed that the limits of exhaustiveness could be in principle pushed very far, in any case beyond those where the statement of such a rule remained relevant in practice for human persons (see above 2.3.1).

In order to clarify this, one may imagine even more complex cases of exceptions that would nevertheless be considered as acceptable: it may be, for example, that the rule to be applied depends on the repetition of n specific A marks in the situation, or on the fact that there are twice as many A marks as B marks, or that the number of A, B, and C marks are in an arithmetical relationship with each other, and so on. One may now see where this discussion is heading: the problem, before applying the right rule, is to *know how to recognise according to rules* which situation one is dealing with. This recognition must itself be exact if the exactness of the whole procedure must be preserved. It must therefore be described itself as a composition of operators defined exactly on a given space. In other words, it must be possible to traverse the space of possible cases, guided by a finite number of rules, for the procedure to remain systematic.

It is clear at this stage that this enrichment of the notion of exhaustiveness lies in the fact that the description of some parts of the situation depends on that of others. Until now, description has been seen, somewhat naively, as an immediate act. However, it often depends on other descriptions. For example, if you want to know which floor the person waving at you is standing on, you have to count them one by one by looking down (or up from the ground floor). In chess, knowing whether a square is under threat requires inspecting its diagonals and lines, to find out whether it is within range of an opponent piece. This property therefore depends on the properties of the surrounding squares. They may be directly observable (occupation or not by an opponent piece), but in other cases they may themselves depend on other properties.

Such properties are defined inductively, *i.e.* from other properties already known. The key point is that these properties are well defined, *i.e.* that one cannot fall into a circle while evaluating them, and that this inspection ends well at the end of a finite series of observations. This is what I call the *principle of inductive definition.*

Such a principle is essential because it provides the description of problematic situations with a rich logical structure, enabling properties to be modified in a chained manner following an action, for example as all the positions in a queue change when a new element is inserted in its midst. This principle, which enables the local and relative definition of the observable properties of a situation, is complementary to that of locality, which concerns the local and relative definition of actions.

Up to now, the discussion has been restricted to situations that can be described by a finite number of propositions. Such a restriction seems necessary (wouldn't an infinite vocabulary be required to describe infinite situations?) and sufficient, because a procedure that has to be terminated can only consider a finite subset. However, when it comes to designing a pure ratiocination that has to decide what to do in a situation where several options are possible, i.e. replace deliberation itself, then the procedure may need a space of indefinite size simply to explore the space of possibilities, like in chess, where many potential chessboard positions need to be explored.

In this case, the aim is to ensure that the description of the situation is completed in a finite amount of time. For example, on an infinite chessboard, the rule forbidding the king from putting himself in check would be unverifiable, as the case could arise of apparently empty rows or diagonals that would have to be explored *ad infinitum* to ensure that no enemy threat was present. Also, transformation rules which may apply simultaneously to an infinite number of elements should be excluded. These restrictions can be grouped under a fourth principle, which Dershowitz and Gurevich (2008) call the *principle of bounded exploration* of a transformation.

It is nothing else than the condition for the termination of an action stated by Knuth (see above 1.2).

The four principles listed here are indeed consistent with those presented by Nachum Dershowitz and Yuri Gurevich in the ambitious paper just quoted, which aims to give a generic framework to the notion of computability. My analysis reaches similar conclusions without using this notion. It suggests that *systematic ratiocinations*, that can be performed without recourse to judgement, are the true objects of computer science, and that *computation* and *algorithm* are only aliases for this concept.

An important remark here is that the possibility of representing the solution of any problem in a systematic way is not obvious. The game of chess shows that the relevant space of conditions may be too *entangled* for this to be practically feasible (see above 2.3.1). It is also conceivable that in some cases the full space of conditions cannot be enumerated *in advance*. The question whether these limits are only practical or essential is metaphysical. It is preferable to turn the question around and note that systematicity is not a necessary condition for problem-solving, but only one of its possible ways.

A "non-systematic" human inquirer is in fact content, most of the time, to compose a solution to their problem from the empirical material of possible conditions and operations that they know how to recognise in experience, and which they can marginally adapt to the present situation. Such an approach is doubly approximate, in that the rules they use are not necessarily extracted from a body of verified rules, and in that they cannot determine in advance the scope of its solution in relation to possible exceptions.

The key point here is the aim of exhaustiveness. As discussed, it is relative to the scope that one wishes to give to the solution of a problem; in chess, for example, it is not obvious that *the same* thinking procedure should solve problems in the middle of the game as well as those at the end of the game, those where the player is in a position of strength as well as those where they are in a position of weakness, those where they have their queen as well as those where they have lost it, and so on. The human player will tend, in all these scenarios, to think differently in each case. So the search for exhaustiveness can very well be limited to a fairly narrow definition of the problem at hand.

Conversely, it is possible, using machine-learning statistical methods, to find global solutions to extremely complex problems, such as the game of chess and the game of go, image recognition, and so on. Such problem-solving is not systematic in that the principle of locality is not respected: each operator (e.g. the weights of an individual neuron in a neural network) depends in principle on all the conditions of the situation.

In a way, then, undertaking to solve *systematically* a problem is a methodological choice, which is not the only one possible, even when computational procedures are mobilised. This choice consists in representing the problem by a space of conditions generated by a finite number of rules, which can be transformed by a finite number of locally operative rules; and in seeking a solution to this problem by the composition of the latter alone. The choice of systematicity offers clear advantages in terms of automation, control and traceability of solutions (see above 3.3), but it is important not to confuse it with the very concept of inquiry and procedure, which are much broader.

8 Machines and problem-solving

The historical survey of the previous chapter shows the difficulty of the very concept of systematic problem-solving. Through the *pons asinorum*, Aristotle has a first intuition of its possibility (see above 4.3.2). While, strictly speaking, this method requires only that it be provided with a problem space to explore exhaustively, it only becomes truly systematic when this space is itself derived or generated by a system. The *pons asinorum* can rigorously solve a problem, as long as it is posed correctly. It is a local guarantee of an outcome that depends on global demands made on its environment. Such demands seem to Aristotle incompatible with the contingencies of the human world. He reserves therefore systematic thought to the realm of scientific inquiry.

In the Renaissance, following the changes in the ideal of knowledge that have taken place since the Middle Ages, this systematic ambition is extended to all fields of knowledge. However, the *pons asinorum* gradually falls into oblivion in the 15th century, even among the Aristotelians themselves (Hamblin 1976; Thomas 1965). The usefulness of systems for the closest concept of problem-solving that the early moderns have, *invention*, is not perceived by La Ramée. In order to organise invention, he relies on the then common idea of a *Topica Universalis* (Schmidt-Biggemann 1983), more pragmatic because more circumstantial. Systems are conceived as *pedagogical, i.e.* as external devices designed to educate the mind to acquire an inner systematicity, in other words to make it capable of self-orientation in knowledge.

This ambition is confronted with a formal problem: because many fields of knowledge cannot obviously exhibit principles, it is a "cartographic" approach to knowledge that is favoured, of which tree diagrams are the most emblematic forms. Such an approach very quickly shows its limits. Trees and tables can serve to accumulate knowledge once it has been constituted, but do not generate it. They can give a solution that has been recorded beforehand; their sole value is to classify and map these solutions properly so that they may be searched efficiently, in the same way that a library must serve efficiently the books it keeps.

The fidelity to the spirit of the *pons asinorum*, and to the belief in the usefulness of systems for invention, is probably best found in the Lullist tradition, to which Alsted and Leibniz participate. For these representatives of what has been called *baroque rationality* (Knecht 1981), there must indeed exist a unique method of procedure for deriving all truths from a finite number of concepts and axioms, which together would constitute the *key* to universal knowledge. It is the search for these first principles that concentrates the efforts of these thinkers, rather than reflection on the articulation between the systems of knowledge and the procedures of discovery.

https://doi.org/10.1515/9783111494500-008

It is through mathematics, and in particular through geometry, that the appropriate model of the articulation between system and problem-solving is eventually found: geometry does not claim to give solutions to particular problems (except as paradigmatic examples) but to provide the concepts essential to the description and rigorous resolution of any problem in its field. In geometry, the function of the system is not to enumerate all possible problems by means of some combinatorial or tree structure, for which it would immediately provide the solution – but simply to construct an open space for the indeterminate formulation of any problem and for its efficient inquiry. This is no longer the arborescent scheme of the humanists which must branch out in the direction of all possible problems, nor the combinatorial scheme of the Baroque philosophers, which multiplies the tables of symbols at the risk of becoming an immense *tarot* card deck. In contrast, the geometrical model of systematicity favours the strict divisions of labour in problem-solving that inspire military engineering in the 17th century.

Geometry makes it possible to list instructions for the *construction* of objects and to *demonstrate* consequently their properties. In this it echoes the perfect knowledge by which God knows things *insofar as he created them.* It is the double relationship of construction / demonstration that is essential here. From the 17th century onwards, military engineering makes decisive use of geometry to implement a new organisation of construction projects, involving the standardisation of their operations and the introduction of a threefold division of work, firstly between trades, then between project owner, engineer and contractor, and finally between scientist and engineer. These divisions of labour clearly reveal two decouplings that are essential to modern technology: firstly, between the design and execution of the work, and secondly, between systematic (global) inquiry and (local) problem-solving.

The purpose of this final chapter is to understand how these two decouplings were made to work beyond the limited domain of the building crafts. How is it possible that systematic problem-solving applies to generic technical questions and to human affairs? Buildings are static and structural artefacts which design may obviously benefit from geometry. It is not obvious that problems of transport, production, health, law, business management could also be tackled within generative mathematical systems.

The general thesis of the chapter is that *machines* are the handles through which systematic problem-solving gets its leverage in the material world and in the world in general. What is meant here is not merely practical, *i.e.* that machines are agents perfectly suited to the execution of effective procedures. It is not obvious how this could be possible at all. If procedures are summaries of successful inquiries, how could machines ever come to *understand* such sequences of meanings and have the capabilities to act upon them?

Answering that machines are designed and built in order to have this property, of obeying rules, only pushes the question further: how is such a design possible? One could answer that the design leverages the natural laws to which the material parts of the machine conform, and in particular the laws of mechanics. For example, the lever machine uses the law of mechanical moment to multiply a force, the mill concentrates a force at a point of application, and so on.

However, in these machines there are no other rules than those for the transmission of forces, and nothing resembling rule-following by humans. It is thus not at all obvious that procedures *in general*, established by reasonable inquiry, could thus be transferred to a material object.

Section 8.1 explores the origins of the modern meanings of *machine* in the 17th century, as the geometric object of movement, and as the primary conceptual scheme of scientific explanation; it argues that the core template of mechanical explanations is still very much constitutive of contemporary science. Section 8.2 goes further to characterize machines as inferential devices regarding the dynamic phenomena of experience: mechanisms are nothing else than procedures for reasoning well about an object presenting some regular behaviour. Section 8.3 explains how these networks of inferences can be leveraged to intervene on reality and to control it: the specification of a machine to be built is a logical contract between a principal who would like to change a given situation and a designer who must deliver a machine to support this goal. The specification is the place where both agree on the articulation between the problematic situation and the machine that will deliver or enable the required changes. Finally, section 8.4 reflects on this interpretation and identifies the *coherence of intentions* as a core principle driving the progress of practical rationality.

8.1 Geometric machines

The metaphorical association of machines with reasoning is ancient, and probably stems from the fascination that always captured the public at the sight of *automata*. It may have roots in the medieval tradition of the *Timaeus*. Ficino, one of the first to associate the old expression *machina mundi* with an automaton, expresses thus his fascination with a festive machine demonstrated in Florence in 1475:

> We recently saw in Florence a small cabinet made by a German craftsman in which statues of different animals were all connected to a single ball and kept in balance by it. When the ball moved, they moved too, but in different ways: some ran to the right, others to the left, upwards or downwards, some sat up, others stood down, some crowned others, and the latter injured others. There was also the sound of trumpets and horns and the singing of birds; and other things happened there simultaneously and a host of similar events took place, and all

> this simply from the movement of a ball. So God by his own being, which is really the same as his understanding and will, or which is something entirely simple – the universal centre from which (as we have stated elsewhere) the rest of things proceed like lines – has only to nod and all that depends on him trembles. (*Platonic theology*, 2001: 201)

The practical reason of God by which he knows his creation, according to Thomas (see 5.1.1), is here identified as the principle of the mechanism. This metaphor develops in the Renaissance, but its ties to theological considerations loosen gradually. Later, in the Enlightenment, Diderot's uses it to express his fascination with a manufacturing machine:

> The stocking loom is one of the most complicated and consequential machines we have: it can be seen as a single piece of reasoning, of which the manufacture of the work is the conclusion; there is also such a great dependence between its parts that to cut out a single one, or to alter the form of those that are deemed less important, is to harm the whole mechanism. (article "Bas", E 2: 98a-113b.)

Since the apparition of computers, the image of "electronic brains" has become so pervasive that it barely causes a stir, except at times of artificial intelligence innovation.

This section explores the basis of this analogy, which is much more than a metaphor. In the 1930s, mechanism was set up as a norm for the very definition of an effective procedure (8.1.1), as a ratiocination that forbids any judgement and initiative to the agent. *This* idea can be traced back to Galileo and the emergence of mechanics as the geometry of motion. It is thus in the very model of systematic inquiry, geometry, that the connection between machine and ratiocination can be elucidated (8.1.2). Machines had the core function, in the early modern scientific enterprise, to provide the template for the explanation of dynamic phenomena (8.1.3). This epistemic privilege is related to their *constructible* character, in the geometric sense. Such a schema still remains active in current scientific explanation. The decisive shift is that, while the explanation is no longer required to *construct* the machines underlying the phenomenon in question, it has now to specify which laws, already established by the theory, regulate their behaviour. The object of explanation is no longer a machine, but a *system* whose behaviour, governed by theoretical laws, is sufficient to account for the phenomena. This shift in the ideal of explanation, from the *constructibility* to the *instrumentality* of the entities posited by scientific inquiry, is not a rupture that separates pre- and post-Newtonians, but on the contrary the maturation of the same project of knowledge (8.1.4).

8.1.1 What is a "mechanical" behaviour?

In the 1930s, solving Hilbert's famous *Entscheidungsproblem* ("to know a procedure that allows us to decide, for each formula presented, for which domains of individuals it is universally valid (or can be satisfied) and for which it is not" (Hilbert and Ackermann 1972: 91)) required to agree on exact criteria that this procedure should meet. Very rapidly, references to the "mechanical execution" of such a procedure crippled in, as mysterious as they were recurrent in the fundamental texts of computability theory. Turing's famous analysis presents machinery, strictly speaking, as a metaphor: "A man computing a real number can be compared to a machine that is only capable of a finite number of conditions..." (Turing 1936: 231). It is in particular Gödel, after his *Gibbs Lecture* of 1951 who insists on the importance of Turing's *mechanistic* definition of the notion of effective procedure:

> The most satisfactory way... [to arrive at such a definition] is to reduce the concept of a finite procedure to that of a machine with a finite number of parts, as the British mathematician Turing did. (Gödel [1951] 1995a: 304)[1]

At the 1946 Princeton conference he organised on the subject, Tarski had formulated the decision problem as follows:

> Materially, we can justify the choice of the decision problem by the fact that the task of logic is to mechanise thought [...] Intuitively, the solution of the decision problem consists in determining whether there is a mechanical way of deciding whether a given statement of a formal system is a theorem.[2]

Yet it is indeed in Gödel that we find one of the first occurrences of this idea, in a text from 1933:

> [The rules of inference [...] refer only to the outward structure of the formulas, not to their meaning, so that they could be applied *by someone who knew nothing about mathematics, or by a machine*. This has the consequence that there can never be any doubt as to what cases the rules of inference apply to, and thus the highest possible degree of exactness is obtained. (Gödel [1933] 1995b: 45, my italics)

In this seminal text, the image of the exteriority of rules to the content of symbolic formulas leads to that of the machine, as of a device *closed* to all understanding. This reasoning is fully explained by Minsky, which text we quote in its entirety:

---

1 Cited by Sieg (2006: 196). I rely on the latter article regarding this historical remark on Gödel.

2 Text published by Bénis-Sinaceur (2000: 23-24). Tarsksi also uses the concept of machine in a 1939 text, also cited by Bénis-Sinaceur (2000: 8).

> We will begin by saying that an effective procedure is a set of rules that say, from moment to moment, precisely how to behave.
> This attempted definition is subject to the criticism that the interpretation of rules is left to a person or agent. However, a person's ability to obey instructions depends on their background and intelligence. If their intelligence is too low, they may not understand what we mean. If their intelligence is too high or too foreign, they may invent a coherent interpretation of the rules that was not intended. We know how often a seemingly "simple" explanation turns out to carry unsuspected ambiguity.
> We could avoid the problems of interpreting understanding if we could specify, along with the statement of the rules, the details of the *mechanism* that is to interpret them. That would leave no ambiguity. Of course, it would be very tedious to have to do all this again for each individual procedure; it is desirable to find a reasonably uniform family of mechanisms that obey the rules. The most practical situation would be one in which we set up
> (1) a language in which groups of behavioural rules are expressed, and
> (2) *a single machine* which could interpret instructions in that language and so execute the steps of each specified process (Minsky 1967: 106, my italics)

In this remarkable text, it is the problem of interpretation and interpersonal communication that prompts the appeal to mechanism and machine so as to remove any ambiguity, like the Leibnizian *"Calculemus!"*. But whereas the Leibnizian calculation had to be based on a language and characters that immediately conveyed the primitive meanings of things, here, on the contrary, it is by emptying the symbols of their meaning, and the agents of their ability to understand, that communication is ensured.

What all these texts have in common is that they present the machine as a kind of ideal limit to the human execution of rules, that expresses the certainty that the agent only obeys rules set up by the designer of the procedure, devoid of any ambiguity, without exhibiting any other form of intelligence or intuition. These texts must be interpreted in the light of their Hilbertian inspiration, in the words of Bénis-Sinaceur (2000: 7), *i.e.* of the project to define the criteria that mathematical inferences must satisfy, so that they make as little appeal as possible to the mathematician's intelligence or intuition – since it was this kind of justifications that sparked disagreements concerning the question of infinity. Hilbert takes calculation as a paradigm of exact rules; he therefore proposes as a criterion that inference be defined by finitary symbolic rules, *i.e.* rules that proceed only by recognising and substituting symbols from one proposition to another, in a finite number of steps.

The machine can therefore automate human judgement only because the designed procedure has already suppressed it entirely. In the framework of Dewey's theory of knowledge, this means that the procedure, as reconstructed by the final resolution of the problem, makes it possible to *do without* the inquiry entirely. This is only possible for purely procedural (mathematical) inquiries established on pre-established systems of meanings. It is easy to see what is intuitive about such a possibility: when seeking a solution *systematically, i.e.* by traversing a problem space according to pre-established rules, it is crucial "not to skip any steps" and not to let intuition take over.

At the end of the text quoted above, Minsky adds an interesting remark concerning the particularity of machines as *artefacts*. Their mechanism is not a *description* but a *specification* of their operation. Unlike a theory, if there is disagreement between the representation of the mechanism and the machine, it is the machine that must be replaced. Minsky also adds – and therein lies the important thought:

> How can we be sure, in any particular case, that there is any way at all to build the machine we have in mind? How do we know that the physical world will permit it? In some particular cases, no reasonable person will have any doubts. Can I really build an adding machine? When we look at the structure of a simple adding machine we see that it should work; it cannot help working – this is as clear as it is that 2 + 2 = 4.
> The cliché is illuminating. Some people believe that simple mathematical statements are "self-evident". Other people maintain that they are based on rather obscure but still empirical observations, Either view may be held for machines. That a conventional adding machine will work (if properly constructed) seems on the one hand equivalent to the arithmetic propositions involved, and on the other hand seems derived from our empirical experience about the ways in which objects behave in the world. *Perhaps one could even maintain the view that belief in an arithmetic statement is equivalent to the belief that certain machines, if properly built, will work.* (Minsky 1967: 5, italics in the original)

The question Minsky poses at the beginning of this text is the very one of this inquiry: how is it possible to implement "mental" procedures in "concrete" devices? He suggests, with great subtlety, that the answer might lie in the *superposition* of two regimes of argumentation: an empirical regime, which must ensure that the concrete machine meets the specification, and a logical-mathematical regime, which is nothing more than a reformulation of mathematical theorems. A machine works *because it is constructed* to work that way, in the same way that in constructive logic a judgement is entirely constituted by the proof that allows its assertion. But on the other hand, all descriptions of the machine's constituent parts, from which it is constructed, are *empirically assumed* to be valid.

The superposition of the empirical and mathematical arguments is suggested in the last sentence of the text. The equivalence of the two beliefs is a double implication: on the one hand, there is the "platonist" implication, that the machine works because the mathematical statement is correct, but on the other hand also the opposite implication, that this statement is correct *because the machine, correctly constructed, works*. Any reflective judgement about a phenomenal reality is both the construction of a "logical machinery", so to speak, and the assertion that it applies to that reality. This machinery is nothing more than the indication of a procedure for acting (empirically) on this reality, and the empirical promise that this action will lead to the result predicted by the procedure.

8.1.2 Machines and judgements

The idea that mechanisms work because every parcel of judgement has been exhausted from the procedure in advance is actually quite old. Dubourg Glatigny and Vérin argue that the machine is already the "ideal" of reductions in art:

> [According to] the remark Galileo makes at the beginning of his *Mechanics*, contrary to what the vulgar believes, the machine does not save force but judgement. [...] Mechanical devices avoid all the trial and error, reflection, judgements and decisions involved in the gestures to be made and their succession. The machine is [...] the ideal model of an action that reduces to a minimum the obligation to make choices. (Dubourg Glatigny and Vérin 2008: 62)

Galileo's text referred to by the authors reads:

> The third utility of Machines is very great, because great expense and cost are avoided by using an inanimate force, or one without reason, which does the same things as the force of men animated, and driven by judgement, as happens when mills are made to grind with the water of ponds, or rivers, or with a horse which supplements the strength of 5 or 6 men. And because the horse has great strength, and lacks speech, the necessary reasoning is supplanted by means of wheels and other Machines which are shaken by the strength of the horse, and which fill and carry the vessel from one place to another, and empty it according to the design of the engineer. (Galilei [1634] 1966: 25)

The judgement that the machine saves is that of the workman, who is assigned to menial tasks. This idea points to the division of labour between design and execution that will occur later on in the 17th century (see above 7.4). The residual judgement found in the execution is transferred to the design, *i.e.* to the judgement of the supervisor who has to build the machine or configure it for the work to be done.

What allows Galileo to make such a statement is a new conception of mechanics that defines it as a species of geometry, the geometry of movement. It stems from the rediscovery, in the middle of the 16th century, of the treatises by Archimedes and Heron of Alexandria, both mathematicians and engineers, and the treatise by pseudo-Aristotle. The early modern engineers, such as Leonardo, Brunelleschi, Galileo, revived this tradition and "sought in the works of Euclid, Archimedes, Vitruvius or Heron of Alexandria an answer to their questions" (Rossi 1996: 19, 34–42). A profound reflection begins concerning the status of motion studied by mechanics: is it "against nature", as the Aristotelian doctrine maintained? or is it natural itself, since it can be explained by mathematical principles? The formation of a consensus in favour of the second hypothesis, thanks in particular to Galileo, enables mechanics to become a legitimate explanation of natural movements. The rejection of final causes and substantial forms puts mechanics at the centre of scientific inquiries into natural phenomena.[3]

Just as geometry is the mathematics that constructs its objects and demonstrates their properties, likewise mechanics is the geometry of machines (see below, next section). The properties of the machine are its movements, which carry out its function, just as the rotation of the millstone performs the grinding of the wheat. Problem-solving here is rigorous because it becomes possible to *demonstrate* that the requested function is accomplished by the structure presented, and to *demonstrate* how to construct it. It can be shown that the rotation of the millstone follows that of the gears driven by the wings, provided the mill is built in a certain way. The construction of machines, in classical mechanics, is conceived as composition of simple machines, which are postulated just as points, lines and circles are in plane geometry. The classical list of simple machines in the Renaissance is the pulley, screw, wheel and axle, inclined plane, wedge, and lever – simple geometric figures whose kinematic properties are considered basic. Descartes himself writes in a letter describing the properties of simple machines:

> And it would be useful for those who meddle with inventing new machines, if they knew nothing more of this matter than what little I have just written; for they would not be in danger of making a mistake in their account, as they often do by supposing other principles. (*Letter to Huyghens, 5 October 1637*, AT 1: 447)

The constructibility and function of machines are therefore not empirical matter, based for example on certain strength of materials, certain chemical properties, and so on. It is an *a priori* constructibility which allows the demonstration of its properties, *i.e.* of its movement and function, because its constitutive principles are entirely clear to the mind.

In this conception of mechanics one can already see the fundamental reason of the applicability of *systematic* problem-solving to practical and technical problems. Machines mediate and articulate the relationship between two moments of planned human action, *use* on the one hand (or more generally the execution of a plan), and *construction* on the other (or more generally *planning*) – The new idea here, which would have been preposterous for Aristotle, is that there exists "geometric"

3 See on this subject Roux and Festa 2001. On the place of mechanics in the scientific revolution in the early modern period, see Dear 2019; Dijksterhuis 1986; Garber and Roux 2013; Koyré 1961; Palmerino et al. 2004; Rossi 1996; Westfall and Torchin 1977.

relationships between these two series of phenomena, which make it possible to pass from one to the other by means of exact and exhaustive reasoning.

This thesis may be found attractive and suggestive but unconvincing. Firstly, it is known that the pre-Newtonian conception of mechanics proved largely inadequate to build machines beyond simple mechanics and optics, as it didn't account for forces and empirical factors such as friction. Secondly, even if it worked, its scope would be limited to procedures reducible to spatial movements, and would miss all the others. A more generic explanation is required.

8.1.3 The mechanistic interpretation

To meet this challenge, it is useful to examine the way scientific explanation evolved from the "mechanical philosophy" of the 17th century to Newtonian science. Machines and mechanisms apparently lost their role as privileged explanatory models against natural laws expressed in equations, but they did not disappear in this transformation. The focus shifted from providing mere analogies to actually studying the behaviour and interactions of elementary mechanisms (e.g. moving particles) under given conditions. By becoming invisible, they revealed their true nature as inferential machines.

If artefacts are defined on the one hand by their function and on the other by their structure (Kroes 2012), then machines are artefacts whose structure is dynamic and "sufficiently composed" (Roux 2013: 71). In Roux's view, it is precisely this duality that, for many philosophers of the early modern period, makes the machine, much more than a metaphor, a model for the intelligibility of phenomena, giving rise to what she calls "machinic explanations". Descartes, of course, is one of the great originators of this kind of explanation: in his eponymous treatises, he reconstructs the world and the human being as machines operating solely by "size, figure and motion". As Wagner (1998: 152) says, for Descartes the machine is "an explanatory model [...] that expresses an ideal of intelligibility of the phenomena of nature", notably in opposition to the occult qualities of scholastic physics, in that its explanation is always clearly reduced to geometry. Séris remarks that

> the (artificial) machine has no privilege over the (natural) machine. It is rather the opposite, it is *less* machine than the living being, for example; and it is inserted into a "mechanistic" world where everything is already machine. (Séris 1994b: 153)

This type of explanation has been much criticised since Newton.[4] At best, they allow the construction of a *model* of the phenomenon to be explained, which can be solved mathematically and verified in experiment – such as Descartes's use of the bounce of a ball to model the optical laws of reflection and refraction. In most cases, however, the explanations merely propose analogies between the phenomenon to be explained and structural relationships observed in everyday life – which themselves, however, remain unexplained. For example, Boyle compares the "spring of air" (by which he means the force of expansion that air exerts after being compressed) to that of a ball of wool, made up of little hairs each of which has a certain elasticity. One can speak here of a "sketch of a mechanism", because its modelling according to known laws of matter and motion is still a long way off. Finally, in the worst case, one remains stranded in the realm of abstract metaphors that explain nothing, or of preposterous assumptions of figures or properties that produce the very phenomenon under investigation. For example, the sense of smell is explained by the specific shape of the olfactory molecules, which can only be related to those of the olfactory organ. Historian of science Alan Gabbey comments these explanatory excesses in this way:

> In a sense, the explanatory difficulties in the case of sensations would seem to be of the mechanists' own making. They tried to explain everything, which was too much by a long chalk, and their critique of traditional modes of explanation failed to alert them to the dangers of trying to replace Peripateticism *en bloc*. The physics of forms and qualities was tailor-made to explain everything: the mechanical philosophy was not. Explanatory effectiveness was to lie not in comprehensiveness, but in a selective delimitation of the problem domain […] The ontology and explanatory ideals of the mechanists were new, but very often the "problem domain" was largely Peripatetic. (Gabbey 1984: 14)

More positively, one may bring the *intention* of early modern philosophers closer to a legitimate mode of contemporary scientific explanation, which for a time was called *structural*,[5] before the term *mechanistic* took over again, with the emergence of the "new mechanists" movement in the early 2000s, which aims to give right to the dynamic structural explanation of phenomena. Unlike the "machinic" explanations of modern philosophers, however, the constituent parts of the explanatory mechanism as well as their causal relationships are not worked out by analogy or metaphor: they are based on scientific theories already established. Thus, the explanation of a certain behaviour of a cell by certain actions of the mitochondria, the

4 See a summary of critical positions on this subject in Roux 2017: 31–33. I follow here the classification of mechanistic explanations in Gabbey 2001 and the detailed case studies in Roux 2013. Examples are taken from the first article.

5 McMullin 1978; this paper is discussed below 8.1.4.

nucleus, the Golgi apparatus, etc. is not a matter of analogy, or even of modelling: it is the resolution of a global phenomenon into elementary behavioural components which are themselves already established or observed. When hypotheses have to be made about an element or action that is as yet poorly understood, it must be possible to explore and falsify them through experimentation.

From the point of view of the *intentions* of modern philosophers, this is a difference of relevance more than of essence. Today's scientists are simply much better equipped with explanatory instruments than those of four hundred years ago. Sophie Roux highlights two essential points in common between early modern and contemporary mechanistic explanations. Firstly, these explanations operate at "mid-scale": describing a machine or mechanism does not require resolving it into its most elementary components and movements. The ones that the explanation manipulates can be taken as stepping stones in the analysis of the phenomenon towards its reduction to elementary laws of nature. As it was just said, the new mechanists admit the idea of hypotheses and therefore of *sketches of mechanisms* – just like some of the best early modern scientific explanations.

The second common feature of these explanations is that they are primarily addressed to the *understanding* of the inquirer. They aim to reveal the internal structure of the phenomenon in question, as well as the causal relationships between the parts of this structure. This unveiling does not naturally involve propositions and demonstrative syllogisms, but rather diagrams that enable to grasp them more intuitively, in particular through the mental simulation of how they work. According to Sophie Roux (2017: 36–41), Descartes may have been inspired in his physical treatises by the illustrated explanations that could be found in theatres of machines.[6] He was the first to include diagrams in works of natural philosophy.

This interpretation of the role of diagrams in the mechanical philosophy echoes Koyré's interpretation of the role of machines in the birth of the modern scientific spirit, as *invitations* to apply the mathematical method to dynamic phenomena:

> [This] possibility finds its expression and guarantee in the very fact that the act of intelligence which, in breaking down and recomposing a machine, *understands* its arrangement, as well as the structure and operation of its multiple workings, is exactly analogous to that by which, in breaking down an equation into its factors, it understands its structure and composition. (Koyré 1961: 316)

Machines invite to an *active* understanding. Their functioning must be put in relation with their structure, and this structure with its process of construction. In other words, it is the possibility of this *active knowledge* of natural phenomena that

6 On theatres of machines (or catalog books of machines), see below 8.2.1.

is central in mechanical philosophy. It is not in essence limited to geometry; but geometry is its first ideal model, because of the exact (demonstrable) correspondence it afforded, between principles of design and observable properties. Geometric machines are transparent objects, in the sense that their function is a strict consequence of their movements, and these movements are in turn strict consequences of their design.

8.1.4 Scientific problem-solving

The transformations brought by Newtonian science must be put in the perspective outlined in the former discussion. While Newton's famous *Hypotheses Non Fingo* ("I feign no hypotheses"), criticized quite harshly the "machinic" explanations of mechanical philosophers, particularly Cartesian ones, it vindicated and, in many ways, brought to reality their aim at active knowledge, in the sense just outlined.[7]

What Newton in particular does not hypothesise about is the *cause* of universal gravitation, which implies that it must be seen as a *fact* or *phenomenon* that can be measured and for which a mathematical theory may be established inductively, without this theory saying anything about the reasons for this behaviour.

This position has long been interpreted in the direction of logical positivism. For Ernst Mach ([1883] 1919: 193, 246), Newton would have understood that the real object of science was to give a mathematical description of the phenomenal regularities of nature, where the metaphysical concept of *cause* had no place.

In such a vision, which makes the establishment and application of the laws of nature the essential content of science, any machinic explanation may seem conjectural or at best metaphorical, serving perhaps only to set the inquiry on the right track towards a good formulation or application of these laws.

Against such a view, McMullin (1978), in a famous article that anticipates the "new mechanists" movement in philosophy of science[8] contrasts what he calls *structural explanation* with the *nomothetic explanation* promoted by logical empiricists, and declares the former preferable to the latter. Thus, explaining the elongation of an iron bar under the effect of heat is more satisfactory when one exhibits the behaviour of its molecular structure than when one merely evokes the law of expansion of metals. This empirical law can only be established on the basis of reasoning of the

7 This epistemological shift has already been the subject of numerous analyses and interpretations. I have consulted in particular: Brading 1999; Cohen 1969; Gabbey 1992, 1996; Levitin 2021; McMullin 2001; Walsh 2012.

8 On the new mechanists, see notably Bechtel 2008; Levy 2013; Machamer, Darden and Craver 2000; Piccinini 2015.

first type. McMullin (2001) thus challenges the positivist interpretation of Newton and attempts to show that his physics does offer structural explanations.

This opposition seems forced. The two forms of explanation, nomothetic and structural, are compatible and complementary, although they are situated at different levels of discourse. A problem can be solved simply by appealing to the law of metal expansion, if it has already been established in a body of theory. But this means that it has been derived from other more elementary laws, for example concerning the effects of increasing the kinetic energy of a rigid structure and, as McMullin says, such a derivation requires a structural explanation. All problem-solving is partly structural, and this is its constructive part, and partly nomothetic, and this is its descriptive part, since the relations and objects underlying this structural interaction must be well described.

An explanation may seem merely nomothetic when the explanatory problem posed is very close to a typical question that has already been incorporated into the theoretical corpus for reasons of convenience or illustration. This is the case with the law of expansion of metals, a common and important phenomenon. For example, if one asks why a bar of iron expands when it is heated, all one has to do is recognise that this is a question that has already been answered. However, this procedure can only provide a satisfactory explanation because we already possess a derivation of this law from the fundamental laws of science. As long as it was only an empirical law that could not yet be integrated into the systematic corpus of science, the explanations that drew on it were only provisional.

The opposition between structural and nomothetic explanations thus originates in that between two forms of inquiry, the resolution of particular problems and the construction of a systematic theoretical corpus. Problem-solving makes it possible to establish new scientific laws that can be added to theories, but these theories in turn form the basis of all problem-solving.

Newton's *hypotheses non fingo* forbids structural explanations only when they cannot be warranted by experience,[9] which is the case of gravity, and this is why he poses it as an elementary principle. But then he develops structural explanations throughout the rest of his work. The law of universal attraction was considered admirable *because* of the number of such explanations that could be derived from it, concerning the speed of falling bodies, the tides, the motion of the planets, and so on.

Therefore – and this is the main point – machines and their construction are by no means banished from Newtonian physics; it is only from the "machinery" of the elementary laws of nature that nothing is to be conjectured. In a way, this

9 I owe this remark to Sophie Roux.

remark merely repeats the Aristotelian assertion that the principles of a science cannot be demonstrated: for if one could give a structural explanation of these laws, then they would no longer be elementary. This is in fact what Einstein did when he explained the law of gravitation as a simple effect of the law of inertia in a space curved by bodily masses. But when Newton explains the phenomena of the tides, then indeed Earth, Moon, and bodies of water do behave like machines, but it is no longer required to understand *why* this is so. It is enough to show that their interaction produces the phenomenon to be explained.

Apart from this difference, scientific explanation still proceeds by highlighting global structural behaviour from the laws of behaviour of its parts. The demonstration may involve mechanistic explanations, which are in the majority in the life sciences, where the "machines" are the organisms, tissues, cells or their components, both objects and means of the explanation. In the physical sciences the range of explanations is wider. For example, it often involves statistical considerations. Without going into complex epistemological considerations on this subject, it can be pointed out that here again the idea of an interlocking of structural behaviours, which can be related according to precise rules, remains central. It is through such explanations that the various descriptive levels of phenomena proposed by (for example) chemistry, thermodynamics, atomic physics and nuclear physics can be linked together, and that their respective laws can hope to be integrated (at least ideally) into a unified system of knowledge.

This elementary fact of scientific explanation is in direct continuity with the "machinic explanations" attempted by the early moderns. In order to *apply* a scientific law, one must clearly circumscribe the objects of its application and therefore distinguish a *system of interest* from an environment with which it interacts. If, for example, one wants to apply the law of conservation of energy or the Newtonian second law of dynamics, one must identify the bodies whose energy, mass, force balance and acceleration one is going to calculate, in order to highlight the phenomenon targeted by the explanation. *System*, here, is a methodological term used by scientists to designate *what they are talking about* and whose effects they want to demonstrate. It will be argued in the next discussion that *machine* or *mechanism* would be equally legitimate terms.

The essential difference lies elsewhere. Whereas early moderns often make do with analogies drawn from everyday life to describe their mechanisms, scientific explanation since Newton requires one to explicit which precise law, proven or provisional, is *used* to describe the behaviour of the system of interest or its components. The focus now is on the structural design of the system of interest (e.g. the relative distance of two masses connected by a rod), from which kinematic properties can be derived by applying universal laws at each of its components. It is, in a certain sense, much more faithful to the spirit of the mechanical philosophy than the early machinic analogies, as it keeps the discipline of rigorously composing and decomposing the machine to analyse into elementary ones whose laws are well known.

This is the possible source of the instrumental flavour that has often been associated with the Newtonian way of doing science. If the explanation of a phenomenon is given by rules describing directly the interactive behaviour of underlying machines without the need to describe these machines themselves, then the same rule may also be used for other purposes, such as *predicting* the phenomenon in question, and in some cases technically taking *control* of it to put it at the service of our purposes – since a different arrangement of the machines underlying the phenomenon will generate new interactions between them, the consequences of which can be derived by the same rule. As Peter Dear says:

> The universe according to a natural philosophy like Newton's, which consisted of lifeless matter bouncing around according to mechanical laws, was clearly exploitable for human ends: it lent itself to instrumentalisation. (Dear 2006: 10)

Instrumentality is in fact merely the mirror image of constructibility, in that it results directly from the division of work that the machine makes possible, between its construction and its use.

8.2 Inference machines

This section proposes a general interpretation of machines as dynamic objects of experience which interactions with their environment seems regulated (8.2.1). This definition exposes the nature of the concept of *machine*, that it is a principle of intelligibility of experience, *i.e.* a category (8.2.2). A Kantian interpretation of this thesis is attempted (8.2.3).

8.2.1 Machine as regular dynamic objects

The difficulty to define the term *machine* is not recent: Marx ([1867] 1976: 492) already noted it, and Séris (1994b: 167–172) remarks more generally that the history of machines lends itself to constant reconceptualisations. When confronted with machines, there is an irresistible tendency, he says, to define their scope and meaning in terms of their most recent developments, as if technical progress always consisted, beyond the apparent increase in their complexity, in better revealing their simple essence, from which the meaning of their successive sketches would be illuminated in retrospect.

A historiographical study by Popplow (2007) of the term *machina* shows that it was used almost exclusively in Latin antiquity to designate either the simple machines (the wedge, the lever, the inclined plane, etc.) studied by mechanics as a science, or scaffolding and war machines, *i.e.* structures composed in a complex way that connoted stability. It was not until the beginning of the 16th century that its meaning was enriched to also refer to motion-transmitting machines (or *cinematics* as Sophie Roux (2013: 8) calls them), such as mills and automata, principally clocks (Popplow 2007: 65–68). All these objects were in fact invented or extensively perfected and disseminated during the Middle Ages, and it was initially through similarity to the admiration that Vitruvius found in ancient machines that the engineers of the early modern period gave them the same name.

These new objects were fascinating to all sections of the public, whether learned, noble or popular. This is evident in the widespread popularity of *theatres of machines*, lavishly illustrated works written by engineers keen to showcase their expertise.[10] In the 17th century, the term was so popular that it became a catch-all phrase, as one contemporary, Charles Sorel, remarked:

> There is currently great pleasure in these fashionable words, and there are even some that, on their own, mean anything we want. There are a quantity of people who, when they cannot express something by a proper word, use the word *Machine*, they say, *Machines must be made for that*, and *these are Machines*.[11]

The concept of *machine* thus extends far beyond the realm of technology. The 1873 edition of the French dictionary Littré devotes about half of its large article to the term[12] to uses that do not designate material artefacts, but qualify things as diverse as governments, animals, intrigues, heavens, and literature. This diversity can be traced back to Antiquity, where the machine already referred both to mechanical devices and to any ingenious invention or means implemented to achieve some ends.

This is not an accidental lexical phenomenon. The hypothesis put forward here is that *machine* is a term which, throughout its complex history, has gradually acquired a generic function: that of designating phenomena whose regular behaviour should, in principle, be subject to a stable system of inference, even if this system is imperfectly known. To say that an organ, say a kidney, is a biological machine is not

---

10 Vérin 1993: ch. 1, § 2, and more generally Dolza and Vérin 2004; Lefèvre, Renn and Schoepflin 2012.

11 Charles Sorel, *De la connaissance des bons livres* (1671), quoted by Roux (2013: 74).

12 Littré 1873–1874: 3: 368. 9 entries out of 21 do not concern material artefacts, those from no. 12 onwards and except no.16, which deals with theatrical machines.

to say that it is an artefact or that it consists of pistons and gears, but that, given certain observations about its state and its interactions with its environment, it should in principle be possible to explain other observations by known inferences about it.

A machine is always situated because its inferential function is to give an account of a dynamic situation, *i.e.* to explain it or to predict its evolution, usually partially. The machine is then a focal point of the modelling activity mentioned above.

Modelling, as already discussed, means setting up a problem space, usually based on a theory, that adequately describes a situation in order to solve a problem. Within a dynamic situation, machines are those phenomena whose behaviour can be guaranteed under verifiable environmental conditions. The construction of a more or less complex procedure using them should ideally provide the solution, *i.e.* the transformation of the situation from an initial state to a final state, whether this final state is to be explained, predicted or brought about.

For example, I know that if the satellite and the meteorite are in this position and at this speed at this moment of observation, they will collide at this other moment. I also know that if I activate a certain amount of thrust on the auxiliary engine, the satellite will veer slightly and avoid impact. Or again: I know that a certain type of cancer cell reproduces thanks to a certain enzyme, theta polymerase. So I know that if I inhibit the function of this enzyme, the spread of this cancer will be reduced. In these examples, the satellite, the meteorite, the cell and the enzyme can be considered as machines, *from the point of view of the problem at hand*, because they allow given inferences to be made about them that are considered certain.

A machine can thus, in plain terms, be described as *any dynamic object whose interaction with its environment seems regulated.* This definition makes it possible to account for the diversity of its uses in everyday as well as scholarly language, without making them appear as a mere multiplication of metaphors. In this sense, the heavens are a natural machine, as well as rivers and tides. Crops and livestock are also natural machines, in the sense that we can describe, at a certain level of abstraction, their regular behaviour as the stable production of cereals, fruit, milk, eggs, etc. – as long as their environment and their care also follow certain rules and procedures. The *machine of the State* usually refers to all the administrative organs of government, insofar as they are perceived as operating in a coordinated and regulated manner. A plot (political or theatrical) is a machine in that it aims to govern all the meaningful interactions between its participants. To return to technology, bio-technologies today invite us to describe synthetic organisms as "biological machines", without this use sounding metaphorical (Bongard and Levin 2021). Finally, according to this definition, abstract machines such as Turing machines have full right to this name. They are no less machines than "concrete" ones, even if their environment is symbolic and their dynamics take place in a logical temporality.

The operating principle of a machine is called *mechanism*, and this term also has a very wide usage. For example, one talks about the mechanism of a chemical battery or the reproduction of a virus, even though these explanations are based on chemical or biological notions rather than those described by the physical science called *mechanics*, and which is now associated fairly widely with Newtonian physics.

8.2.2 Machines as devices of inquiry

It is clear from the above examples that there is no such thing as a machine *per se*. Some phenomena can be considered as machines, depending on the problem at hand and on the aspect of their behaviour for which rules of inference are available. A machine is a category in the Kantian sense, *i.e.* a principle of intelligibility of experience, mobilised especially for the systematic solution of dynamic problems.

This observation is important for understanding why machines often fail us: it is that we have unduly projected certain rules of inference onto phenomena that don't qualify for the concept of those rules, or that we have misjudged the environmental conditions for applying them. Reality is not mechanical *per se* (or that is a metaphysical claim from which this inquiry is entirely independent), but we often make sense of and intervene in reality by understanding it *mechanically*.

When the projection is successful, machines act as pockets of relative certainty in situations that at first glance may seem highly problematic. But as one gradually populates it with machines and begins to reason about their interactions, things gradually become clearer and a global understanding of the situation may be achieved.

The workings of machines may not be fully understood or certain or certified. They are devices of inquiry, which stand as hypotheses about a situation, and which have useful consequences for the problem at hand. The consequences are only as certain as the hypotheses themselves, *i.e.* that such and such a phenomenon denotes *this machine* (this set of rules of inference) and that its observed state in the situation is such or such.

It is thus not necessary to know *all* the laws of interaction of an object with its environment in order to treat it as a machine, but only the ones that are currently of interest for the problem at hand. Thus, I may not understand the reasons for a hen's behaviour in general (for example, why and how it communicates with its fellow hens), but know very well what conditions are necessary for it to produce such and such a quantity of eggs. In this case, the hen is a machine for laying eggs, and not for its behaviour in general. In the same way, I may not understand *exactly* what is going on inside the washing machine drum, which stirs the washing and the water in chaotic trajectories; but I know that, whatever these precise trajectories, the washing will come out clean, because this transformation depends only on the intensity of this stirring, the temperature of the water, the soap, and so on.

Weather forecasting is the opposite case: climate scientists have put a lot of effort into understanding the micro-behaviour of low-pressure systems, anticyclones, hot and cold air columns, etc., but still: the global interaction of these phenomena is complex and can't be reduced to a simple mechanism. It is therefore necessary to calculate the climatic conditions with infinite precision in order to predict how the weather will change in a given place. It is a phenomenon whose rules cannot be described systematically, although its parts can be. One speaks then of *open system*, or *complex, chaotic* system, rather than of a machine. All of these terms, which refer to the *system theory* introduced by Bertalanffy ([1968] 2015), are quite telling. An open system refers to a phenomenon that is not a machine because it is difficult to describe its behaviour systematically: it is a *problem*, an object of study for systematisation.

Note that not all systematic problem-solving needs to posit machines. Linear ratiocinations, *i.e.* that go from initial premises to conclusions are in this case. Machines are mostly necessary in the case of dynamic problems, where multiple objects interact according to their own rules. But even then they may not be necessary, as can be shown from some forms of differential equations which can accommodate the descriptions of these situations, provided they are not too complex.

Machines are, however, flexible vectors for the exposition of a theory aimed at dynamic problems. As in geometry, the theory simply presents these elementary inference engines and their laws of combination, leaving the investigator to solve their own problems. This is where the original inspiration of classical geometry is most evident and still very much alive: as an object of the geometry of motion, the machine represents the ideal of a part of material nature to which a geometrical construction would correspond, itself capable of demonstration and certain knowledge.

This perspective in some ways recovers one the fundamental insights of cybernetics, as they were understood by Ashby, one of its founders: machines are not specific things in the furniture of the world but a category that can be mobilised to describe reality and structure experience. Ashby proposes the idea of a theory of machines that would be similar to geometry:

> Many books have been titled "Machine Theory", but they usually contain information about *mechanical* things, such as levers or gears. Cybernetics too, is a "theory of machines", however it does not deal with things, but with *ways of behaving*, [...] Cybernetics has to do with the actual machine – electronic, mechanical, neural, or economic – just as geometry has to do with objects in our terrestrial space. [...]

> A determinate machine is defined as that which behaves in the same way as a univalent closed transformation. (Ashby [1956] 1957: 1–2, 24)[13]

8.2.3 The machine as a category

It is possible to go further in this direction and to give a transcendental meaning to the assertion that the machine is a category or a principle of intelligibility of phenomena in a Kantian sense. Machines appear in some complex inquiries as ways to organize the elements in the situation that allow to settle the problem at hand.

In the course of an inquiry, the requirement of the problem is the end sought. It may be a simple goal or a complex set of constraints to be satisfied. These goals or constraints may concern phenomena taking the form of an event, of a permanent state, or of a dynamic process. The problem itself may be about their explanation, prediction or production. The situation may be located in experience itself (material phenomena) or in the realm of meanings (symbolic phenomena). Indeed, it has been shown that none of such distinctions are essential to the work of inquiry itself, and that the construction of machines could be relevant in all cases: machines can be explicative or productive, abstract or concrete, and whatever else: they are a generic device of inquiry. For the sake of simplicity the discussion will focus on the two simple cases of a problem requiring the production or explanation of some concrete (physical) phenomena.

One remembers that, in Kant's terms, the technical understanding settles the end (here the target situation) and then looks up for mechanical means – *i.e.* phenomena which natural causality is known, or supposed to be known (see above 5.1.3). A machine, then, is only what is conceived as the subject of these phenomena. Kant makes this precision in the context of his distinction between material and spiritual automatas, already evoked:

> For this reason, all necessity of events in time in accordance with the natural law of causality can be called the mechanism of nature, although it is not meant by this that the things which are subject to it must be really material machines. Here one looks only to the necessity of the connection of events (*Verknüpfung der Begebenheiten*) in a time series as it develops in accordance with natural law, whether the subject (*Subjekt*) in which this development takes place is called *automaton materiale*, when the machinery (*Maschinenwesen*) is driven by matter, or with Leibniz *spirituale*, when it is driven by representation. (*Critique of Practical Reason*, AA 05: 97, transl. Gregor 1997)

13 By *univalent closed transformation* Ashby means a total function of a set of states on itself, *i.e.* nothing else than a state transition system.

The term *subject* is significant. In the *Prolegomena to any future metaphysics* it designates, for things thought under the category of substance, the problematic ultimate substrate of the predicates. It is a regulative idea of reason (§ 46, AA 04: 333). This means that the concept of machine derives from that of substance. Its predicates are the natural "connections of events" evoked in the quotation.

This idea can be made more precise thanks to Kant's qualification of the machine as a product where "each part is conceived as existing only through all the others, and also as existing for the sake of the others and the whole, that is, as a tool (organ)", taking here the traditional example of the clock (*Critique of Power of Judgement*, § 65, AA 05: 373). Machines are substances which predicates or partial concepts *cohere* with one another, giving the whole its stability across time.

This idea can be compared to the one by which Kant characterises phenomena thought under the category of substance. It is useful here to follow Longuenesse (1998)'s interpretation of the meaning of the tripartition of possible relations made in judgements: categorical, hypothetical and disjunctive.

It has been already noted (5.1.2) that a categorical judgement could be expressed as hypothetical by emphasising the object in question (*Bodies are divisible* = *If an object x is a body, then x is divisible*). Kant takes this convertibility, as Longuenesse (1998: 95) reminds us, from Wolffian logic. But in this case, one may wonder all the more why Kant distinguishes sharply one from the other in the *Logic* (§ 25, AA 09: 106). The answer is to be found, according to Longuenesse (1998: 140–147), in the *Amphibology of the Concepts of Reflection*, which closes the *Transcendental Analytic*. This text associates the categories of relations to the pair *internal / external*. Longuenesse highlights their logical role in the formation of judgements; they are the essential concepts by which we compare phenomena with each other in order to delimit the perimeter of the object. Thus, a condition that is judged to be internal to the object gives rise to a categorical judgement, like: "*If x is a tree, then x has a trunk*", because the concept of *trunk* is part of the synthesis by which I think that of *tree*. On the contrary, the idea of *freezing* water presupposes a condition external to water, that of a low temperature. In "*If the temperature is low, the water will freeze*", the judgement is naturally expressed in hypothetical form.

The fact that in our judgements the cursor between internal and external is mobile responds to the convertibility of any hypothetical judgement into a categorical judgement. If the judgement considers not only the water, but also the freezer in which it has been placed, then it can state categorically that the water is frozen there. What ultimately fixes the perimeter of our concepts is the stability or instability of these settled relations through time, which we evaluate *through* our pairs of *a priori* concepts substance / predicate and cause / consequence. They are rules for forming rules, or the general form of rules. The texts of the *Amphibology* regarding phenomenal substances give a meaningful picture of their transcendental characteristic:

> The inner determinations of a *substantia phaenomenon* in space [...] are nothing but relations, and it is itself entirely a sum total of mere relations. We know substance in space only through forces that are efficacious in it, whether in drawing others to it (attraction) or in preventing penetration of it (repulsion and impenetrability). (*Critique of Pure Reason*, A 265 / B 321, AA 03: 217, transl. Guyer and Wood 1998)
> Whatever we can cognize only in matter is pure relations (that which we call their inner determinations is only comparatively internal); but there are among these some self-sufficient and persistent ones (*selbständige und beharrliche*), through which a determinate object is given to us. [...] It is certainly startling to hear that a thing should consist entirely of relations, but such a thing is also mere appearance, and cannot be thought at all through pure categories; it itself consists in the mere relation of something in general to the senses. (*Critique of Pure Reason*, A 285 / B 341, AA 03: 229, transl. Guyer and Wood 1998)

These texts make clear that an essential role of the understanding is to discriminate between "self-subsistent and permanent" relations in phenomena from transient ones; the first ones will be conceived as internal to some substance, and the other ones, when they have some influence upon it, as external conditions.

From the comparisons of these texts, it appears that a machine is a specific way of attributing substance to a set of dynamic phenomena, which individual rules of change appear to cohere strongly together, *i.e.* driven themselves by some rules. What is "self-subsistent and permanent" in these machines are not static qualities, but these dynamic relationships, such as the ones between the movements of the clock's parts.

When these relationships appear modified by some external element (e.g. new inputs, parameter changes which modify its behaviour or its outputs), this element is understood as a condition or cause of its modification. Although some machines seem to be autonomous, such as clocks, most of them work in interaction with their environment, as noted in the previous discussion.

It is now possible to understand the function of machines in the course of an inquiry about the explanation or the production of some complex phenomenon. Here the machine is not what is observed – indeed, any layman may say *"This is a machine!"*, as Sorel noted in his time (see above 8.2.1), without knowing how to put this remark to any use – but what is to be constructed, either physically (for a problem of production) or symbolically (for a problem of explanation). The machine, then, is a coherent set of dynamic phenomena which the technician or scientist can give an account of. They can make categorical judgements about the consequences that derive from their machine, because they have constructed it – they are merely stating what they intended to do. For example, when the technician asserts that the water will flow into the irrigation canals with such or such flow rate, it is because they have built them in this intention. and when the scientist says that this flow rate is sufficient for crops to grow, it is because he has a model of crops as natural machines requiring such and such amount of water, sun and nutrients to grow.

These machines may also need external conditions to meet the problem requirement, which do not depend on its design, *i.e. hypotheses* to be made. It may be an external constraint that is necessary but cannot be controlled (or whose natural law is unknown), or an external interaction that is deliberately kept loose to give the machine some versatility (e.g. to produce different outputs depending on different inputs). In the case of the irrigation canal, the water source may experience fluctuations, and the technician must work to mitigate such effects, e.g. by building retention basins. The crops may suffer from excessive heat, and require therefore more water to be irrigated at specific times of day, and so on.

This remark leads us to consider the role of disjunctive judgements in this kind of problem-solving. Generally speaking, they have a different role from the categorical and hypothetical ones which, as we have seen, are strongly coupled through the internal / external distinction. Kant describes it this way:

> The disjunctive judgment contains the relations of two or more propositions to one another, though not the relation of sequence, but rather that of logical opposition, insofar as the sphere of one judgment excludes that of the other, yet at the same time the relation of community, insofar as the judgments together exhaust the sphere of knowledge proper; it is therefore a relation of the parts of the sphere of some knowledge where the sphere of each part is the complement of that of the others in the sum total of the divided knowledge. (*Critique of Pure Reason*, A 74/ B 99, AA 03: 89, transl. Guyer and Wood 1998 slightly modified)

By *sphere of knowledge* is meant what would be called today the extension of the proposition, the *cases* in experience where it happens to be valid (Longuenesse 1998: 378). The disjunctive judgement aims to determine the set of conditions relevant to the exhaustive description of a phenomenon. Thus, to describe the possible physical states of water, one needs to distinguish between its critical temperatures of freezing and vaporisation. Or, to describe all the possible trajectories of a massive body, one needs to consider it as part of a larger system including all the other bodies exerting an attraction on it. In this second case, it must be recognised that there is *co-determination* of the conditions between them, since the body under study also exerts forces on the other parts of the system. Here, as is often the case when systematicity is sought, there is a re-conceptualisation of the problem: the conditions are no longer represented by the bodies acting on the body under study, but by the initial conditions of the system as a whole, which determine all its possible trajectories. The search for disjunctive rules in the formation of a concept thus makes it possible, in a way, to calibrate its very perimeter and to reformulate which of its rules are valid categorically and which hypothetically, *i.e.* under external conditions.

Disjunctive judgements perform therefore a fundamental work in the scientific process. If the conditions of a rule can, *after analysis*, only be thought of as internal to the concept in question, then this rule must be qualified as analytic, *i.e.* as logically constitutive of the concept. Thus, the judgement that every tree has a trunk, even if it was initially formed synthetically and empirically, can be considered analytical when scientific reflection has judged, after examining a wide variety of trees, and especially after modelling their internal functions and biological structure, that this element is an integral part of this model. A trunk is nothing else than the stem of the tree, botany tell us, and if there are stemless plants, their order is exclusive of those under which the various tree species are classified.[14] That the tree has a trunk is therefore a result of the conventions by which we have decided to describe plant species. By contrast, there are trees without leaves (conifers) and the judgement *"The tree has leaves"* is therefore not analytical. This is not to say that these conventions are arbitrary and subjective. What is objective are the rules that the understanding has discerned through the mass of phenomena. Their grouping into the "packages" that are models and concepts is the business of the exhaustive ordering that the disjunctive rules aim at, and it is these concepts that ultimately determine analyticity.

The exhaustiveness sought by disjunctive judgements has a critical role in problem-solving through machines; it is what makes it *systematic*. When a machine has been constructed and shown to solve the requirement of the problem with the help of some external conditions, it remains to be demonstrated that this solution will work for all external conditions that must be envisaged, whether uncontrollable constraints that may happen or designed interactions that one wants to allow. For systematic problem-solving to be complete, all the cases of the problem must be verified, *i.e.* proven to yield the required solution. This means that there must be a way to traverse the set of possible cases, a condition explicated by the principles of inductive definition and bounded exploration (see above 7.4.3).

This derivation of the categoricity of the *machine* concept, in a Kantian frame, shows its very elementary character, and explains that it may apply to a large set of problems which satisfy only minimal criteria of regularity. Inserting machines in a given environment to solve a problem – either symbolically or physically – is a basic way of the understanding, to separate rules of phenomena which are known and can be controlled, from those which are unknown or cannot be controlled; the first ones can be thought as mere machinery, or problems already solved, the second ones are challenges that must be met by adjusting or specifying the correct

---

14 I base myself here, very modestly, on Wikipedia (www.wikipedia.com), articles *trunk (botany)*, *stem* (last visited March 4, 2025).

machines. Then, the solution found must be verified to hold in all conditions. This method is a basic division in the work of inquiry.

8.3 Programmable machines

This penultimate section lays the ground for answering the core question of this inquiry: why are computers so versatile and effective in problem-solving? The former section has given a conceptual or transcendental account of the general concept of *machine* and of its use in problem-solving. It must now be concretely shown how this happens in real life.

The first step is to show how machines are inserted in the course of an inquiry. The core concept here is the one of *specification*. It is a logical contract between the owner of a problem (who will be called the *principal* in this discussion) and the designer of a machine which will support the resolution of this problem (8.3.1). A second step is to make sense of this investigation's claim that computers are incomplete machines whose mechanisms can be reconfigured through software. To this end, the concept of *programmability* is introduced (8.3.2). A third step is to understand in which sense some programmable machines can be universal and made to execute *any* effective procedure, in whichever system of meanings it is defined. Only a sketch of an answer is provided here, this problem being the subject of a forthcoming book *Reasoning with machines* (8.3.3).

8.3.1 A contractual definition of programming

The key point of a specification, according to philosopher and computer scientist Ray Turner (2011: 147), does not lie in its content or format, but in its normative role: "Something is a specification when it has the jurisdiction of correctness over an artefact". The specification thus plays a logical role similar to that of a stipulative definition: it describes a concept of what must or can exist, without necessarily making explicit how to construct it; it makes it possible to judge the adequacy to this concept of objects or artefacts that are candidates for its standard (Turner 2011: 139).

The specification is entirely in this stipulative move. It establishes a standard against which an artefact can be judged. It should however not be compared to a law or regulation. Rather, the appropriate legal concept for understanding the role of specifications is that of *contract*, which shares responsibilities between two mutually consenting parties. *De facto*, specifications are essential contractual objects as soon as a product or service is supplied.

As Gunter et al. (2000: 40) say, one of the important motivations for the specification is "the need to divide responsibilities in a contract between the needs of the user and supplier. They build their deal around *S* [the specification], which serves as their basis for communication." This mediating role of specifications is central to the framework developed by computer scientist Michael Jackson with his colleague Pamela Zave in several papers, which are partially collected and put into perspective in Nuseibeh and Zave (2010). The starting point for any agreement between a principal and a machine designer (or programmer) is to make a clear distinction between the *problem domain* and the *machine domain*:

> The first and most fundamental principle of description is to identify clearly what is being described. In software development the primary distinction is between the *Machine* and the *Problem Domain*. The Machine is the software we build and the computer that executes it. The Problem Domain is that part of the world in which the Machine is required to bring about some effect desired by the *Customer* for whom the system is developed. For example, [...] in a lift control system the Problem Domain is the lift shaft, the lift car, the motor and winding gear, the buttons and lights, the floors served by the lift, and the people who use it. (Jackson 2002: 5)

This formulation makes it immediately clear that the expression of requirements is not primarily about the machine itself, but firstly about a "state of the world" that the principal wishes to change. The problem domain is the logical place where both the requirements that the machine must fulfil and the relevant properties of the world that it must take into account are expressed, in particular the constraints and the methods of action of the various means available. M. Jackson, who is wary of the ambiguities associated with the descriptive / prescriptive opposition often used in these discussions, prefers simply to indicate that invariant properties must be expressed by propositions with an *indicative* modality: they express what is, quite simply, without condition. Needs, on the other hand, must be expressed by propositions with an *optative* modality: they express what one would like to see happen.

It is important to clearly identify the contractual character of the specification. Jackson calls *E* (for *environment*) the set of properties of the situation that are relevant for the problem at hand, and by *R* the requirements that one would like to see realised there. The specifications *S* are then a group of propositions that relate to possible observations in the two domains, that of the problem and that of the machine. If the propositions *S* are verified, as the properties *E* are by assumption, then the requirements *R* must be satisfied. M. Jackson therefore writes the following relation, which asserts that it must be possible to prove *R* if *E* and *S* are recognized as valid:

(7) $E, S \vdash R$

One can see here that R and S play an entirely different logical role. Specifications *are not* the rewriting of requirements into a more precise language. They describe a set of phenomena which, if realised, lead to the realisation of the phenomena described by R. Here it is not said how to interpret this relationship, whether it is about causal, volitional, logical, mathematical phenomena, and so on. It all depends on the problem at hand. For example the following specification: *"A person with a valid badge must be able to pass through the automatic gates"* realises the following requirement: *"Only accredited persons must be able to enter the building"*. One can clearly see that the implication of the second proposition by the first presupposes properties of the situation that have nothing to do with the program: for example that only accredited persons have a valid badge and that the gates are the only ways to get into the building. These are obligations that are the responsibility of the building's manager, and which must also be fulfilled so that their requirements are met.

A second difference between specifications and requirements is that the former can only relate to phenomena that are accessible to the machine. *Accessible* here means observable or controllable – because the machine must be able to act on the current situation. Conversely, the requirements that describe a problem are not necessarily located at this interface – for example, the fact that the gates give access to the building (and not for example, to the parking lot) is not a phenomenon observable by the machine that controls the gates.

One may rightly recognize here the logical pattern of dynamic logic, or the idea, in Hoare's logic (see above 5.2.3), that the function of a program is to guarantee the fulfilment of a post-condition *if* a pre-condition is guaranteed by the execution environment prior to its execution. What Hoare says about a program is also true about a machine in general. Jackson's equation may be rewritten in this way:

(8) $E\,\{\pi\}\,R$

to mean that if the precondition E is satisfied before the action of the machine π satisfying the specification S, then the postcondition R will be satisfied on its completion (Hoare 1969: 577). Another way of understanding this formula is that the π machine guarantees the effectiveness of the *rule "If E, then R"*.

This is exactly what the machine designer must guarantee the principal, and what they should ideally be able to demonstrate logically: that such and such phenomena will occur in the situation if such and such initial conditions are realised there. In this respect, a machine is nothing more than *a guarantee of the effectiveness of an inference rule* in a given situation. The machine *connects, within the situation,* the properties that the principal can control with those that they wish to see happen. From the point of view of the principal, the machine is a *guarantee of a solution* to their problem.

Of course, one might object that this is not true in practice, and that the history of engineering and computing could be told as much by its "crashes" as by its innovations. Even today, computers remain open to the vagaries of software updates, undetected incompatibilities and design flaws. However, these breakdowns and failures have always been perceived as anomalies to be eradicated, so as to create a space of certainty in which users should have their expectations met. The machine is a space where uncertainties can be patiently brought into line. What ultimately counts is the *commitment* of the designer to the principal, and the principal's *belief* in that commitment. Engineering, as a profession, was built on this promise of operational certainty.

There is therefore a fundamental dissymmetry between the environment and the machine. The environment is open to the indeterminate horizon of the real situation, full of uncertainty, novelty, change, even contradiction (between people), subject to influences which, appearing initially remote, may suddenly turn out to be of burning relevance and proximity. The machine, on the other hand, presents itself as a closed space, prepared in advance, ideally controlled in all its components. In a problematic situation, it represents an island of certainty within which inquiry can safely make useful inferences.

8.3.2 Programmability

It now remains to make sense of computers' versatility in solving problems. How is it possible for one single machine to implement systematic procedures from fields as different as business, science, healthcare, communications, graphics, and so on? Firstly, does it mean that a universal language of action is possible? Secondly, what does it mean to *implement* a procedure on a machine, given the definition just given? This discussion starts with this second difficulty.

At the start of this inquiry, the hypothesis was advanced that computers were *incomplete machines*, that required programs in order to run. In the sense of the definition given above (a dynamic phenomena which interactions with its environment seems to be regulated), no machine is incomplete. What could incompleteness mean for natural machines, such as the movement of heavens, tides and rivers, or the regular behaviour of plants and animals?

This expression could be understood relatively to a problem which the machine must solve, a requirement R. In that case, the machine may be incomplete relatively to the environmental conditions E that are necessary for it to operate properly. There are some cases where no environmental conditions are required at all, like basic automata, such as clocks or music boxes. Beyond an occasional winding or calibration, a clock operates through a fixed schedule, and a music box just needs opening the lid to let the carousel turn gracefully. In most cases however, a machine is supposed to solve a problem over a given range of environmental conditions. These conditions may be understood as inputs. Think of an industrial injection machine, which sets plastics into precise shapes by injecting them into a mould under high pressure and temperature, or a pocker calculator, which computes input numbers into sums or divisions. However, it is not this trivial sense of *incompleteness* – that many machines need inputs to operate – that is of interest to understand the possibility of general-purpose machines.

As Wiggershaus (2025) remarks, some basic automata may be reconfigured. For example some music boxes allow for the music cylinder to be replaced so that a new tune may be played. It is said that Vaucanson's flute player (1738) was able to play twelve different tunes. Also, the injection machine may be reconfigured to produce new shapes by a change of the mould. In the case of a washing machine, buttons allow to select the duration of the washing-cycle or its intensity, and are often packaged into *programs*.

In these trivial cases of programmability, the handles located at the interface between the environment and the machine (the music cylinder, the mould, the buttons of the washing program) act as *second-order interventions* with respect to the basic specification of the machine (playing a musical tune, producing a plastic artefact or washing dishes), which concern a variation in the output or in the modality of the process. They (re-)configure in some degree the inner mechanism of the machine, or its "selected operations", in the words of Wiggershaus. This "degree", according to him, must be understood as degrees of freedom in the underlying mechanism.

There has been several accounts on the complexity of mechanisms, lending always more degrees of freedom for programmability, some of them historical, and others conceptual, drawing on computability theory.[15] From our perspective, it is useful to draw on the analysis of Blackwell (2002), a psychologist of human-machine interactions, regarding the distinction between computer programs and mere data inputs.

Blackwell clearly sees the relative nature of any distinction between using and programming a computer. Considering, among others, spreadsheets, word processors, scripting languages, HTML, he asks:

> Which of [these] can unambiguously be categorised as a programming language or another form of software? [...] The distinctions between these technologies are not at all clear-cut. All are capable of producing dynamically modified text documents, and several can potentially

15 For a review, see Wiggershaus 2025.

> be used to create apparently identical results. Yet some are classified in professional contexts as programming languages, and some are not. (Blackwell 2002: 2)

The difficulty in making these distinctions stems from the fact that it is not the technologies used that characterise programming, but the *experience* of the user, who has to carry out tasks that are more or less cognitively complex in order to achieve their ends – such as specifying what they want to achieve as a result, characterising the favourable situation for this, identifying exceptions, understanding the rules of the elementary operation of the machine they are trying to modify, and so on. The first characteristic of programming relative to simple computer use is the loss of *direct* manipulation of the elements of the situation. The programmer has to predict *what will happen later* in the situation, and this may involve looking at many different situations, the possibility of which is not immediately apparent. This *indirect* nature of the control of machine behaviour implies the use of notations to describe its rules. They allow indirect references to be made at several levels, like a three-cushion billiard shot – a carefully calculated strategy that hits all the right angles before reaching the target.

Blackwell (2002: 4) thus contrasts the example of programming a video recorder, which requires simply specifying the time and channel (simple programming, close to simple use), with that of a boiler, for which several operating modes can be defined once for all, and then each mode set to work at a particular time of day. Here an abstraction has been created (the mode) that the user can now manipulate as if it were an object, allowing them not to repeat similar instructions.

What essentially distinguishes computer programming from simpler programming, then, is this unbounded flexibility to create *indirect* control of action at as many levels as desirable. Blackwell's analysis supports the epistemological view of machines taken here, that they are primarily trustworthy systems of inference about dynamic phenomena. Computers are machines that allow for the construction of complex networks of inference about their potential behaviour, and for the nesting of these networks one within the other at unbounded levels of reflexivity.[16]

While washing-machine programs could still be considered as inputs, because they are part of the machine's environment, a program that completes the proper working of a machine is something different altogether. It is part of the design process to meet the machine's specification. In Jackson's framework, it can be described by a set of invariant properties M of the machine which, together with the configur-

16 My other book, forthcoming, *Reasoning with machines*, explains how such subtle mechanisms may be engineered in material phenomena, and in which sense one can say that there are maximally programmable machines (universal computers).

ations established by the program P, realise S. There is therefore a second formula complementary to the previous one 7:

(9) $P, M \vdash S$

M and P relate to phenomena which are "hidden" from the environment, *i.e.* which do not matter to the principal. For example, the program may, in order to solve a task, store and modify certain data in the machine's memory; the correct operation of the program may be expressed by the values taken by this data, but this is irrelevant to the specification and even more so to the requirements, since other solutions might be conceivable, for example other machines, running other programs, or "hard-wired" machines altogether, with no program at all. These phenomena are internal to the machine.

Here, a second difference appears between inputs and programs. It is not only a matter of degree of complexity in the mechanism, or of indirectness in the inferences, but a logical stipulation that cuts through the main two logical roles in the specification contract 7, that of the principal, who owns the problem and controls its environment, and that of the designer, who must deliver the machine. Programming activities may sit at either side – for example, the designer may let the principal do a lot of configuration work, or let them even write some macros; but professional programming clearly sits on the designer's side. Note that nothing prevents the principal and the programmer from being one and the same person who is thinking alternately about their problem and about how to program it, and is thus addressing conditions and commitments to themselves: this internal dialogue is part of the programmer's job to weave the nested webs of inference that characterize computer programs.

8.3.3 Versatility

This account of programmability is not yet sufficient to explain the versatility of computers in so many different domains of human practice. A machine, as defined above, is a dynamic phenomenon of experience and as such is rooted in a given materiality. One may well understand that there are programmable industrial machines, programmable scientific machines, programmable surgical machines, programmable business machines, and so on, but not that one generic programmable machine is sufficient to perform all their tasks.

Explaining this possibility is the core topic of my forthcoming book, *Reasoning with machines*. Computer scientist Peter Denning calls this problem the "information paradox". An easy answer indeed is to say that problems are all about information processing, and that information can be coded into numbers. Dennning points out that this condition is not sufficient. Information coding has been commonplace since at least the telegraph, but until the advent of the computer, codes were only used to transmit information, but in no way to solve various problems using these codes:

> Shannon's classical theory of information demonstrates that information can be accurately transmitted and received by processes that do not depend on the meaning of the information. Computers extend this theory, not only transmitting but transforming information without reference to meaning. How can machines that operate independently of meaning generate meaning for observers? Where does new information come from? (Denning and Bell 2012: 470)

Denning's answer to this paradox is as follows:

> The [program] design process transfers the idea of addition from our heads to instruction patterns that perform addition. The meaning of addition is preserved in the design of the machine and its algorithms.
> This applies to any other calculable function. We transfer our idea of what it means for this function to produce its result to a program that controls the machine to do just that. We transfer our idea of what the function means to the design of the machine.
> From this point of view, the idea that machines and communication systems carry out processing without taking into account the meaning of binary data is fragile. Algorithms and machines have meanings implanted in them by engineers and programmers. We design machines so that the meaning of each incremental step, up to the result, matches our intention, if the input data has equally matching meanings. (Denning, Martell and Cerf 2015: 50)

In other words, what is far more fundamental than the coding of information is the programming of mechanical operations that correspond to the signified operations. This explanation depends on the fact that all systematic procedures have a deep structure, which corresponds to Dewey's pure procedural means (see above 6.3.1). This deep structure may be used as a pivot to transpose inferences from one system of meanings to another. In other words, a sequence of inferences in, say, a legal system, may *correspond* to another sequence regarding a mechanism.

It is no place here to develop and justify this sketch of a solution. However, in order to make it more concrete, it may be illustrated by the canonical example of addition used by Denning. This example also allows to make sense of Minsky's enigmatic view that "the belief in an arithmetic statement is equivalent to the belief that certain machines, if properly built, will work" (see above 8.1.1).

So let us imagine a simple machine, initially made up of water flowing regularly from a source to a river through a single canal. Programming this machine involves diverting additional canals and placing manual or automatic sluices on them. Putting this network into operation (running the program) involves activating one or more manual sluices on canals connected directly to the source, which are the binary inputs to the circuit and, once the flows have stabilised, recording the presence or absence of water at points located downstream, which are the outputs.

This circuit can be built using two operations. The first is to dig a canal from an existing point in the network to a new point. The second is to build a sluice blocking an existing canal, depending on whether it is actuated by a flow from a third point. A classic mechanism for achieving this second action is the *piston – rack – pinion – shaft* assembly: the flow actuates a piston which sets in rectilinear motion a rack linked to a pinion which transforms this motion into circular motion. This in turn turns a shaft on its axis, raising or lowering the sluice.

A simple illustration putting this machine to work is the basic assembly in figure 8.1, where blue represents the flow of water, and ochre its absence.

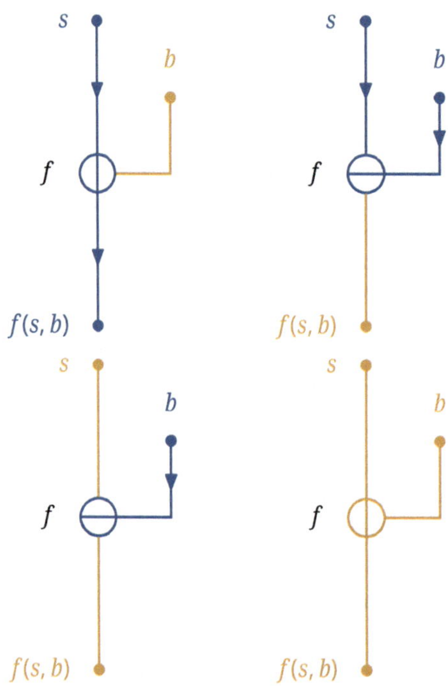

Figure 8.1: The f function, and its four configurations

In the diagram, s and b refer to two points upstream in the irrigation network, to which water may or may not flow. $f(s, b)$ refers to the endpoint of a newly constructed canal downstream. There is an evaluation function e applying to any point, which can take a binary value T or F, depending on the presence of water on points s and b:

Table 8.1: The truth table of function f

$e(s)$	$e(b)$	$e(f(s,b))$
T	T	F
T	F	T
F	T	F
F	F	F

In other words, function f *corresponds* to the logical proposition $s \wedge \neg b$ (S and (not B)). This construction allows to *delegate* the execution of this rule of inference to the flow of water.

Modelling addition in this irrigation network is barely more complex. For example, one-bit addition has the following table:

$$0 + 0 = 00$$
$$0 + 1 = 01$$
$$1 + 0 = 01$$
$$1 + 1 = 10$$

If $b1$ and $b2$ are the two input bits, and t and u the two output bits corresponding to the ten-bit and unit-bit respectively, this table is captured by the following logical equations:

$$u = xor(b1, b2)$$
$$t = and(b1, b2)$$

where *xor* and *and* are the usual *exclusive or* and conjunction logical operators. They can each be written in terms of function f defined above, yielding:

$$u = f(b1, b2) \; or \; f(b2, b1)$$
$$t = f(b1, f(S, b2))$$

where S is a source of constant water. These two equations are translated in the figures 8.2 and 8.3 illustrating for $1 + 0 = 01$ and $1 + 1 = 10$, respectively.

In these examples, the basic rules of logic and arithmetic are made congruent with the rules of water flow in irrigation canals. Minsky's claim is now clear: thanks

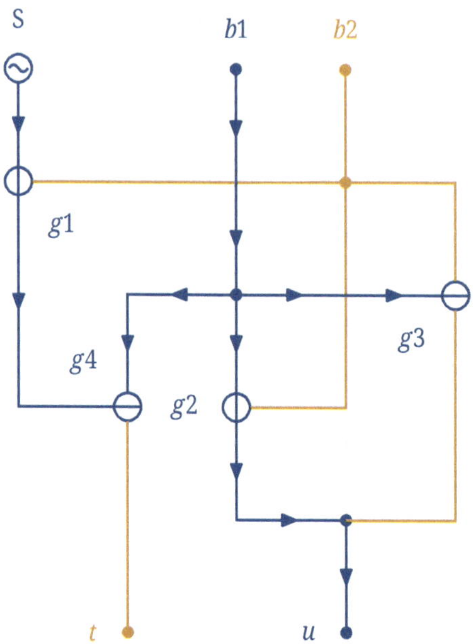

Figure 8.2: Addition of $1+0$

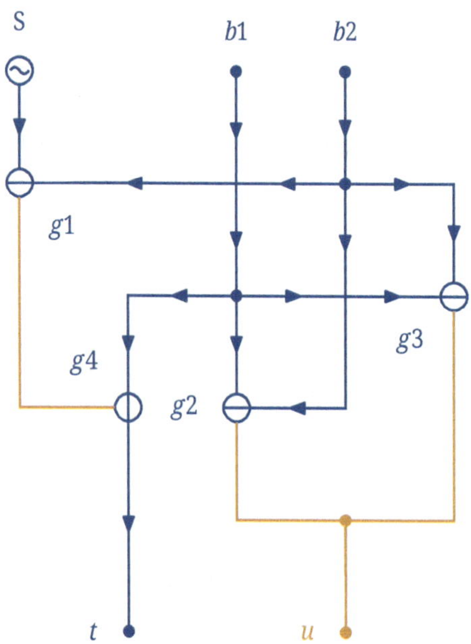

Figure 8.3: Addition of $1+1$

to this congruence, the belief that 1 + 1 = 2 is made equivalent to the belief that the irrigation circuit thus described will work *if* its canals are properly dug and its sluices rise and fall as intended.

Of course, it has not been shown that it is possible to construct such equivalences for any systematic procedure in any system of meanings. Understanding this surprising possibility is the subject of my forthcoming book.

8.4 The effectiveness of rules, machines, systems

This section analyses the *meaning* of this deduction of machine effectiveness. Systematicity itself is a search for the maximal coherence of inquiry, or for the *coherence of intent* (8.4.1). With the automatic execution of ratiocinations, this coherence is partly *pledged on the machine*, since it must be assumed to be constructed in accordance with the program's intention (8.4.2). The effectiveness of machines is a form of *productive detour* whose pattern characterises any kind of practical knowledge, consisting in the *construction* of means, symbolic or material, to solve general classes of problems (8.4.3).

8.4.1 The coherence of intentions

Whatever the epistemic differences that may exist between deliberation, technical ratiocination and systematic problem-solving, they are only modalities of the same effectiveness of knowledge; and the effectiveness of machines stems entirely from the third one.

This thesis can also be expressed as follows: what makes a machine effective is not the power of its energy source, the solidity of its materials, the sophistication of its technologies, or the quality of its operation. All these principles only become effective because they are integrated *in a coherent way* into the machine's intention.

This *coherence of intent* is at work in systematic computation as much as in technical ratiocination and deliberation.

1. When a computation error is made, we inspect which rule may have been applied incorrectly (how we may not have been consistent with the system of rules we had given ourselves), or we search which figure may have been misread or miscopied (how we may have made an "interface" error between the system of rules and the situation to which we were applying it).
2. In the same way, when a technical ratiocination fails, we check whether the procedure has been followed correctly, or we inspect the situation to see whether it was in a conformation that the procedure had not foreseen.

3. Finally, when a deliberation leads to an unexpected result, we first make sure that we were not the victim of an error in reasoning (such as not having perceived the two possible meanings of the same word), or we incriminate a fact that we did not know, or that we had misinterpreted; in other words, once again, not the reasoning itself, but its inadequacy to the situation to which we had applied it.

There is a *similar pattern* that repeats itself at the three levels of inquiry that we have identified. On the one hand, it is possible to contradict oneself, particularly when one has not proceeded in accordance with the injunctions of exact meanings, *i.e.* previously verified in a system. On the other hand, errors of judgement may arise from *misreading* the situation, *i.e.* from *applying* to it an ill-suited reasoning or system of rules, *i.e.* that do not lead to the situation one hoped to obtain. This error may be not only a *local* one, as in the examples above, but also a *global* one. When, for example, I apply the rules of decimal calculation to hexadecimal numbers, I do not achieve the desired result, not because the rules are wrong, but because they are not adapted to the problematic situation I am trying to solve.

What changes in the shift from deliberation to technical ratiocination, and from the latter to systematic calculation, is the *breadth* of the compact of meanings attributed to the situation, and therefore also that of the *bet* that is performed. To the set of *facts* that deliberation deems valid in order to make them the subject of its reasoning, the technical procedure adds a sequence of operations associated to expected reactions of the situation; the system of rules, for its part, projects onto the situation a whole "model" or a whole "theory" as to its behaviour. To this difference in *amplitude* are directly correlated those of effectiveness, but also of risk and difficulty that are commonly observed between deliberative, technical, and systematic judgment.

Everything here is well known since the beginnings of logic. From a simple syllogism to a chain of reasoning, and from this chain to a system based on first principles, it is each time the amplitude of the possible consequences, but also the difficulty and the risk that increase with the number of premises that must be predicated of experience. There is also an operational and pragmatic way to look at this, which may be illustrated by the *pons asinorum* (see above 4.3.2). This logical machinery asserts that the procedure for finding the syllogism will surely succeed if the lists of terms describing the problem are exhaustive. One might object to Aristotle that nothing is gained by this procedure, which replaces the search for the right term by a search for *all* possible terms. But this would be wrong: for the problem of reasoning well has been transformed into a problem of describing well. A small nucleus of certainty has been formed within the situation in exchange for its careful description. Once this movement has been initiated, it may be gradually expanded by composition. The share taken by these cells of certainty in the problematic environment may gradually form *chains* and soon *networks* of possible procedures. Through this movement, the instability of a given class of problems is gradually circumscribed and described by common conditions that are *extracted* from the problems themselves. These conditions constitute a problem space that an *inferential machine* can traverse freely without worrying about agreement with experience, because it regards these conditions as mere parameters. Its criterion of truth is only that it correctly applies the rules that define it.

There is thus a division of intellectual labour. First, there is the work, which applies once and for all to a given class of problem, of preparing systems of rules in which exact reasoning procedures can be developed and examined. Once this work has been done, it remains for the judgement, in each situation, to transcribe the terms of the problem into those of the system under consideration, and then to carry out a ratiocination according to the prescriptions of the solution. This idea corresponds to the concept of *notation* that emerges among English algebraists at the beginning of the 19th century, *i.e.* the idea that reasoning can progress by manipulating uninterpreted symbols according to rules (Durand-Richard 1997). According to Michael Detlefsen (2007: 263–277), this "instrumentalist" conception of notations prepares the Hilbertian idea of systems of formal calculus.

This division of labour helps to explain why systematic procedures are called *effective* and what is the link between the computational meaning of this adjective and the more general sense in which it has in common language. Effective procedures are of course not the only ones to meet their goals. All concepts, in so far as they are packages of rules (see above 5.1.2), have a certain degree of effectiveness. Simply, effective procedures are granted by the system which expresses them a *guarantee of effectiveness*, thanks to the preliminary work that has been done in its construction – looking for exact terms and exhaustive ranges of situations. This guarantee is the very one expected from the division of labour between global inquiry, which verifies rules of inference and assembles them into a system, and local inquiry, which mobilises them within such a framework. Of course, such a guarantee is only as good as the credit given to its issuer. This is why it is possible to doubt the deductions of astrological systems, however coherent and elaborate they may be.

The effectiveness of machines is of this same nature. Machines "in themselves" are no more or less effective than any other natural phenomenon in the world; what is effective *in them* are the intentions we delegate *to them*; and so they never give us more than a *hypothetical guarantee of effectiveness*. This is what the next discussion aims to clarify.

8.4.2 The effectiveness of machines and the coherence of intentions

The instrumentalist conception of notations has encouraged the assimilation of programming to a syntactic or formal activity, unconcerned with semantics and meanings.[17] This understandng of programming is very partial. The *coding* phase of programming may indeed be called formal because the meanings of the situation are already prepared to constitute a problem space for the machine, which then has only to be directed through this space according to valid rules. However, coding only reflects the sharing of responsibility that results from a negotiated agreement between a principal and a programmer. The phase of specification phase, which precedes coding, and the phases of implementation, testing and production, which follow it, are eminently "semantic".

In the contractual interpretation of programming, it is indeed the *negotiation* between the programmer and the principal that sets the boundary between the machine and the situation. The specification separates the *inside* of the machine, as a place governed by certain and binding rules, and its *outside*, governed only by intentional hypotheses. Specification allows the principal to infer: *"If the situation (e.g. the user) behaves this way relative to the machine, then the machine will do this or that thing that will change the situation accordingly"*.

For the principal, therefore, the machine has the same role, *in experience*, as (human) ratiocination in inquiry. The machine is for the principal a *guarantee of solution* and represents a space governed by rules of *superior* certainty to those of the situation. *To reason with machines* is thus to carve out, within the situation, sets of phenomena that are recognised to be governed by systems of rules. Within these systems, reflection can develop and make inferences in an assured manner. This possibility is *the same*, whether the machine is natural or the result of a prior human construction. This form of reasoning is therefore not specific to "technical" knowledge. On the contrary, it has such great effectiveness because it has the very form of scientific knowledge (systematicity). It is no accident that science and technical thinking met in *mechanics* and in the idea of *machine*; it is perhaps possible to argue that it is from this common concept, which is first of all a very special way of apprehending experience, that they began their fruitful collaboration.

It is important to note that mathematical certainty is not opposed here to empirical certainty; for the machine is itself an empirical device, and it is within its scope that the division of labour takes place between the material verification of meanings and the mathematical (procedural) reflection.

17 See my forthcoming book *Reasoning with machines*.

The modal character of the definition of the machine, as a phenomenon whose dynamic interactions with the situation *seem* regulated, does not mean that it would only be a matter of appearances or conjectures. *To judge* that something is a machine is to give oneself the licence to carry out all the compositions of inferences described by the system of meanings predicated of the machine. This initial judgement is always empirical, whatever the checks that may have been made beforehand, and whatever their possible systematic rigour. It is not a question of probability, but only of the status of this judgement, which is not about pure procedural meanings, but about a subject-matter situated in experience (section 6.2).

Thus, the principal's assurance that the machine will work must be divided into two parts. First, they must *trust* the designer, who gives them a *guarantee* of the validity of the machine. Such a condition can be removed, because the principal could very well be the designer themselves, or could check the entire design by themselves. But they must also trust that the machine is properly built according to the design. Let us come back again to Minsky's statement: "The belief in an arithmetical statement if equivalent to the belief that certain machines, if properly built, will work". The analysis in the previous section suggests that the core reason of this equivalence lies in the fact that these beliefs fall under the same principle: *"If you understand and accept the meaning of the rule, then given its condition, you must admit its consequence"*. In the case of $2 + 2 = 4$ and $10_b + 10_b = 100_b$, these are the rules of addition in arabic decimal or binary numbers. In the case of the irrigation circuit, these are the rules for the flow of water from one canal to another and for stopping it by a sluice.

It may be objected that these rules *are not of the same nature*: whereas the former are simply *stipulative* – they define what it is *to add*, the latter are empirical, or in any case depend on natural laws such as those of gravitation and fluid flow. The former are *applied* by the acting subject, the latter *observed* by them, and therefore depend on an external condition.

Things are not as simple as that, however. It is not possible to understand this opposition as one that would exist between *procedural rules* and *material rules*. Performing an addition is a material action that can fail like any other device. Babbage declared that the advantage of a machine would be to reduce the error rate of human computers. On the other hand, the operating rules of the irrigation system are not empirical in the sense that they would depend on assumptions to be made about the flow of water, that it will pass through the canals, that the sluices will stop it, and so on. Such assumptions do not have to be made, because the flow of water *already behaves in such a way*; building a water flow circuit is merely modifying another, more elementary one. Programming merely modifies the control *of a machine that is already running*, or even more precisely: it modifies the machine according to the very rules by which it was described *in the first place* – whether these prove correct or not when experience is summoned. Whether we are talking about building a canal or making an addition, we can say that these actions conform to the exact meaning of what *are* an addition and an irrigation canal (a passage of water); they both have the same level of purity and materiality. Whether water evaporates from the surface of the earth, or a brain injury affects my numbering skills, these facts have nothing to do with the correct definition of these acts.

It is therefore not in these kinds of oppositions that lies the difference between calculative ratiocinations and ratiocinations about an irrigation circuit. The relevant difference lies rather in the agentivity of the rule.

Let us return to the principle just stated: *"If you understand and accept the meaning of the rule, then given its condition, you must admit its consequence"*. One can recognise here the *modus ponens*, which seems to require only a form of *consistency with oneself*: for to accept a rule is to agree to apply it. Yet it is useful here to recall the difficulties the tortoise makes with Achilles on this subject: while she agrees to *accept* this *meta-rule*, at Achilles' request, she nevertheless refuses to *apply* it without the help of another one (Carroll 1895), and she thus sets in motion an infinite regress.

The case of the machine seems to circumvent this difficulty. The rule of addition, transcribed into an irrigation circuit, needs no longer be applied; for the flow of water gives its result directly; it substitutes for the effort of the inquiry the simple act of "pushing" the symbols one after another towards their next transformation. It seems that Achilles would quickly get rid of the tortoise with such a machine.

However, another condition appears under Minsky's pen: "if the machine is properly built".

This condition is necessary because the act of the inquirer that enables the object-rule (here addition) to be made effective in this way is *indirect*, and external to it; building or modifying a machine obeys rules that are not those of the object-rule; one must know how to dig a canal correctly and build an automatic sluice mechanism in order for the modification of the circuit to be effective.

This is where the tortoise could again give Achilles a hard time, in a much more direct and obvious way than with the *modus ponens*: because there is an effective and visible decoupling here between the construction of the machine and its operation. If it is not built according to rules, there is no reason to trust it. It is a contingent natural device to which one may or may not entrust one's ratiocinations, depending on one's empirical trust in its designer. If it is, then one must accept these rules and recognise their correct application. It could be that the application of these rules is entirely automated; in the irrigation example, one may imagine drills digging canals according to a program. But if the program can be validated, one still need to have confidence in the drills themselves, and so on. Now this interweaving of machines is very real and to a large extent circular, since today's computers are *always* built from other computers. This fact is sometimes used by opponents of formal methods:[18] if checking the correctness of one program requires another program, then one is caught in an infinite regress. This is nothing else than the tortoise's argument.

The difference between ratiocinative rules and mechanised rules is now clear. The effectiveness of the former appeals to the *modus ponens*, which can be seen as a developed form of the principle of coherence, applied to the intentions of the inquirer; if they admit the rule, they must admit to applying it, because that is *what it means.*

Something different happens with mechanised rules; they still have to do with the internal coherence of the investigator's intentions, but these rules now *enter into a contract* with an external regime of facts or rules, *which promises* them the correct application of the object-rule. The inquirer must accept the machine's conclusions – because that is indeed why *they made the contract* in the first place – however a complementary condition appears: *as long as they trust the contractor.* It is not so much the skills of the programmer that may be questioned here, for the program is *what may be verified in the machine*, what is *mathematical* in it, but rather the reliability of the machine itself. In a way, the distinction between program and machine only indicates that this verification is not absolute, and that there remains a background of facts and rules that must again be verified. The machine appears here, from yet another perspective, as the boundary limit of the enterprise of mathematising experience.

The principal must therefore believe in the effectiveness of the application of the rule *insofar as they believe in that of the machine.* This is nothing else than the contractual interpretation of programming and of Hoare's formula $E\,\{\pi\}\,R$ as asserting that the program π *guarantees* the inference from E to R (see above 8.3.1). It is in this division of labour that the effectiveness of mechanised rules is constituted, *superior* to that of ratiocinative rules. Through this additional detour, and the *additional uncertainty* that results from it, the inquirer gains in effect the automation of ratiocination, with all the advantages formerly indicated (see above 3.3.3). The effectiveness of the robotic arm in building the car results from that of thousands of small contracts between programmers, which have enabled each one to implement their own rules on the machines of their colleagues, thus building this sophisticated machine through *trust* in the effectiveness of ever more elementary machines.

The effectiveness of machines is thus nothing else than that of a network of meanings on the basis of which it is designed. Insofar as this network is *external* to the concept of the machine, its effectiveness may indeed seem *empirical*, and one may be tempted to evoke the energy, the engineer, the technology necessary for its

18 See the discussion of Bringsjord 2015: 275.

realisation as sources of this effectiveness. However, if one bears in mind that elaborate machines can only be designed within *technical systems*, it is clear that their effectiveness responds entirely to that of a *system of meanings* that organises part of the experience. Machines are effective because they are designed on the basis of an extremely complex system of controlled inferences about experience, which was built up gradually and cumulatively.

8.4.3 The productive detours of knowledge

Mechanisation appears therefore as a move by which an agent accepts to delegate certain conditions of the coherence of their intentions to third parties – e.g., that this car they are riding in will not break down –, and by which an industrial world can appear. It is rooted in the same principle of coherence of intentions that characterises problem-solving as a whole, whether deliberative, ratiocinative or systematic.

This idea is paradoxical, in that it seems to associate the effectiveness of knowledge (its capacity to generate effects in the world) with tautology, which is often considered as *empty*, even by logicians themselves. This final discussion delves deeper into the reasons why the simple coherence of intent may be *productive*.

When, instead of reacting directly to a situation, I state a rule that describes my action as implied by the situation, and then apply that rule, it seems that I have only made a detour "which amounts to the same thing". The same seems to be true when, instead of simply stating a procedure, I describe a whole system of rules in which this procedure appears as a particular composition. Why take such detours?

It should first be noted that the entire discursive inquiry may be described as such a detour: faced with a problematic situation, instead of giving in to the impulse in the face of some of its salient marks – whether this impulse is organic or acquired through habit – human beings have the capacity to use instituted signs to distance themselves from the situation and *to control* this impulse in order to examine its consequences, as well as other possible options (see above 6.1.4). *This is* the fundamental detour of which the other two are mere outgrowths. To state a rule is to "rehearse" the action mentally before carrying it out, and it is therefore to give oneself the opportunity to check beforehand a whole series of possible inferences about it, which will confirm or deny it. To state a rule within a system of rules is, moreover, to give oneself from the outset a whole set of already established verifications, and if that is not enough, a whole arsenal of *a priori* verification procedures further increasing the effectiveness of deliberation.

These detours, which are the statement of rules and the construction of systems, are thus "investments" that allow the inquirer to solve a problem once and for all, and then content themselves with ensuring that the conditions of that problem apply to the situation. Ryle writes, in his characteristic style:

> If a farmer has made a path, he is able to saunter easily up and down it. That is what the path was made for. But the work of making the path was not a process of sauntering easily, but one of marking the ground, digging, fetching loads of gravel, rolling and draining. [...] Similarly a person who has a theory can, among other things, expound to himself, or the world, the whole theory or any part of it; he can, so to speak, saunter in prose from any part to any other part of it. But the work of building the theory was a job of making paths where as yet there were none. (Ryle [1949] 2002: 289)

It is the same idea that enables Aristotle to define *technè* as "a productive disposition, accompanied by true reason, of contingent things" (text quoted in 4.2.1). To act technically is to construct a general solution to problems in which a certain variability is inscribed; it is to conceive of the possible generalisation of certain deliberations, and it is in this generality that its *truth* consists.

Thus, the principle of coherence that guides the application of procedures and systems to problematic situations is quite simply that *they were designed for that purpose.* To recognise a certain problem in a situation, and to refuse to apply the solutions specially designed for it, would be (all other things being equal) to contradict oneself, like the tortoise refusing to apply the *modus ponens.*

Mechanisation, which has just been characterised as the exchange of part of our ability to check the coherence of our intentions for the benefit of effectiveness provided by machines, appears here as the particular case of a more general pattern, which is already present in deliberation. As Dewey says, the meanings of ordinary language are themselves the first tools and the first inferential machines on which our deliberations rely. As for our procedures, they also rely on the network of tools (technical, institutional, etc.) available in our world.[19] Mechanisation, thanks to which the capacity to carry out systematic research develops, appears in this perspective as the third modality of an equal and unique relationship of exchange between inquiry and the world, within which networks of signs, tools and machines gradually grow and, in turn, allow inquiry to develop and gain in effectiveness.

What makes these detours "productive"? First, it should be noted that they are not always so. There are cases in which we fail to state rules that are sufficiently "productive" to outwit direct deliberation. This is what Aristotle observed of situations concerning human life, and which is still very largely true. What makes these detours productive is therefore *that reality lends itself to them.* In what sense to in-

---

19 For the idea of a network of tools, see Heidegger [1927] 1993: § 15.

terpret such a formula comes down to metaphysics. It can be understood, for example, realistically (since the situation must have a certain structure for this productivity to take place), or idealistically, or transcendentally like Kant. But, more importantly, it is compatible with the tradition of scepticism – the minimal philosophy that a positive epistemology must seek to convince. Scepticism, as expounded by Sextus, quite admits the truths of common sense, "useful to life", in the extension of which lies technical knowledge.[20]

There ends this research. It can be useful to metaphysics if it helps to clear a vast *positive* epistemological field, that of problem-solving, which is thus *subtracted* from its inquiry. In the same way that metaphysics does not discuss the *content* of logic, mathematics, sciences, technology, but only the *nature* of the truth they reach, this subtraction is not a loss, but on the contrary a clarification of its question.

20 On this subject, see Marchand 2015.

Conclusion

The primary aim of this inquiry was to dissipate a perplexity regarding the versatility and effectiveness of computers in our human world. This goal is partially fulfilled. Some core concepts necessary for a successful account of the pervasiveness of digital technology, such as *problem-solving, procedure, ratiocination, systematicity* and *machine* are now grounded in a coherent epistemological framework, expanding John Dewey's theory of inquiry, and drawing on specific readings of Kant and Martin-Löf. Its background is the Aristotelian tradition of inquiry as well as early modern reflections on the notions of *machine* and *system*.

The work to be done remains important though. It has by no means been explained how universal machines were conceptually possible, nor how they could be concretely built. Furthermore, it must be shown how this framework provides a convincing account of the theoretical and practical reality of computer science, preferably to its more traditional groundings in the concepts of information or computation. Finally, it must help to explain the wide-range transformations of the institutions and infrastructures of our human world. On this topic, the fruitfulness of the perspective defended here may be glimpsed if one remarks that institutions and infrastructures can be viewed as *systems of meanings* at par with scientific theories: they provide frameworks for humans to solve their problems, and are therefore deeply challenged by the transformation of inquiry. All these topics are the object of a forthcoming book *Reasoning with Machines*.

The relevance of the philosophy of knowledge sketched here goes beyond its ability to account for the digital transition. It allows to precisely articulate the traditional ways by which people learn to solve practical problems (habit, craftsmanship, recipes) and the systematic ways that develop outside the realm of mathematics from the 17th century onwards. They all rely on the same reflective moment of inquiry, when it tries to generalise its successful deliberations. However, systematic thinking operates a huge break in the modality of this reflection, as it tries to ground it in exact meanings which allow for an exhaustive description of all problems in a given field – what I have called *global inquiry*, taking inspiration from Dewey. This articulation explains why traditional and systematic problem-solving only overlap. Systematicity allows old problems to be solved differently or new problems to be solved altogether (see above 3.3.3). Conversely, problematic situations which do not have any regularity do not lend themselves to systematization, and are best solved by the human capabilities that Aristotle called cleverness or prudence (see above 4.1.1). This gap translates in the difference, many times commented in this research, between canonic and Procrustean rules, in Ryle's terms. The first ones support inquiry by guiding judgment, the second ones replace inquiry by exact ratiocinations

https://doi.org/10.1515/9783111494500-009

where judgement has no place. Exactness here has nothing to do with precision. Judgement may be guided by an exquisitely precise decision tree, making it ever more sensitive to circumstances by further distinctions, but still remain necessary. Exactness is a matter of *method*, which consists in agreeing in advance on a set of material meanings – however rough their definition may be, as long as they are agreed upon – and then expressing all problems and procedures in terms of these meanings alone. Exactness is nothing else than the logical status of Euclid's postulates (see above 7.4.1).

This attitude to problem-solving can apply to any field of human practice, as long as its meanings can be organized in verified systems of inferences. The traditional distinctions between science and technology, or between theory and practice, become here irrelevant. Martin-Löf's theory of type shows that knowing-how and knowing-what are conjoint in any problem-solving, whatever its domain. There are descriptions and explanations in the most practical disciplines, and prescriptions and constructions in the most theoretical ones. What is fundamentally relevant is rather the division of work between global inquiry and local inquiries, or to say it differently, between system-building and problem-solving. Global inquiry is not a cousin of local inquiries, that could be distinguished by its size, duration or methods. It is rather like their step-mother; it articulates local inquiries into research programs, interprets their results, and redefines occasionally its own scope, adopting new inquiries as its own or refocusing on some core ones – thus the constant flux and crossing-over of disciplines.

Finally, this research explains why the two views outlined in the introduction regarding the very possibility of systematization are unsatisfactory. As it was immediately underlined against the intellectualists, it is simply not true, from a phenomenological perspective, that deliberation proceeds by applying rules. Rules are rather artefacts consciously constructed after the fact, in order to facilitate future inquiries. When they make up a fully effective procedure, deliberation simply disappears. This is why it is also wrong to present effective procedures as gracefully emerging from forms of of life. That may be true of procedures in the broad sense, but systematic thinking lies entirely in the conscious break with habitual forms of life and in the project to submit them to a critical reconstruction and verification.

This is where new questions arise. Where does the expansion of systematic problem-solving come from? Does it appear in time, or is it inherent in the human spirit? Are the conditions of its development historical? Kant, in a formula that has remained famous, had interpreted it as a tension peculiar to the human understanding, "at all times busy scanning phenomena in search of some rule in them".[1] The German term used by Kant, *durchspähen*, evokes the idea of a gaze that carefully scans the horizon in search of the slightest trace of regularity, and *jederzeit* (*at all times*), preferred to the more banal *immer* (*always*), evokes being on alert.

I do not, of course, intend to deal with these questions here, nor to enter into the rich debate among historians of ideas and culture concerning the much criticised notion of *European miracle*, its singularity, and its possible causes.[2] Nor is there any question of discussing the *interpretation* of this fact, particularly that of Heidegger, who closely associates it with the *project* of metaphysics.

This research, however, provides three highlights for the discussion of why the idea of *systematic* problem-solving, which does not seem to depend on material technical progress, appears so late in our intellectual history.

First, Plato and Aristotle clearly formulate this systematic idea, and they *do* have the idea of applying it to human affairs: According to Martha Nussbaum, technique (τέχνη) is conceived from the outset by the Greeks as the set of undertakings mobilising reason to control the effectiveness of action and remove it from the vagaries of fortune (τύχη) (see above 4.1.3). However Aristotle, more than Plato, is sceptical about the possible systematisation of techniques, because they appear to him as endowed with a *weak* universality. Without even mentioning the Aristotelian metaphysics of contingency, the observation of the diversity and historically situated character of human cognitive ecologies (see 3.1) gives this view a very strong plausibility. The difficulties that reductions in art had to formulate universal problem-solving methods confirm the validity of this common-sense position.

Here lies the second point, which is, quite simply, that this idea is *difficult*, and that difficult ideas take time. One needs only think of the time taken by mathematicians to articulate the idea of algebra in all its generality. Yet this idea was necessary for Lovelace to seriously consider that it may be possible to algebrize plans of action, and to glimpse the possibility of a *science of operations*. The attitude towards human practices had to change, in order for engineers, administrators and entrepreneurs to begin to re-imagine them as absolutely and exhaustively controllable, and to undertake to rationally reconfigure, not only their methods, but also the environments and situations in which they must be applied.

It is certainly not indifferent, in this respect (and this is the third remark), that this evolution of the practical stance took place at the same time as that of the scientific one, and that the decisive contact between science and technology occurred

1 *"Dieser ist jederzeit geschäftig, die Erscheinungen in der Absicht durchzuspähen, um an ihnen irgendeine Regel aufzufinden"*, *Critique of Pure Reason*, A 126, AA 04: 92.

2 See, on this subject, the discussions in the collective work by Roberts, Schaffer and Dear (2007); another thesis is by Mokyr (2016).

around the idea of the *machine*. This is where the possibility of mathematising natural phenomena as well as artefacts is conceived in all its clarity, *i.e.* of systematising the understanding of natural phenomena as well as human practices. It is therefore not accurate to say, with Knuth, that computer science should have existed before the invention of computers. The conception of the general possibility of systematic problem-solving certainly does not depend on the *empirical* existence of machines capable of carrying out any procedure, it requires however their concept to be worked out in all its clarity and universality.

The last remark concerns the very terms *problem* and *problem-solving*. I pointed out in the introduction how late (mid-20th century) was the emergence of this phrase to denote a large and quite generic domain of human inquiry. This expansion was inspired to psychological educationalists by mathematical problem-solving in a classroom setting, and their runaway success shows that they evidently filled a conceptual need, that of giving an account of epistemological practices that had spread across many theoretical and practical disciplines.

People confront much broader problematic situations – intrigues, passions, ethical puzzles, aesthetic knots, or spiritual questions, to name but a few. If it is obvious to the layman that such situations cannot be reduced to systematic treatment, it is much less obvious to philosophers to say exactly why – and some still try to provide such a reduction. One could, following Aristotle, point to the irregularity and complexity of human affairs. A bolder hypothesis is also possible: that the instrumentalist flavour of the concepts manipulated in this research shows that they are only adequate to a very specific form of inquiry; and that entirely different sets of concepts are required to make sense of problems and inquiries in other regimes of truth.

Appendices

A The General Problem Solver

A.1 Introduction and illustration of the algorithm

The GPS is an interesting algorithm for at least three reasons. Firstly, it is, if not the first, at least the most emblematic of the first artificial intelligence programs. Published in 1959, it caused a sensation, no doubt because of its claim to generality (the G of GPS). Its abstraction allows it to deal with a whole class of problems with a minimal structure. So in 1959, GPS could really give the impression that the machine was capable of reasoning.

The second reason for its interest is that the GPS implements the famous heuristic of ends and means which Aristotle presents in *Nic. Eth.* III.3 (see above 4.1.3). The procedures composed in this way correspond to the intuitive idea of a plan transforming a situation in sequential steps from an initial state to a final one. The interest of the algorithm lies in its model of the situation, which it represents as a collection of atomic attributes that are independent and indifferent to each other. Actions, on the other hand, can only remove or add attributes to the situation. This model is admittedly crude, but it is flexible enough to allow the representation of a large number of simple problems. Its interest lies in showing that it is precisely in the representation of the situation that an essential part of problem-solving is played. It thus illustrates in a simple manner two important theses defended in the book:

1. The four principles identified for systematic problem-solving (see above 7.4.3) are clearly visible: the principle of locality (each operator depends on a finite number of conditions in the situation), the principle of composition (actions transform the same situation), the principle of induction (since the combinatorial range of situations is finite), and the principle of bounded exploration (as the tree of candidate sequences is finite in length and cannot loop on itself).
2. The heuristics proceeds according to a logic entirely different from the resulting procedure, illustrating the difference between inquiry and its procedural reconstruction (see above 6.2.2).

The third reason for its interest is that, despite such simple situations, the difficulty of achieving systematicity is obvious. For example, the program we are about to present can easily be fooled by the order in which the objectives are presented.

Eckroth (2014) explains very well how Newell, Shaw and Simon (1959)'s GPS works, taking the classic example of the chimpanzee and the banana. The chimpanzee wants to stop being hungry (the desired end state). In the initial situation, the bananas are hanging from the ceiling in the middle of the room; they are accessible

https://doi.org/10.1515/9783111494500-010

to the chimpanzee if it climbs onto a chair and has its hands free. The chair and the chimpanzee are near the door and the chimpanzee has a ball in his hand.

The algorithm takes three sets of information:

1. A series of valid initial states, here:
   - "The chair is near the door".
   - "The chimpanzee is near the door".
   - "The chimpanzee is on the floor".
   - "The chimpanzee has a ball in its hands".
2. A series of target states, here simply: "not hungry".
3. A series of operators, each defined by four elements:
   - the name of the action, e.g. "catch bananas";
   - its preconditions, *i.e.* the states which must be valid for it to take place, here: "empty handed", "at bananas";
   - the states that become valid after the action has been performed, here: "has bananas";
   - the states that become invalid after the action is performed, here: "empty handed".

A.2 Running the algorithm

The result of the algorithm is illustrated graphically in figure A.1. Each card describes an action with its preconditions and its consequences. The arrows show the solution found by the GPS algorithm, going from the desired end state to its preconditions, until it finds some which are all already met.

The execution of the program of (Eckroth 2014) gives the result displayed in listing A.1. Line 22 displays the desired result, *i.e.* the sequence of actions taken to solve the problem.

Listing A.1: GPS program run

```
0 Achieving: not hungry
0 Consider: eat bananas
    1 Achieving: has bananas
    1 Consider: grasp bananas
        2 Achieving: at bananas
        2 Consider: climb on chair
            3 Achieving: chair at middle room
            3 Consider: push chair from door to middle room
                4 Achieving: chair at door
                4 Achieving: at door
```

```
        3 Action: push chair from door to middle room
        3 Achieving: at middle room
        3 Achieving: on floor
    2 Action: climb on chair
    2 Achieving: empty handed
    2 Consider: drop ball
        3 Achieving: has ball
    2 Action: drop ball
  1 Action: grasp bananas
0 Action: eat bananas
```

```
['Executing␣push␣chair␣from␣door␣to␣middle␣room', 'Executing␣climb␣
    on␣chair', 'Executing␣drop␣ball', 'Executing␣grasp␣bananas', '
    Executing␣eat␣bananas']
```

With these two visualisations, one understands intuitively how this algorithm works. It starts with the series of target states, which it tries to achieve one after the other. The program then displays `Achieving: xxx`, preceded by the number of target states to be achieved.

For each one, it tries to successively apply the operators which make one of these states valid. The program then displays `Considering: xxx`. The operator preconditions are added to the series of target states to be achieved. If they are all part of the set of current states of the system, then the program executes the action (it displays `Achieving: xxx`), modifies the set of current states accordingly, and moves on to the next target state. Otherwise, it proceeds recursively and tries to achieve the missing preconditions. If an impossibility occurs (no more operators to apply, or circularity) the branch being explored is abandoned.[1]

1 The interested reader can explore the code of Eckroth 2014, available on GitHub and archived by Software Heritage.

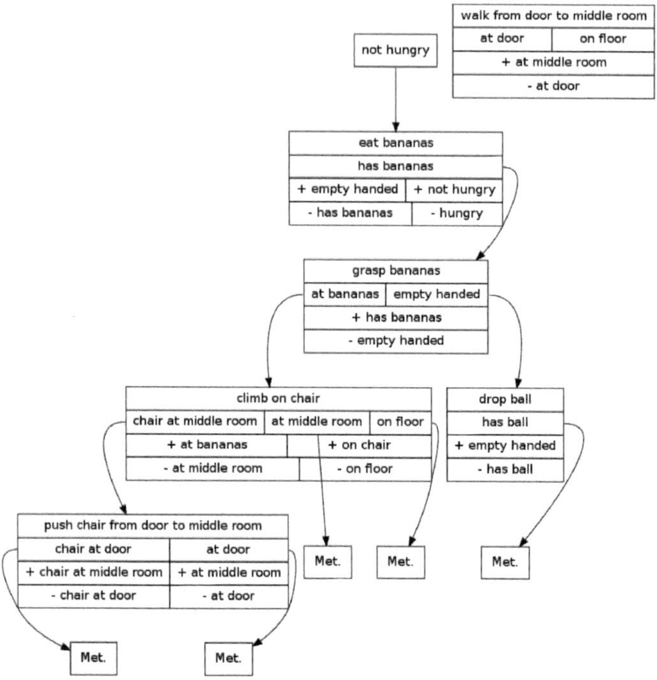

Figure A.1: GPS illustration
Source: (Eckroth 2014)

B A constructive interpretation of geometry

B.1 Interpretations of geometry

Is the justification of the identity of the rules of judgement and of the transformation of judgement, as two moments of the same *act*, suitable for mathematical inquiry in general (see above section 5.3)? In this appendix, this thesis is put to the test of a particular theory, Euclidean geometry. There are many reasons for this choice. Firstly, geometry was historically decisive in the advent of the form of systematic rationality. Secondly (in its classical form), its propositions take a dual form, that of problems that require to construct figures, and that of theorems that state truths about them. This duality, which was a cause of debate from the outset, lends credence to the dichotomy proposed by Pólya, between problems-to-find and problems-to-prove. It poses therefore a challenge to my interpretation. Finally, Euclidean geometry has been the subject of active research by contemporary logicians. Jean-Yves Girard has thus developed a *geometry of interaction* to obtain a mathematical definition of the notion of algorithm; in homotopy type theory (whose common acronym is "HoTT"), the geometric inspiration is also explicit, in the constructive sense of the term:

> The homotopic theory of types provides a synthetic description of spaces, in the following sense. Synthetic geometry is geometry in the style of Euclid: one starts from certain basic notions (points and lines), constructions (a line connecting any two points) and axioms (all right angles are equal), and logically deduces the consequences. This is in contrast to analytic geometry, where notions such as points and lines are represented concretely using Cartesian coordinates in $\mathbb{R}^n$ – lines are sets of points – and where basic constructions and axioms are derived from this representation. (Univalent Foundations Program 2013: 61)

All these projects are rooted in the tradition of intuitionist logic which, as early as the 1930s, proposed to consider the notions of *problems* and *acts of knowledge* as fundamental in mathematics. It found in the 1970s an important revival of interest through the constructive theory of types, a major figure of which is the logician and philosopher Per Martin-Löf. Exploring these highly technical projects would take this research too far. On the other hand, comparing classical geometry and type theory, as simply as possible, promises to bring out a general idea of constructivity, rooted in the original practice of mathematics but also made accurate by recent progress in logic.

Attempts to express constructively Euclidean geometry was begun by Mäenpää and Von Plato (1990) and continued by Von Plato (1995), Beeson (2010), Naibo (2015) and Naibo and Panza (2021) among others. This research is still active, which shows that Euclidean geometry "resists", as it were, these formalisms.

https://doi.org/10.1515/9783111494500-011

This section starts with the description of the first attempt at a constructive re-expression of geometry due to Mäenpää and Von Plato (B.1.1), and details one of the examples they propose, concerning the construction of a parallelogram (B.1.2). The next discussion of this section (B.1.3) is devoted to an examination of its weaknesses, which will lead, in the next section, to propose a new interpretation concerning the duality of the rules of judgement and action.

B.1.1 The proposal by Mäenpää and Von Plato

Mäenpää (1997) denounces in these terms the axiomatic interpretation of geometry based on predicate logic:

> Hilbert's notion of abstract axiomatization, as illustrated in his *Grundlagen der Geometrie* (1899), reduces geometry to the proof of theorems, reducing the existence of mathematical objects to the coherence of the axiomatic system that implicitly defines them. [...] Hilbert's model was generalised throughout the mathematics of this century, and reduced it to the proof of theorems. Problem-solving, which was the primary concern of Greek mathematicians (Knorr 1986, ch. 8), has been discarded. (Mäenpää 1997: 210)

The constructive enterprise, on the contrary, aims in his view to restore the very content of Euclidean geometry, *i.e.* its constructions and demonstrations by faithfully following their arguments, rather than to retrieve its results by other axiomatic and inferential means. In this development we will follow this project that he develops with his colleague Von Plato in a 1990 article, and which the latter takes up again in 1995 (Mäenpää and Von Plato 1990; Von Plato 1995). Beeson's much more precise formalisations are placed from the outset in a Hilbertian (2010) or Tarskian (2015) framework, which makes them less relevant for analysing the distinction between theorem and problem.

The idea of Mäenpää and Von Plato is the following:

> Problems, propositions and sets are identified conceptually in type theory, they are equal types. [...] Accordingly, elements of a set, proofs of a proposition, and solutions to a problem are identified in type theory (Martin-Löf 1984, pp. 5–13). Solving geometric problems involves a phase of demonstration (*apodeixis*) in addition to a phase of construction. One first constructs an object and then demonstrates that it really is of the required kind. Hence a logic of problems must be suitable for proving propositions as well as carrying out constructions. Intuitionistic type theory treats them as formally identical activities. It also expresses constructions explicitly, which allows the division of geometric proofs into the phases of construction and demonstration to be reflected formally. (Mäenpää and Von Plato 1990: 280)

The first and third postulates can be written, for example, with the following introduction rules:

$$\text{Post.1}: \frac{a : \text{Point} \quad b : \text{Point}}{\text{ln}(a, b) : \text{Line}} \qquad \text{Post.3}: \frac{a : \text{Point} \quad r : \text{Line}}{\text{c}(a, r) : \text{Circle}}$$

The first introduction rule says that from two Point objects a and b we can introduce a Line object ln(a, b), and the second says that from a Point a and a Line r we can construct a Circle object c(a, r). All these objects are associated with very specific acts that generate them: *mark* a point, *draw with the ruler* a straight line, *draw with the compass* a circle. These are the acts designated by the introduction rules. The second means, for example: *"If you have a point and a line, then it is assumed that you know how to draw with the compass the circle having this point as its centre and this line as its radius".*

The ambition of Mäenpää and Von Plato is to show, following Euclid, that all other acts of construction derive from these, such as tracing an equilateral triangle or a parallelogram.

Von Plato models properties of figures, such as that of distinction between two points DiPt(a, b) or between two lines DiLn(l, n), of apartness of a line and a point *Apt*(a, l), which means that a point is not located on a line, or that of convergence between two lines *Con*(l, m). He defines about twenty axioms, such as (Von Plato 1995: 172):

(1) $\vdash \text{DiPt}(a, b) \rightarrow \vdash \text{DiPt}(a, c) \vee \text{DiPt}(b, c)$;

(2) $\vdash \text{DiLn}(l, m) \rightarrow \vdash \text{DiLn}(l, n) \vee \text{DiLn}(m, n)$;

(3) $\vdash \text{Con}(l, m) \rightarrow \vdash \text{Con}(l, n) \vee \text{Con}(m, n)$.

These are separation axioms; (1) says that if two points are distinct, then a third point is distinct from at least one of them, (2) that the same is true with lines, and finally (3) that if two lines are convergent, then a third is convergent with at least one of them. It is important to note that these relationships are types, in the same way as Point, Line objects. They can enter as conditions in the construction of new objects, as here, to mark a point of intersection between two converging lines:

$$\frac{l : \text{Line} \quad m : \text{Line} \quad \vdash \text{Con}(l, m)}{\text{pt}(l, m) : \text{Point}}$$

Note that in his 1995 paper, Von Plato rewrites Euclid's postulate 1 by adding the condition that a and b are distinct:

$$\text{Post.1}: \frac{a : \text{Point} \quad b : \text{Point} \quad \vdash \text{Dipt}(a, b)}{\text{ln}(a, b) : \text{Line}}$$

I come back to this point later.

This choice allows Mäenpää and Von Plato to propose the unification between theorem and problem in a single form of judgment, as seen in 5.3.4. In order to fully understand what this formalisation means, we will follow an example of a problem that Von Plato details, that of constructing a parallelogram (Von Plato 1995: 186–188).

B.1.2 The construction of the parallelogram

The input data for the parallelogram problem (the x of Mäenpää) are three unaligned points, which are written thus in Von Plato's formalism:

(1) $(a, b, c : \text{Point};$

(2) $\vdash \text{DiPt}(a, b);$

(3) $\vdash \text{Apt}(c, \text{ln}(a, b))).$

– where (1) states that a, b, c are points, (2) that a and b are two distinct points, and (3) that the point c does not lie on the straight line (ab) (Apt = "apart"). In other words, these three formulae invite us to consider three distinct non-aligned a, b, c points. Von Plato calls this the *initial context* of the problem. The desired object $y : B(x)$ is a

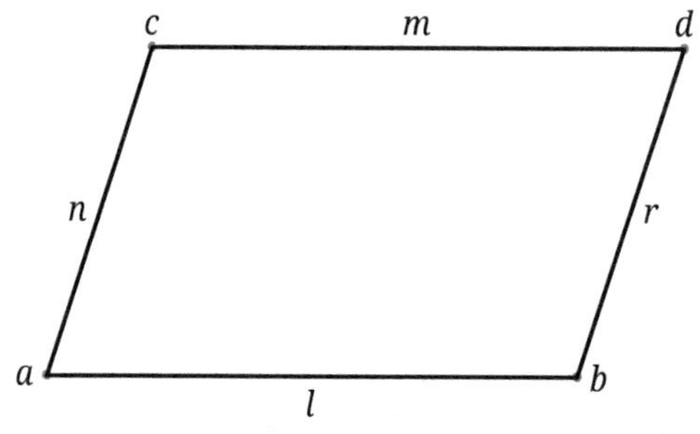

Figure B.1: Constructing a parallelogram

parallelogram, which can be defined as a construction of four points and four lines:

(1) $a, b, c, d : \text{Point};$

(2) $l, m, n, r : \text{Line};$

The $B(x)$ object must satisfy the $C(x)$ conditions, which are as follows:

(3) $\vdash \mathrm{DiPt}(a, b); \vdash \mathrm{DiPt}(c, d); \vdash \mathrm{DiPt}(a.b, c.d);$

(4) $\vdash \mathrm{Inc}(a.b, l); \vdash \mathrm{Inc}(c.d, m);$

(5) $\vdash \mathrm{Inc}(a.c, n), \vdash \mathrm{Inc}(b.d, r);$

(6) $\vdash \mathrm{Par}(l, m); \vdash \mathrm{Par}(n, r);$

(7) $\vdash \mathrm{DiLn}(l, m); \vdash \mathrm{DiLn}(n, r).$

(3) requires that the points are distinct two by two, (4) and (5) that the four straight lines are indeed those connecting the four points together (Inc = "incidental"), and finally (6) and (7) that the straight lines are parallel and distinct two by two.[1]

Solving the problem proceeds as follows:

1. Starting from a, b, c the lines $\mathrm{ln}(a, b)$ and $\mathrm{ln}(a, c)$ can be constructed by postulate 1 mentioned above.
2. From the point c and the line $\mathrm{ln}(a, b)$, a line parallel to the latter, passing through c, can be constructed, and another from the point b and the line $\mathrm{ln}(a, c)$. This procedure is defined by the following construction rule:

$$\frac{l : \mathrm{Line} \quad a : \mathrm{Point}}{\mathrm{par}(l, a) : \mathrm{Line}}$$

   Von Plato makes this a postulate in his axiomatics, for foundational reasons, but this procedure is well constructed by proposition I.31 of the Elements, which uses I.23, and which requires no less than ten tracings of circles!
3. The point d is finally constructed as the intersection of these last two straight lines. To do this, it must be shown that they are convergent. Here Von Plato uses the following separation axiom:

$$\vdash \mathrm{Con}(l, m) \rightarrow \vdash \mathrm{Con}(l, n) \vee \mathrm{Con}(m, n)$$

   Which says that a third line intersects at least one of two converging lines. Since the lines $par(ln(a, b), c)$ and $ln(a, c)$ are convergent at c, according to this axiom the line parallel to $ln(a, c)$ passing through b is necessarily convergent with the first, since it is not convergent with the second.
4. The properties requested concerning the points and lines thus obtained must be verified.

1 Von Plato uses the following abbreviated notation, which holds for any property Ppt: $\vdash \mathrm{Ppt}(x.y, z) \equiv \vdash \mathrm{Ppt}(x, z) \wedge \mathrm{Ppt}(y, z)$.

This proof is all contained in the final context, where we find the four moments of the construction:

(1.1) $(a, b, c : \text{Point},$

(2.1) $\text{ln}(a, b), \text{ln}(a, c) : \text{Line},$

(2.2) $\text{par}(\text{ln}(a, b), c), \text{par}(\text{ln}(a, c), b) : \text{Line},$

(1.1) $d \equiv_{def} \text{pt}(\text{par}(ln(a, b), c), \text{par}(ln(a, c), b)) : \text{Point};$

(3) $\vdash \text{DiPt}(a, b), \vdash \text{DiPt}(c, d), \vdash \text{DiPt}(a.b, c.d);$

(4) $\vdash \text{Inc}(a.b, \text{ln}(a, b)), \vdash \text{Inc}(c.d, \text{par}(\text{ln}(a, b), c));$

(5) $\vdash \text{Inc}(a.c, \text{ln}(a, c)), \vdash \text{Inc}(b.d, \text{par}(\text{ln}(a, c), b));$

(6) $\vdash \text{par}(\text{ln}(a, b), \text{par}(\text{ln}(a, b), c)), \vdash \text{par}(\text{ln}(a, c), \text{by}(\text{ln}(a, c), b));$

(7) $\vdash \text{DiLn}(\text{ln}(a, b), \text{par}(\text{ln}(a, b), c)), \vdash \text{DiLn}(\text{ln}(a, c), \text{par}(\text{ln}(a, c), b)).$

We can check that all the expressions in this context correspond line by line to the specification given above, and that they are relative to the starting elements a, b, c. For brevity, we have introduced the term d, but it can be replaced everywhere by the right-hand side term of the equation (1.1). The proofs of these assertions about relations between these objects, simply introduced by $\vdash$ (which means "that can be proved"), are omitted for brevity, but they follow from the axioms posited by Von Plato, as we have seen for the case of the convergent of two parallel lines.

In this new reading, the object $y = B(x) : C(x, y)$ appears in all its complexity: it is not just the procedure that constructs the object specified by the problem, but the one that constructs it *from* an initial context, and exhibits it in a final context where the properties necessary for its verification are immediately readable.

This final object can be presented either as a theorem or as a problem, depending on whether one chooses to highlight a figure or a relation (for example that the lines *par*(*ln*(*a*, *b*) and *par*(*ln*(*a*, *c*) are simply convergent, and whether or not one asks for a proof.

Here would appear the profound identity of theorems and problems in the interpretation of Mäenpää and Von Plato: whether it is a proof or a construction, it is a matter of transforming an initial context into a final context, which constructs new objects from those initially given, whether these objects are elementary objects (points, triangles, etc.), magnitudes, or proofs of relations between these objects or these magnitudes.As Von Plato himself says:

> From the above parallelogram construction, a general way of looking at solutions of geometric problems is suggested: The solutions are mappings from the *data context* of the problem into its *goal context*. We can give explicit expressions to these mappings in type theory, as functions between contexts. (Von Plato 1995: 188)

This is how all the proofs of geometry should be represented, whether they are theorems or problems: as functions of transformations of geometrical contexts, which include plots, observations, and measurements.

B.1.3 Insufficiencies of this interpretation

The formalisation of Mäenpää and Von Plato is not entirely satisfactory. Firstly, the unification of geometric objects (points, lines, circles) and their properties (distinction, separation, convergence, etc.) under a single title called "type" appears rather artificial. It certainly makes it possible to formally present the operations of construction and demonstration within the unified framework of the single formalism of natural deduction, but the convergence stops there. Properties are types clearly marked by $\vdash$, their objects are always omitted, and their derivation is carried out according to the classical rules of propositional logic, starting from axioms. Now the many axioms introduced by Von Plato do not appear in Euclidean geometry, which has only definitions, postulates and common notions.

Moreover, propositions and geometric objects remain distinguished implicitly in the very formulation of the notion of problem as $\forall x, \exists (y : B(x)) : C(x, y)$, where C is a proposition and B a geometric object. If the types were really treated equivalently, this distinction would fall away, and there would be nothing to stop the complex types B and C being represented by a single type that aggregates them, such as $D(x) \equiv_{def} \exists (y : B(x)) : C(x, y)$.

This is precisely what Von Plato arrives at with his notion of *context*, which is, however, unsatisfactory. Imported from logic, it presents the complex geometric object that the mathematician constructs as a disordered and incomplete list of objects and propositions. This list is disordered, since one could just as easily have grouped together, as Von Plato does, the points with their properties, the lines with theirs, and kept the incidence relations between points and lines at the end. This list is also incomplete, as many more properties could be deduced from those presented, such as $\vdash \mathrm{Apt}(d, n)$ or $\vdash \mathrm{Apt}(d, l)$. The fragmentation of the constructed geometric object into a multiplicity of types that must then be brought together in the artificial unity of a context does not translate the unity of the inquiry.

Secondly, it is not very clear what justifies the introduction of propositions. When one translates the starting hypothesis of a geometric statement, for example *"Let a and b be two distinct points"*, by the position of two objects a, b of a type Point and the assertion that a parametric type $\vdash \mathrm{Dipt}(a, b)$ is valid, this assertion must, in all rigour, also have a proof or object. However it is not said what it might be. It seems to be an act of "reading" the properties of the object consisting of the two points, but Von Plato does not explain what these elementary acts are.

One could propose introducing elementary acts of observation, such as:

$$\mathrm{obsdi}(a, b) : \mathrm{DiPt}(a, b)$$

which would denote an act of observing that a and b are indeed two distinct points. However such an act cannot be valid as a proof in Euclidean geometry. It is indeed *obvious*, for example, that the orthocentre and barycentre of a triangle are in general distinct, but geometry does not accept unproven observations. The answer will be: the canonical acts $\mathrm{DiPt}(a, b)$ are not observations, but conditions – for example they could be constructed from a segment (a, b). However in Von Plato's logic, it is the segments that are constructed from two distinct points, not the other way round. One could consider the segments as the original objects of our logic, and define the points as extremities of the segments. This is the solution chosen by Naibo and Panza, however it has the defect of inverting the intuitive priority one generally attributes to points over segments, which can be seen in the order of definitions given by Euclid himself. Another solution, finally, is to make an elementary act of the *choice* of one point distinct from another. But in this case, what is constructed is not this relation but the point itself.

Finally, Mäenpää and Von Plato do not deal with the essential properties of figures which are magnitudes (distances, angles, areas) on which an algebraic calculation, a congruence relation and an order relation operate. Magnitudes thus seem to behave as *distinct* types of figures, but neither their rules of introduction nor their rules of use are defined.

These reflections lead us to a negative conclusion: in Mäenpää and Von Plato's formalisation, the proofs of theorems are not special cases of geometrical problems, except in the very superficial sense that one can find a logical formalism unifying their presentation. For the same reason, the properties of figures are not types in the same way as a point, a line or a circle. Rather, they concern the equality of figures ("this point, constructed in such a way, is *the same* as this one, constructed in another way"), the order of points relative to each other, and the comparative magnitude (of lines, angles, closed figures). But Mäenpää and Von Plato's formalisation says nothing about all this.

B.2 A new constructive interpretation of geometry

This section sketches a new constructive interpretation of geometry, which significantly complements the interpretation of the common nature of the rules of judgment and action, itself based on the complementarity of the notions of *proof* or *construction* on the one hand, and *proposition* or *object* on the other, the former being the acts that produce or make manifest the latter. We show that the complementarity of the constructive and theoretical contents of geometry cannot be interpreted in this way.

I begin by highlighting the problematic nature of the *properties* by which the figures of geometry are specified, in particular the magnitudes (distances, angles, areas) that play an essential role in the logical framework of the Elements (B.2.1). The detailed study of a theorem (I.13) shows that the discourse on magnitudes proceeds through acts of "reading" the figure. It is thus possible to state relations concerning them, which can then be transformed thanks to the rules of inference constituted by common notions (B.2.2). The study of a sequence of theorems deriving from I.13 shows that their purpose is to make licit more and more sophisticated acts of reading, in the same way that problems learn to construct more and more complex objects (B.2.3).

B.2.1 Questions and leads

Let us start with the central idea that the main objects of geometry are the figures – points, segments and circles – that it can construct and compose in the plane. There are elementary construction methods for these objects. For segments and circles, they are postulates 1 to 3. As for points, their elementary acts are to be *marked* at particular situations, which may be fully determined (intersection of two converging lines), finitely determined (there are two intersections of two incident and non-tangent circles) or infinitely indeterminate (any point on a segment or circle, or distinct from any other object already constructed). The indeterminacy of these points represents the inputs to the problem, with respect to which the latter behaves like a function. By this I make a first proposition which departs from Mäenpää and Von Plato's theory, in that I consider it possible to mark a point *distinct* from a given figure, and in particular of another point.

These initial constructions may be composed together to form methods for constructing elementary objects (such as drawing the line parallel to another, passing through a given point) or complex figures, such as a parallelogram: this is the purpose of *problems*. The exact form of these constructive principles for expressing Euclidean geometry may be written in a formal language such as Coq.

Two difficulties immediately arise: firstly, the constructions cannot be composed arbitrarily, and certain sequences of instructions may be incorrect. For example, one cannot be requested to mark two distinct points at the intersection of two straight lines (since there is only one), or to mark the intersection of two circles whose centres are further apart than the sum of their radii (since there are none). There seem to be objective properties of Euclidean space that prohibit certain constructions.

On the other hand, how can *interesting* problems be specified? It doesn't seem possible, as with natural integers, to proceed by recursion to define interesting properties and functions like **even()** or **double()**, as was done above (5.3.1).

When one looks more closely at what the *Elements* are talking about, it appears that it is about *relations* between figures. Thus there are relations of coincidence (if three equal segments, each starting from a point on a circle, have a common end, then *that end is the centre of the circle*, III.9) of distinction (if two circles intersect then they do not have the same centre, III.5), intersection (two circles intersect at no more than two points, III.10), parallelism (if a transversal to two lines makes their opposite angles equal, then these two lines are parallel, I.27), relative situation (the chord between two points of a circle is inside this circle, III.2).

However, the relations most present in the Elements, which form the bulk of its conceptual framework, are comparisons of *magnitudes* between figures, mainly distances between points (or lengths of segments), angles and areas. There is almost no proposition that does not appeal to them, either in terms of specification (constructing an *equilateral* triangle, I.1) or in terms of premises (triangles that have the same base and are equal have their vertex on the same parallel, I.39). Finally, even when no magnitude is mentioned in the wording of a theorem, as in proposition I.30: two lines parallel to a third are parallel to each other, the proof, itself, calls on it (in this case, Euclid uses the property of equality of opposite angles). Magnitudes are therefore used, in particular, to indicate which constructions are possible or impossible, especially concerning the intersections of figures. They are therefore an essential part of the logical framework of the Elements.

But what magnitudes are is problematic from a constructive perspective. The relations of coincidence or distinction between figures may be interpreted, at a pinch, as the elementary form of judgement $a = b$ that the theory of types admits. The relations of intersection, parallelism, or situation may be interpreted, for their part, in terms of the possibility or impossibility of constructions of points or lines. But relations between magnitudes seem to escape these interpretations. Two angles located at two distant vertices on a figure can be said to be equal by virtue of an algebraic reasoning that turns away from construction and seems to abstract from it.

So what are all these relations, and in particular the magnitudes? If they are types, what are their rules of introduction and elimination? And above all, how is one to understand that they *constrain* the constructions of figures, which one thought were the primitive types of geometry?

It is tempting here to agree with Plato, and interpret these properties as objects in their own right, characteristic of space, accessible by pure acts of observation –

by properly "theoretical" acts. Theorems, in particular, would play the essential role of bringing them to light. This would lead to the recognition that there are indeed two forms of intellectual acts, of construction and of observation, and therefore also, distinct rules for each of these forms. The analyses that follow confirm this hypothesis; however, far from calling into question a constructive interpretation of geometry, it reinforces and enriches it.

B.2.2 An angular theorem

In order to answer these questions, it is useful to study in detail one of the first theorems of the Elements, Proposition I.13, which has the advantage of appealing only to common notions and to an intermediate construction (see figure B.2):

Theorem. *If a straight line lies on another straight line, it forms either two right angles or angles whose sum is equal to two right angles.*

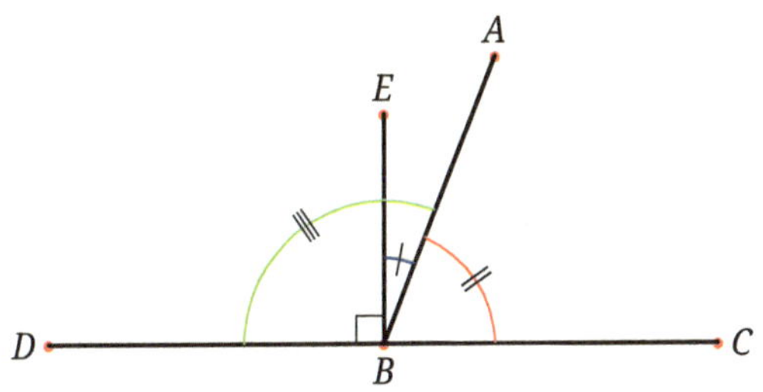

Figure B.2: Theorem I.13 of the *Elements*: sum of supplementary angles

Here is the text of Euclid's demonstration:

Proof. Let any straight line AB lie on a straight line CD, and form angles CBA and ABD. I say that angles CBA and ABD are two right angles, or that their sum is equal to two right angles.

1. If angle CBA is equal to angle ABD, then they are both right angles. (definition 10)
2. If this is not the case, then BE must be drawn from point B forming a right angle with CD. (I.11) Angles CBE and EBD are therefore two right angles.
3. Since angle CBE is equal to the sum of the two angles CBA and ABE, add angle EBD to each of them. The sum of angles CBE and EBD is equal to the sum of the three angles CBA, ABE and EBD (common concept 2).

4. Similarly, Since angle DBA is equal to the sum of the two angles DBE and EBA, add angle ABC to each of them. The sum of angles DBA and ABC is equal to the sum of the three angles DBE, EBA and ABC (common concept 2).
5. But the sum of angles CBE and EBD was also found to be equal to the sum of the same three angles, and things that are equal to the same thing are also equal to each other, so the sum of angles CBE and EBD is also equal to the sum of angles DBA and ABC. But angles CBE and EBD are two right angles, so the sum of angles DBA and ABC is also equal to two right angles. (common concept 1)

Therefore, if a straight line is laid on a straight line, it forms either two right angles or angles whose sum is equal to two right angles. □

Leaving aside the trivial case where AB is perpendicular to CD (point no. 1), the demonstration comprises three forms of acts: first, the construction of an auxiliary line BE (point no. 2), then the "reading" of an additive relationship between angles ABC, EBA and EBC on the one hand, and between DBE, EBA and DBA on the other (points no. 3 and 4), and finally inferences drawn from these relationships thanks to the common notions (points 3–5). We can detail these last two forms of acts in the form of the following reasoning (here $2d$ is "two right angles", and CN = common notion):

$CBA + ABE = CBE$ ("basic additive reading") (1)

$CBA + ABE + EBD = CBE + EBD$ (from 1 and CN 2) (2)

$ABE + EBD = ABD$ ("basic additive reading") (3)

$CBA + ABE + EBD = CBA + ABD$ (from 3 and CN 2) (4)

$CBE + EBD = CBA + ABD = 2d$ (from 2, 4 and CN 1) (5)

What, then, is the nature of these inferences and acts of "reading" the figure? Concerning the inferences, it is clear that we are dealing with a propositional calculus whose common notions form the rules of inference. This propositional calculus includes a few elementary forms, concerning the equality or inequality of two angles, or the additive relationship between three or more angles. We thus find ourselves in the case predicted by type theory, where propositions are themselves taken as objects of a type subject to specific rules (see above 5.3.2).

It therefore remains to explain the nature of the acts of "reading" that one finds in points no. 3 and no. 4 of the demonstration. They concern the additive relationship between three angles, *i.e.* given three segments having one same vertex, the sum of the two angles formed by the inner segment with the two outer segments is equal to the angle formed by the latter two (in the diagram, for example BA is the inner segment to the two outer segments BE and BC).

The possibility of such a reading is not justified by Euclid, as it has been noted for example by Joyce (1998). It seems to be rooted in the obvious principle that the whole is the sum of its parts. In any case, it can be formalised in type theory as a rule forming a proposition from three lines of the same vertex. It is therefore an act in its own right, of a nature entirely distinct from that of acts of construction, which makes it possible, as it were, to form propositional objects from certain figures.

It is important here to note that, like elementary constructions, the "acts of reading" are reduced to the strict minimum by Euclid. Thus equality between angles is reduced to the case of right angles, and is the subject of postulate 4 ("all right angles are equal to each other"). All other equalities between angles result from inferences using common notions.

Theorem I.13 thus makes it possible to relate two acts of a different type: that, on the one hand, of constructing a line incident to another, and that, on the other hand, deduced by acts of reading and inference, of judging that the angles formed by this line are supplementary (*i.e.* of a sum equal to two rights). It thus forms a kind of "coupling of meanings" linking two judgements, one concerning a construction, and the other concerning magnitudes. Theorem I.14, moreover, closes this "capsule of meaning", by showing the reciprocal proposition: if the sum of two adjacent angles are equal to two rights, then the two outer segments of these angles lie on the same line.

The theorems are used in two ways. On the one hand, they allow us to develop other theorems associating figures and relations of magnitudes; on the other hand, they allow us to demonstrate the possibility or impossibility of certain constructions. As far as the second point is concerned, an example is proposition I.27, which shows that if a transversal with two lines makes equal alternate angles, then *it is not possible* to mark a point of intersection between them (*i.e.* they are parallel). We will develop the first point in more detail, since we are seeking to elucidate the nature of theorems.

B.2.3 A sequence of theorems

I am now interested in the logical progression, schematised in figure B.3, of some theorems arising from I.13. Starting from the bottom, I.32, which establishes the famous theorem that the sum of the angles of a triangle is equal to two rights. The proof calls only on auxiliary constructions, common notions and Theorems I.13 and I.29. I.29, in turn, which establishes the angular properties relating to parallels intersected by a transversal, calls only on common notions, postulate 5 and Theorems I.13 and I.15. Finally, I.15, which establishes the equality of opposite angles, depends only on the common notions, postulate 4 and theorem I.13 (the common notions and auxiliary constructions are not shown on the diagram). One can see here that Proposition I.13, just studied, plays an essential role in the angular inferences of Book I of the Elements. A quick look at these theorems is useful, summarising their proofs

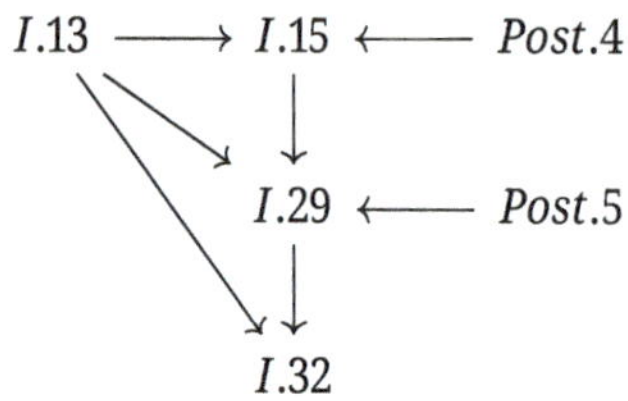

Figure B.3: Some developments of Theorem I.13

for the sake of brevity.

Theorem I.15 is a good illustration of how new angular relationships can be derived from the slightly expanded figure in Theorem I.13:

Theorem. *If two straight lines intersect each other, they make the angles at the vertex equal to each other.*

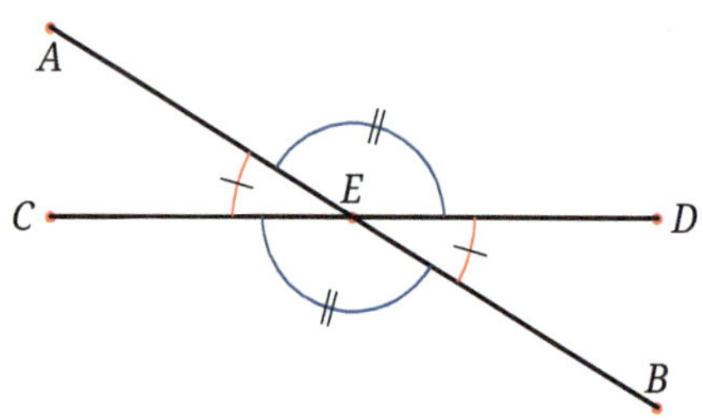

Figure B.4: Theorem I.15 of the *Elements*: equality of opposite angles

The proof operates in the same way, through reading and inference. Only this time, the acts of reading relate to the configuration already studied in proposition I.13, where one segment is only placed on top of another (without cutting it). On the one hand, this concerns AE, which is laid on CD in E, and on the other hand DE, which is laid on AB in E. It is very important to note the active reading that is required here: each time it is a matter of *selecting* a subpart of the complete figure, by "forgoing" the segment EB first, and afterwards the segment EC. These selections *make visible* the property already proved by I.13, which becomes available for propositional calculus:

$CEA + AED = 2d$ ("reading" prop. 13) (1)

$AED + DEB = 2d$ ("reading" prop. 13) (2)

$CEA + AED = AED + DEB$ (1, 2, Post. 4 and CN 1) (3)

$CEA = DEB$ (3 and CN 3) (4)

Theorem I.29, meanwhile, enriches Figure I.15 in the following way (see Figure B.5):

Theorem. *A straight line falling on parallel straight lines makes the alternate angles equal to each other, the outer angle equal to the inner and opposite angle, and the sum of the inner angles of the same side equal to two right angles.*

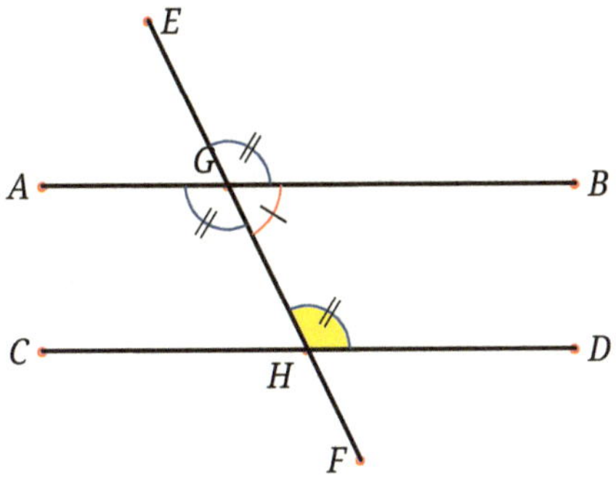

Figure B.5: Theorem I.29 of the *Elements*: equality of angles of a transversal with two parallel lines

I omit the demonstration, as it proceeds as for I.15, by acts of selective readings and by inferences following common notions. Angle *GHD* is central here: the theorem shows its equality on the one hand with the alternate angle *AGH*, on the other with the interior opposite angle *BGE*, and finally that its sum with the same-sided interior angle *HGB* is equal to two right angles. Two acts of reading are mobilised here, one from the configuration of type I.13 formed by the segment GH placed on the segment AB at G, and the other from that of type I.15 formed by the intersection of segments AB and EH at G.

There is finally Theorem I.32, which establishes the famous equality of the sum of the angles of a two-right triangle (see figure B.6).

Theorem. *In any triangle, if one of the sides is produced, the exterior angle is equal to the sum of the two interior and opposite angles, and the sum of the three interior angles of the triangle is equal to two right angles.*

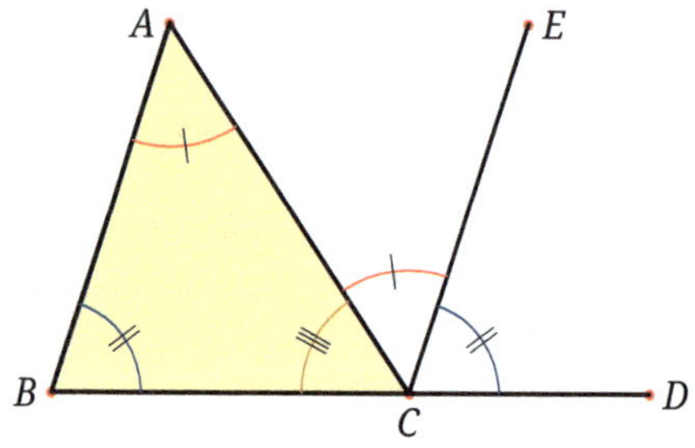

Figure B.6: Theorem I.32 of the *Elements*: equality of the interior angles of a triangle to two right angles

It is thanks to the construction of *CE*, the line parallel to *BA*, that this equality can be demonstrated. It should be noted that this construction plays no role in the final result. It is therefore possible that a property of a figure can only be made visible by "extraction" from a more complex figure, which is what gives Euclidean geometry all its subtlety. As in the previous examples, the reflection proceeds by four successive *selective readings* of the figure, which allows only sub-parts to be considered. A configuration of type I.29 is thus read on the *AB*, *AC*, *CE* lines, then another on the *AB*, *CE*, *BD* lines. An elementary "additive" reading is then performed on the three segments *CA*, *CE*, *CD*. Finally, a configuration of type I.13 is read on the *AC* segment laid on the *BD* segment in *C*. Angular equalities are extracted from all these readings and transformed through the arithmetic logic of common notions (step 4). Algebraic reasoning can then be deployed in steps 5 and 6:

$$CAB = ACE \qquad \text{(“reading” prop. 29)} \quad (1)$$

$$ABC = ECD \qquad \text{(“reading” prop. 29)} \quad (2)$$

$$ACD = ACE + ECD \qquad \text{“basic additive reading”} \quad (3)$$

$$ACD = CAB + ABC \qquad \text{(1, 2 and 3)} \quad (4)$$

$$BCA + ACD = BCA + CAB + ABC \qquad \text{(4 and CN 2)} \quad (5)$$

$$2d = BCA + CAB + ABC. \qquad \text{(“reading” prop. 13 and CN 1)} \quad (6)$$

Therefore, this sequence of theorems progressively enriches by demonstration the acts of reading angular equalities so that they can be applied to increasingly richer configurations; they also become less "obvious", if obviousness is related to their adjacency – as suggested by the initial act of reading which is content to add adjacent angles and which is, it must be admitted, postulated. These acts of reading therefore behave exactly like acts of construction, which, starting from initial postulates, build on each other in increasingly complex acts.

List of Figures

https://doi.org/10.1515/9783111494500-012

Bibliography

Académie Française [1694] (2001), *Dictionnaire de l'Académie Française (Première Édition)*, ARTFL, Chicago: University of Chicago, URL: https://artfl.atilf.fr/dictionnaires/ACADEMIE/index.htm (visited on 05/04/2025).

Alexander, Christopher (1964), *Notes on the Synthesis of Form*, Harvard University Press.

Allen, James V. (2015), "Practical and Theoretical Knowledge in Aristotle", in: *Bridging the Gap between Aristotle's Science and Ethics*, ed. by Henry Devin and Karen Margrethe Nielsen, Cambridge: Cambridge University Press, pp. 49–70.

Angier, Tom P. S. (2010), *Technē in Aristotle's Ethics: Crafting the Moral Life*, Continuum Studies in Ancient Philosophy, London: Continuum.

Anscombe, Gertrude Elizabeth Margaret (1981), *The Collected Philosophical Papers*, vol. 2. Metaphysics and the Philosophy of Mind, Minneapolis: University of Minnesota Press.

Anscombe, Gertrude Elizabeth Margaret [1957] (2000), *Intention*, 2nd ed., Cambridge, Mass: Harvard University Press.

Arendt, Hannah [1958] (2012), "Condition de l'homme moderne", in: *L'humaine condition*, ed. by Philippe Raynaud, trans. by Georges Fradier, Quarto, Paris: Gallimard, pp. 53–323.

Aristotle (1985 –), *Aristoteteles et Corpus Aristotelicum*, ed. by Maria C. Pantelia, Thesaurus Linguae Grecae Digital Library, Irvine, CA: University of California, URL: http://www.tlg.uci.edu (visited on 23/09/2023).

Aristotle (1984), *Complete Works of Aristotle: The Revised Oxford Translation*, ed. and trans. by Jonathan Barnes, 2 vol., Bollingen Series LXXI, Princeton: Princeton University Press.

Aristotle (2014a), *Œuvres: éthiques, politique, rhétorique, poétique, métaphysique*, ed. by Richard Bodéüs, Bibliothèque de la Pléiade 601, Paris: Gallimard.

Aristotle (1950–1962), *Organon*, trans. by Jules Tricot, 5 vol., Bibliothèque des textes philosophiques, Paris: Vrin.

Aristotle (2014b), *Premiers analytiques*, ed. and trans. by Michel Crubellier, GF, Paris: Flammarion.

Aristotle (2005), *Seconds Analytiques*, ed. and trans. by Pierre Pellegrin, GF, Paris: Flammarion.

Aristotle (2015), *Topiques ; Réfutations sophistiques*, ed. by Pierre Pellegrin, trans. by Jacques Brunschwig and Myriam Hecquet-Devienne, GF, Paris: Flammarion.

Aristotle (1960), *Posterior Analytics ; Topica*, trans. by E. S. Forster and Hugh Trennedick, Loeb Classical Library 391, Cambridge, Mass: Harvard University Press.

Aristotle [1973] (1994), *Posterior Analytics*, ed. and trans. by Jonathan Barnes, 2nd ed., Clarendon Aristotle Series, Oxford: Clarendon Press.

Aristotle (1993), *Analytica Posteriora*, ed. by Ernst Grumach, trans. by Wolfgang Detel, 2 vol., Aristoteles Werke in deutscher Übersetzung, Berlin: Akademie-Verlag.

Aristotle (1987), *Ethique à Nicomaque*, trans. by Jules Tricot, 6th ed., 2 vol., Bibliothèque des textes philosophiques, Paris: Vrin.

Aristotle [1926] (1968), *Nicomachean Ethics*, trans. by Henry Rackham, Loeb Classical Library, Cambridge, Mass: Harvard University Press.

Aristotle (2014c), *Nicomachean Ethics*, trans. by Charles David Chanel Reeve, Indianapolis: Hackett Publishing Co.

Aristotle (1948), *Métaphysique*, trans. by Jules Tricot, 2nd ed., 2 vol., Bibliothèque des textes philosophiques, Paris: Vrin.

Aristotle [1926] (1990), *Physique I-IV*, ed. and trans. by Henri Carteron, Collection des universités de France, Paris: Les Belles Lettres.

https://doi.org/10.1515/9783111494500-013

Aristotle (2013), *Le Mouvement des animaux - La Locomotion des animaux*, ed. by Pierre-Marie Morel, GF, Paris: Flammarion.

Arnim, Hans von, ed. [1905] (1964), *Stoicorum Veterum Fragmenta*, Stuttgart: B. G. Teubner.

Ashby, William Ross [1956] (1957), *Introduction à la cybernétique*, trans. by Marcel Pillon, Paris: Dunod.

Asper, Markus (2009), "The Two Cultures of Mathematics in Ancient Greece", in: *The Oxford Handbook of the History of Mathematics*, ed. by Eleanor Robson and Jacqueline Stedall, Oxford: Oxford University Press, pp. 107–132.

ATILF - CNRS and Université de Lorraine (1994), *Trésor de La Langue Française*, TLFi, Nancy, URL: http://atilf.atilf.fr/tlf.htm (visited on 04/03/2025).

Aubenque, Pierre [1962] (1972), *Le Problème de l'être chez Aristote: Essai sur la problématique aristotélicienne*, 3rd ed., Bibliothèque de philosophie contemporaine, Paris: Presses universitaires de France.

Ayer, Alfred Jules [1936] (1952), *Language, Truth and Logic*, 2nd edition, reprint, New York: Dover.

Babbage, Charles (1989), *The Works of Charles Babbage. 3: The Analytical Engine and Mechanical Notation*, ed. by Martin Campbell-Kelly, The Pickering Masters, London: Pickering.

Babbage, Charles [1832] (2010), *On the Economy of Machinery and Manufactures*, Cambridge: Cambridge University Press, (visited on 02/12/2022).

Bachimont, Bruno (2010), *Le Sens de la technique : le numérique et le calcul*, Paris: Encre Marine.

Backus, John (1978), "Can Programming Be Liberated from the von Neumann Style? A Functional Style and Its Algebra of Programs", in: *Communications of the ACM* 21.8, pp. 613–641.

Bacon, Francis [1620] (2004), *Novum Organum and Associated Texts*, ed. by Graham Rees and Maria Wakely, The Oxford Francis Bacon 11, Oxford: Oxford University Press, URL: http://ebookcentral.proquest.com/lib/ensparis-ebooks/detail.action?docID=3055170 (visited on 03/04/2025).

Bailer-Jones, Daniela M. (2002), "Scientists' Thoughts on Scientific Models", in: *Perspectives on Science* 10.3, pp. 275–301.

Barendregt, Henk P. and Erik Barendsen [1984] (2000), *Introduction to Lambda Calculus*, URL: https://www.cse.chalmers.se/research/group/logic/TypesSS05/Extra/geuvers.pdf (visited on 22/09/2023).

Barnes, Jonathan (1969), "Aristotle's Theory of Demonstration", in: *Phronesis* 14.2, pp. 123–152.

Barnes, Jonathan (1981), "Proof and Syllogism", in: *Aristotle on Science: The Posterior Analytics*, ed. by Enrico Berti, Università Di Padova, Padova: Editrice Antenore, pp. 17–59.

Barnes, Jonathan (2000), *Aristotle: A Very Short Introduction*, Oxford: Oxford University Press.

Beaujouan, Guy (1957), *L'interdépendance entre la science scolastique et les techniques utilitaires (XIIe, XIIIe et XIVe siècles)*, Les conférences du Palais de la Découverte, Paris: Palais de la Découverte.

Bechtel, William (2008), *Mental Mechanisms: Philosophical Perspectives on Cognitive Neuroscience*, New York: Routledge.

Beckmann, Johann [1806] (2017), *La technologie générale*, ed. by Guillaume Carnino, Liliane Hilaire-Pérez and Jochen Hoock, trans. by Joost Mertens, Histoire, Rennes: Presses universitaires de Rennes.

Beeson, Michael (2010), "Constructive Geometry", in: *Proceedings of the 10th Asian Logic Conference*, World Scientific, pp. 19–84.

Beeson, Michael (2015), "A Constructive Version of Tarski's Geometry", in: *Annals of Pure and Applied Logic* 166.11, pp. 1199–1273.

Bénis-Sinaceur, Hourya (2000), "Address at the Princeton University Bicentennial Conference on Problems of Mathematics (December 17–19, 1946), By Alfred Tarski", in: *Bulletin of Symbolic Logic* 6.1, pp. 1–44.

Bernard, Claude [1865] (2013), *Introduction à l'étude de la médecine expérimentale*, Champs Classiques, Paris: Flammarion.

Berry, Gérard (2017), *L'hyperpuissance de l'informatique: algorithmes, données, machines, réseaux*, Paris: Odile Jacob.

Bertalanffy, Ludwig von [1968] (2015), *General System Theory: Foundations, Development, Applications*, New York: George Braziller.

Blackwell, Alan F. (2002), "First Steps in Programming: A Rationale for Attention Investment Models", in: *Proceedings IEEE 2002 Symposia on Human Centric Computing Languages and Environments*, IEEE, pp. 2–10.

Blair, Ann M. (2010), *Too Much to Know: Managing Scholarly Information before the Modern Age*, Yale University Press.

Blåsjö, Viktor (2022), "Operationalism: An Interpretation of the Philosophy of Ancient Greek Geometry", in: *Foundations of Science* 27.2, pp. 587–708.

Bocconi, Stefania et al. (2016), "Developing Computational Thinking: Approaches and Orientations in K-12 Education", in: *EdMedia + Innovate Learning*, ed. by George Veletsiano, Vancouver, BC, Canada: AACE.

Bongard, Joshua and Michael Levin (2021), "Living Things Are Not (20th Century) Machines: Updating Mechanism Metaphors in Light of the Modern Science of Machine Behavior", in: *Frontiers in Ecology and Evolution* 9.

Bos, Henk J. M. (2001), *Redefining Geometrical Exactness: Descartes' Transformation of the Early Modern Concept of Construction*, New York: Springer.

Boswell, James (1887), *Life Of Johnson*, ed. by George B. Hill, vol. 2, Oxford: Clarendon Press, URL: http://archive.org/details/in.ernet.dli.2015.49034 (visited on 03/04/2025).

Bourdeau, Michel and Jean Mosconi, eds. (2022), *Anthologie de la calculabilité*, Paris: Cassini.

Bourdieu, Pierre (1980), *Le sens pratique*, Le Sens commun, Paris: Éditions de Minuit.

Bourdieu, Pierre [1972] (2000), *Esquisse d'une théorie de la pratique*, éd. revue par l'auteur, Points essais 405, Paris: Seuil.

Bourdieu, Pierre (2015), *Sociologie générale vol. 1: Cours au Collège de France 1981-1983*, Paris: Seuil.

Boyle, Robert (1674), *The Excellency of Theology Compar'd with Natural Philosophy*, London, URL: http://name.umdl.umich.edu/A28966.0001.001 (visited on 04/03/2025).

Brading, Katherine (1999), "The Development of the Concept of Hypothesis from Copernicus to Boyle and Newton", in: *Revista de Filozofie KRISIS* 08-09, pp. 5–16.

Bringsjord, Selmer (2015), "A Vindication of Program Verification", in: *History & Philosophy of Logic* 36.3, pp. 262–277.

Brioist, Pascal (2008), "La réduction en art de l'escrime au XVIe siècle", in: *Réduire en art: la technologie de la Renaissance aux Lumières*, ed. by Hélène Vérin and Pascal Dubourg Glatigny, Paris: Éditions de la Maison des sciences de l'homme, pp. 238–257.

Brooks, Frederick P. (1982), *The Mythical Man-Month: Essays on Software Engineering*, Boston: Addison Wesley.

Brunschwig, Jacques (1981), "L'objet et la structure des Seconds Analytiques d'après Aristote", in: *Aristotle on science: the Posterior analytics*, ed. by Enrico Berti, Università di Padova, Padova: Editrice Antenore, pp. 61–96.

Bruyère, Nelly (1984), *Méthode et dialectique dans l'œuvre de La Ramée: Renaissance et âge classique*, De Pétrarque à Descartes 45, Paris: Vrin.

Bruyère, Nelly (1986), "Le Statut de l'invention dans l'œuvre de La Ramée", in: *Revue des Sciences philosophiques et théologiques* 70.1, pp. 15–24.

Buchanan, Richard (2009), "Thinking about Design: An Historical Perspective", in: *Philosophy of Technology and Engineering Sciences*, ed. by Anthonie W. M. Meijers, Paul Thagard and John Woods Jr., Amsterdam: Elsevier, pp. 409–454.

Bunge, Mario (1966), "Technology as Applied Science", in: *Technology and Culture* 7.3, pp. 329–347.

Burke, Tom (2002), "Mathematizing Dewey's Logical Theory", in: *Dewey's Logical Theory: New Studies and Interpretations*, ed. by Tom Burke, D. Micah Hester and Robert B. Talisse, Vanderbilt University Press, pp. 121–159.

Caliste, Lisa and Guillaume Carnino (2022), "Qu'est-ce que l'industrie ?", in: *Artefact. Techniques, histoire et sciences humaines* 17, pp. 219–242.

Campbell, Murray, A. Joseph Hoane and Feng-hsiung Hsu (2002), "Deep Blue", in: *Artificial Intelligence* 134.1, pp. 57–83.

Campbell-Kelly, Martin et al. [1996] (2014), *Computer : A History of the Information Machine*, 3rd ed., Boulder, Colorado: Westview Press, URL: http://archive.org/details/computerhistoryo0000camp_z9s0 (visited on 05/02/2023).

Carboncini, Sonia (1987), "L'encyclopédie et Christian Wolff: à propos de quelques articles anonymes", in: *Les études philosophiques* 4, pp. 489–504.

Carroll, Lewis (1895), "What the Tortoise Said to Achilles", in: *Mind* 4.14, pp. 278–280.

Cașilli, Antonio A. [2019] (2021), *En attendant les robots. Enquête sur le travail du clic*, Points, Paris: Seuil.

Chang, Hasok (2021), "Operationalism", in: *The Stanford Encyclopedia of Philosophy*, ed. by Edward N. Zalta, URL: https://plato.stanford.edu/archives/fall2021/entries/operationalism/ (visited on 15/05/2023).

Charles, David (2015), "Aristotle on Practical and Theoretical Knowledge", in: *Bridging the Gap between Aristotle's Science and Ethics*, ed. by Henry Devin and Karen Margrethe Nielsen, Cambridge: Cambridge University Press, pp. 71–93.

Châtillon, Jean (1965), "Le Didascalicon de Hugues de Saint-Victor", in: *Cahiers d'histoire mondiale* 9.1, pp. 539–552.

Cicero (2003), *De natura deorum*, ed. by Andrew Roy Dyck, Cambridge Greek and Latin classics, Cambridge: Cambridge University Press.

Clarizio, Emanuele (2021), *La vie technique: une philosophie biologique de la technique*, L'évolution des machines, Paris: Hermann.

Clark, Andy (1997), *Being There: Putting Brain, Body, and World Together Again*, Cambridge, Mass: MIT Press.

Clark, Andy (2010), *Supersizing the Mind. Embodiment, Action, and Cognitive Extension*, Oxford: Oxford University Press.

Clark, Andy (2013), *Mindware: An Introduction to the Philosophy of Cognitive Science*, 2 edition, Oxford: Oxford University Press.

Cohen, I. Bernard (1969), "Hypotheses in Newton's Philosophy", in: *Proceedings of the Boston Colloquium for the Philosophy of Science 1966-1968*, ed. by Robert S. Cohen and Marx W. Wartofsky, Boston Studies in the Philosophy of Science 5, Dordrecht: Springer, pp. 304–326.

Condillac, Étienne Bonnot de [1749] (2010), *Traité Des Systèmes*, Chicoutimi: UQAC, URL: http://dx.doi.org/doi:10.1522/030166753 (visited on 04/03/2025).

Crandall, Richard L. (1988), *The Incorruptible Cashier*, Vestal, NY: Vestal Press, URL: http://archive.org/details/incorruptiblecas0001cran (visited on 06/12/2022).

Crombie, Alistair Cameron (1994), *Styles of Scientific Thinking in the European Tradition: The History of Argument and Explanation Especially in the Mathematical and Biomedical Sciences and Arts*, 3 vol., London: Duckworth.

Cross, Nigel (2006), *Designerly Ways of Knowing*, London: Springer.

Crubellier, Michel (2014), "Premiers Analytiques: Introduction et appareil critique", in: *Premiers analytiques*, GF, Paris: Flammarion, pp. 7–48, 235–397.

Crubellier, Michel and Pierre Pellegrin (2002), *Aristote: le philosophe et les savoirs*, Paris: Seuil.

Cunningham, Andrew (1982), "Two Legacies of the Later Alexandrian School: The Preliminary Questions in Commentaries; the Theory/Practice Division of Medicine", in: *Nihon Ishigaku Zasshi. [Journal of Japanese History of Medicine]* 28.3, pp. 400–424.

Daston, Lorraine (2022), *Rules: A Short History of What We Live By*, Princeton University Press.

De Mol, Liesbeth and Maarten Bullynck (2021), "Roots of 'program' Revisited", in: *Communications of the ACM* 64.4, pp. 35–37.

De Pater, Wilhelmus Ant (1965), *Les Topiques d'Aristote et la dialectique platonicienne: La méthodologie de la définition*, Etudes thomistiques Vol. X, Fribourg, CH: Editions St. Paul.

Dear, Peter Robert (2006), *The Intelligibility of Nature: How Science Makes Sense of the World*, Science.Culture, Chicago: University of Chicago Press.

Dear, Peter Robert (2019), *Revolutionizing the Sciences: European Knowledge in Transition, 1500-1700*, 3rd ed., Oxford: Macmillan.

Dehaene, Stanislas [1995] (2010), *La Bosse des maths*, "Quinze ans après", Paris: Odile Jacob.

Denning, Peter J. and Tim Bell (2012), "The Information Paradox", in: *American Scientist* 100.6, pp. 470–477.

Denning, Peter J., Craig H. Martell and Vint Cerf (2015), *Great Principles of Computing*, Cambridge, Mass: MIT Press.

Dershowitz, Nachum and Yuri Gurevich (2008), "A Natural Axiomatization of Computability and Proof of Church's Thesis", in: *Bulletin of Symbolic Logic* 14.3, pp. 299–350, (visited on 08/02/2024).

Descartes, René [1897–1911] (1996), *Œuvres*, ed. by Charles Adam and Paul Tannery, Paris: Vrin.

Descartes, René (2008), *Meditations on First Philosophy: With Selections from the Objections and Replies*, ed. by Michael Moriarty, Oxford World's Classics, Oxford: Oxford University Press.

Descartes, René and Elisabeth of Bohemia (2007), *The Correspondence between Princess Elisabeth of Bohemia and René Descartes*, ed. by Lisa Shapiro, The Other Voice in Early Modern Europe, Chicago: University of Chicago Press.

Detlefsen, Michael (2007), "Formalism", in: *The Oxford Handbook of Philosophy of Mathematics and Logic*, ed. by Edward Shapiro, Oxford: Oxford University Press, pp. 51–74.

Dewey, John (1967–1972), *The Early Works 1882–1898*, ed. by Jo Ann Boydston, 5 vol., Carbondale: Southern Illinois university press.

Dewey, John (1981–1990), *The Later Works 1925–1953*, ed. by Jo Ann Boydston, 17 vol., Carbondale: Southern Illinois university press.

Dewey, John (1975–1983), *The Middle Works 1899–1924*, ed. by Jo Ann Boydston, 15 vol., Carbondale: Southern Illinois university press.

Dewey, John [1938] (1986), *Logic: The Theory of Inquiry*, ed. by Jo Ann Boydston, vol. 12, The Later Works, 1925-1953, Carbondale: Southern Illinois university press.

Diderot, Denis and Jean Le Rond d'Alembert, eds. [1765] (2017–2022), *Encyclopédie ou Dictionnaire raisonné des sciences, des arts et des métiers*, ENCCRE, Académie des Sciences, URL: https://enccre.academie-sciences.fr/encyclopedie/ (visited on 04/03/2025).

Dijksterhuis, Eduard Jan (1986), *The Mechanization of the World Picture: Phytagoras to Newton*, trans. by Carry Dikshoorn and Dirk Jan Struik, Princeton: Princeton University Press.

Dolza, Luisa and Hélène Vérin (2004), "Figurer la mécanique : l'énigme des théâtres de machines de la Renaissance", in: *Revue dhistoire moderne contemporaine* no51-2.2, pp. 7–37.

Dreyfus, Hubert L. [1972] (1992), *What Computers Still Can't Do: A Critique of Artificial Reason*, Cambridge, Mass: MIT Press.

Dubourg Glatigny, Pascal and Hélène Vérin (2008), "La réduction en art, un phénomène culturel", in: *Réduire en art: la technologie de la Renaissance aux Lumières*, ed. by Hélène Vérin and Pascal Dubourg Glatigny, Paris: Éditions de la Maison des sciences de l'homme.

Duncker, Karl (1945), *On Problem-Solving*, trans. by Lynne S. Lees, Psychological Monographs vol. 58, n°5, Washington DC: American Psychological Association.

Dunne, Joseph [1993] (1997), *Back to the Rough Ground: Practical Judgment and the Lure of Technique*, ed. by Alasdair McIntyre, Paperback, Notre Dame, Ind.: University of Notre Dame Press.

Duns Scotus, Johannes (2017), *Questions sur la métaphysique*, ed. by Olivier Boulnois, trans. by Dan Arbib, Dominique Poirel and Olivier Boulnois, vol. 1, Épiméthée, Paris: Presses universitaires de France.

Duns Scotus, Johannes (1997), *Questions on the "Metaphysics" of Aristotle*, trans. by Girard J. Etzkorn and Allan B. Wolter, Franciscan Institute Publications 19, St Bonaventure (N.Y.): St. Bonaventure University.

Durand, Emmanuel (2012), "La Providence Du Salut Selon Thomas d'Aquin: Un Bénéfice Théologique de La Métaphysique Des Singuliers", in: *Revue des sciences philosophiques et théologiques* 96.3, pp. 451–492.

Durand-Richard, Marie-José (1997), "Logic versus Algebra : English Debates and Boole's Mediation", in: *A Boole Anthology. Recent and Classical Studies in the Logic of George Boole*, ed. by James Gasser.

Durand-Richard, Marie-José (2010), "Le regard français de Charles Babbage sur le « déclin de la science en Angleterre »", in: *Documents pour l'histoire des techniques. Nouvelle série* 19, pp. 287–304.

Džeroski, Sašo et al. (1997), "Using Machine Learning Techniques in the Construction of Models. II. Data Analysis with Rule Induction", in: *Ecological Modelling* 95.1, pp. 95–111.

Ebbersmeyer, Sabrina (1995), "Spekulation", in: *Historisches Wörterbuch der Philosophie* 9, ed. by Joachim Ritter et al., pp. 1355–1371.

Eckroth, Joshua (2014), *General Problem Solver*, URL: https://github.com/joshuaeckroth/cse3521-website/blob/master/gps.org (visited on 18/02/2025).

Ensmenger, Nathan (2016), "The Multiple Meanings of a Flowchart", in: *Information & Culture* 51.3, pp. 321–351.

Evans, Jonathan and Keith Frankish, eds. (2009), *In Two Minds: Dual Processes and Beyond*, Oxford: Oxford University Press.

Fantl, Jeremy (2017), "Knowledge How", in: *The Stanford Encyclopedia of Philosophy*, ed. by Edward N. Zalta, URL: https://plato.stanford.edu/archives/fall2017/entries/knowledge-how/ (visited on 09/03/2018).

Al-Farabi, Abu Nasr Muhammad (2015), *Le recensement des sciences*, ed. and trans. by Amor Cherni, Sagesses musulmanes, Beyrouth: Albouraq.

Ficino, Marsilio [1482] (2001), *Platonic theology*, ed. by James Hankins and William Roy Bowen, trans. by Michael John Bridgman Allen and John Warden, The I Tatti Renaissance library 2, 4, 7, 13, 17, 23, Cambridge, Mass: Harvard University Press.

Finerty, J. Patrick (1979), "Cycles in Canadian Lynx", in: *The American Naturalist* 114.3, pp. 453–455.

Floyd, Juliet (2016), "Chains of Life: Turing, Lebensform, and the Emergence of Wittgenstein's Later Style", in: *Nordic Wittgenstein Review* 5.2, pp. 7–89.

Floyd, Juliet (2018), "Lebensformen: Living Logic", in: *Language, Form(s) of Life, and Logic: Investigations After Wittgenstein*. Ed. by Christian Martin, Berlin: Walter de Gruyter, pp. 59–92.

Floyd, Robert W. (1979), "The Paradigms of Programming", in: *Communications of the ACM* 22.8, pp. 455–460.

Friedman, Walter A. (1998), "John H. Patterson and the Sales Strategy of the National Cash Register Company, 1884 to 1922", in: *Business History Review* 72.4, pp. 552–584.

Furetière, Antoine (1701), *Dictionnaire universel*, 2nd ed., 3 vol., La Haye, URL: https://gallica.bnf.fr/ark:/12148/bpt6k56749155 (visited on 27/08/2021).
Gabbey, Alan (1984), "The Mechanical Philosophy and Its Problems: Mechanical Explanations, Impenetrability, and Perpetual Moti", in: *Change and Progress in Modern Science*, ed. by Joseph C. Pitt, Dordrecht: Springer, pp. 9–84.
Gabbey, Alan (1992), "Newton's Mathematical Principles of Natural Philosophy: A Treatise on 'Mechanics'?", in: *The Investigation of Difficult Things: Essays on Newton and the History of the Exact Sciences in Honour of D.T. Whiteside*, ed. by Peter Michael Harman and Alan E. Shapiro, Cambridge: Cambridge University Press, pp. 305–322.
Gabbey, Alan (1996), "Newton and Natural Philosophy", in: *Companion to the History of Modern Science*, Routledge.
Gabbey, Alan (2001), "Mechanical Philosophies and Their Explanations", in: *Late Medieval and Early Modern Corpuscular Matter Theories*, ed. by Christoph Lüthy, John Emery Murdoch and William Royall Newman, Medieval and Early Modern Science 1, Leiden: Brill, pp. 441–466.
Gabbey, Alan (2004), "What Was "Mechanical" about "The Mechanical Philosophy"?", in: *The Reception of the Galilean Science of Motion in Seventeenth-Century Europe*, ed. by Carla Rita Palmerino et al., Boston Studies in the Philosophy of Science 239, Dordrecht: Springer, pp. 11–23.
Galilei, Galileo [1634] (1966), *Les méchaniques de Galilée, mathématicien et ingénieur du duc de Florence*, ed. by Bernard Rochot, trans. by Marin Mersenne, Le Mouvement des idées au XVIIe siècle 4, Paris: Presses universitaires de France.
Garber, Daniel and Sophie Roux, eds. (2013), *The Mechanization of Natural Philosophy*, Boston Studies in the Philosophy of Science 300, Dordrecht: Springer.
Garçon, Anne-Françoise (2008), "Réduire la mine en science?... Anatomie des De re metallica d'Agricola", in: *Réduire en art: la technologie de la Renaissance aux Lumières*, ed. by Hélène Vérin and Pascal Dubourg Glatigny, Paris: Éditions de la Maison des sciences de l'homme.
Gaudillière-Jami, Nadja (2022), "Automatiser l'Architecture", PhD thesis, Paris: Université Paris-Est.
Giglioni, Guido (2022), "The Age of the New Organons : Ramus, Bacon and Descartes, and the Making of Thinking Tools in the Early Modern Period (1550-1650)", in: *Conférence: Anciens et Nouveaux Organons Techniques de Savoir Entre XVIe et XVIIe Siècles*, Université Paris 1 Panthéon-Sorbonne.
Gilbert, Neal Ward (1960), *Renaissance Concepts of Method*, New York: Columbia University press.
Gödel, Kurt [1951] (1995a), "Some Basic Theorems on the Foundations of Mathematics and Their Implications", in: *Collected Works*, vol. 3, New York: Oxford University Press, pp. 304–323, URL: http://archive.org/details/KurtGdelCollectedWorksVolumeIII1995 (visited on 03/04/2025).
Gödel, Kurt [1933] (1995b), "The Present Situation in the Foundations of Mathematics", in: *Collected Works*, vol. 3, New York: Oxford University Press, pp. 36–53, URL: http://archive.org/details/KurtGdelCollectedWorksVolumeIII1995 (visited on 03/04/2025).
Goldschmidt, Victor [1953] (1989), *Le systeme stoïcien et l'idée de temps*, 4th ed., Bibliothèque d'histoire de la philosophie, Paris: Vrin.
Granström, Johan Georg (2011), *Treatise on Intuitionistic Type Theory*, Berlin: Springer.
Grattan-Guinness, Ian (1990), "Work for the Hairdressers: The Production of de Prony's Logarithmic and Trigonometric Tables", in: *Annals of the History of Computing* 12.3, pp. 177–185.
Grier, David Alan (2005), *When Computers Were Human*, Princeton: Princeton University Press.
Gunter, Carl A. et al. (2000), "A Reference Model for Requirements and Specifications", in: *IEEE Software* 17.3, pp. 37–43.
Gurevich, Yuri (2000), "Sequential Abstract-State Machines Capture Sequential Algorithms", in: *ACM Transactions on Computational Logic* 1.1, pp. 77–111, (visited on 11/04/2024).

Hager, F.-P. and Ch. Strub (1995), "System - Systematik - Systematisch", in: *Historisches Wörterbuch der Philosophie* 10, ed. by Joachim Ritter et al., pp. 824–856.
Hamblin, Charles L. (1976), "An Improved Pons Asinorum ?", in: *Journal of the History of Philosophy* 14.2, pp. 131–136.
Hamou, Philippe (2014), "Sur les origines du concept de méthode à l'âge classique: La Ramée, Bacon et Descartes", in: *LISA* vol. XII-n°5.
Hansson, Sven Ove (2018), *Technology and Mathematics: Philosophical and Historical Investigations*, vol. 30, Philosophy of Engineering and Technology, Berlin: Springer.
Harel, David, Dexter Kozen and Jerzy Tiuryn (2000), *Dynamic Logic*, Cambridge, Mass: MIT Press.
Heidegger, Martin [1927] (1993), *Sein und Zeit*, Tübingen: Max Niemeyer.
Hempel, Carl Gustav (1965), *Aspects of Scientific Explanation: And Other Essays in the Philosophy of Science*, New York: The Free Press Collier-Macmillan.
Hendricks, Vincent Fella, Arne Jakobsen and Stig Andur Pedersen (2000), "Identification of Matrices in Science and Engineering", in: *Journal for General Philosophy of Science* 31.2, pp. 277–305.
Hersh, Reuben (1997), *What Is Mathematics, Really ?*, New York: Oxford University Press.
Herzig, Andreas (2015), "Logics of Knowledge and Action: Critical Analysis and Challenges", in: *Autonomous Agents and Multi-Agent Systems* 29, pp. 719–753.
Heyting, Arend (1931), "Die intuitionistische Grundlegung der Mathematik", in: *Erkenntnis* 2.1, pp. 106–115.
Hilbert, David [1922] (1996), "The New Grounding of Mathematics. First Report", in: *From Kant to Hilbert: A Source Book in the Foundations of Mathematics*, ed. by William Bragg Ewald, vol. 2, Oxford: Clarendon Press, pp. 1115–34.
Hilbert, David and W. Ackermann (1972), *Grundzüge Der Theoretischen Logik*, Berlin: Springer.
Hildebrand, David L. (2008), *Dewey: A Beginner's Guide*, Oneworld Beginner's Guides, Oxford: Oneworld.
Hoare, Tony (1969), "An Axiomatic Basis for Computer Programming", in: *Communications of the ACM* 12.10, pp. 576–583.
Houkes, Wybo (2009), "The Nature of Technological Knowledge", in: *Philosophy of Technology and Engineering Sciences*, ed. by Anthonie W. M. Meijers, Paul Thagard and John Woods Jr., Amsterdam: Elsevier.
Houkes, Wybo and Pieter E. Vermaas (2004), "Actions Versus Functions: A Plea for an Alternative Metaphysics of Artifacts", in: *The Monist* 87.1, pp. 52–71.
Houkes, Wybo, Pieter E. Vermaas et al. (2002), "Design and Use as Plans: An Action-Theoretical Account", in: *Design Studies*, Philosophy of Design 23.3, pp. 303–320.
Hoyningen-Huene, Paul (2013), *Systematicity: The Nature of Science*, Oxford Studies in Philosophy of Science, New York: Oxford University Press.
Hume, David [1740] (1990), *A Treatise of Human Nature*, ed. by Lewis A. Selby-Bigge and P. H. Nidditch, 2nd ed., Oxford: Clarendon Press.
Hutchins, Edwin (1995), *Cognition in the Wild*, Cambridge, Mass: MIT Press.
Jackson, Michael (2002), "Some Basic Tenets of Description", in: *Software and Systems Modeling* 1.1, pp. 5–9.
Jardine, Nicholas (1976), "Galileo's Road to Truth and the Demonstrative Regress", in: *Studies in History and Philosophy of Science Part A* 7.4, pp. 277–318.
Johansen, Thomas Kjeller (2017), "Aristotle on the Logos of the Craftsman", in: *Phronesis* 62.2, pp. 97–135.
Joyce, David (1998), *Euclid's Elements*, URL: http://aleph0.clarku.edu/~djoyce/java/elements/elements.html (visited on 06/10/2021).

Kahneman, Daniel (2012), *Thinking, Fast and Slow*, London: Penguin.

Kant, Immanuel (1902–), *Gesammelte Schriften*, Deutsche Akademie der Wissenschaften, Berlin; Göttingen: G. Reimer; Walter de Gruyter.

Kant, Immanuel (1963), *Introduction to Logic*, trans. by Thomas K. Abbot, New York: Philosophical Library.

Kant, Immanuel [1970] (1991), *Political Writings*, ed. by H. S. Reiss, trans. by H. B. Nisbet, 2nd ed., Cambridge: Cambridge University Press.

Kant, Immanuel (1997), *Critique of Practical Reason*, ed. by Mary J. Gregor, Cambridge Texts in the History of Philosophy, Cambridge: Cambridge University Press.

Kant, Immanuel (1998), *Critique of Pure Reason*, ed. and trans. by Paul Guyer and Allen W. Wood, The Cambridge Edition of the Works of Immanuel Kant, Cambridge: Cambridge University Press.

Kant, Immanuel (2000), *Critique of the Power of Judgment*, ed. by Paul Guyer, trans. by Paul Guyer and Eric Matthews, The Cambridge Edition of the Works of Immanuel Kant, Cambridge: Cambridge University Press.

Kant, Immanuel (2011), *Groundwork of the Metaphysics of Morals: A German–English Edition*, ed. and trans. by Mary Gregor and Jens Timmermann, The Cambridge Kant German-English Edition, Cambridge: Cambridge University Press.

Kaplan, Craig A. and Herbert A. Simon (1990), "In Search of Insight", in: *Cognitive Psychology* 22.3, pp. 374–419.

Kerstiens, Ludwig (1958), "Die Lehre von Der Theoretischen Erkenntnis in Der Lateinischen Tradition", in: *Philosophisches Jahrbuch* 66, pp. 375–424.

Kiran, D. R. (2017), "Seven Traditional Tools of TQM", in: *Total Quality Management*, ed. by D. R. Kiran, Butterworth-Heinemann, pp. 271–290.

Knecht, Herbert H. (1981), *La logique chez Leibniz: essai sur le rationalisme baroque*, Lausanne: L'âge d'homme.

Knorr, Wilbur R. (1983), "Construction as Existence Proof in Ancient Geometry", in: *Ancient Philosophy* 3, pp. 125–148.

Knuth, Donald E. (1974), "Computer Science and Its Relation to Mathematics", in: *The American Mathematical Monthly* 81.4, pp. 323–343.

Knuth, Donald E. (1997), *The Art of Computer Programming*, 3rd ed., vol. 1: Fundamental Algorithms, Reading, Mass: Addison-Wesley.

Kolmogorov, Andreï (1932), "Zur Deutung der intuitionistischen Logik", in: *Mathematische Zeitschrift* 35.1, pp. 58–65.

König, Gert and Helmut Pulte (1998), "Theorie", in: *Historisches Wörterbuch der Philosophie* 10, ed. by Joachim Ritter et al., pp. 1128–1154.

Kostic, Neven (2009), "Topologie Des Champs de Contraintes Pour Le Dimensionnement Des Structures En Beton Arme", PhD thesis, Lausanne: EPFL.

Koyré, Alexandre (1961), *Études d'histoire de la pensée philosophique*, Paris: Armand Colin.

Kroes, Peter A. (2012), *Technical Artefacts: Creations of Mind and Matter : A Philosophy of Engineering Design*, Philosophy of Engineering and Technology 6, Dordrecht: Springer.

La Ramée, Pierre de [1555] (1996), *Dialectique 1555: un manifeste de la Pléiade*, ed. by Nelly Bruyère, De Pétrarque à Descartes 61, Paris: Vrin.

Langley, Pat and Herbert A. Simon (1995), "Applications of Machine Learning and Rule Induction", in: *Communications of the ACM* 38.11, pp. 54–64.

Le Blanc, Gilles (2005), *L'industrie Dans l'économie Française, 1978-2003: Une Étude Comparée*, Institut de l'entreprise.

Le Blond, Jean-Marie [1936] (1970), *Logique et méthode chez Aristote: étude sur la recherche des principes dans la physique aristotélicienne*, 2nd ed., Bibliothèque d'histoire de la philosophie, Paris: Vrin.
Lefèvre, Wolfgang, Jürgen Renn and Urs Schoepflin, eds. (2012), *The Power of Images in Early Modern Science*, Birkhäuser.
Leibniz, Gottfried Wilhelm (1923–), *Sämtliche Schriften und Briefe*, Deutsche Akademie der Wissenschaften, Darmstadt; Leipzig; Berlin: Akademie-Verlag.
Leibniz, Gottfried Wilhelm [1875–1890] (1962), *Philosophische Schriften*, ed. by C. J. Gerhardt, 7 vol., Hildesheim: Olms.
Leibniz, Gottfried Wilhelm (1903), *Opuscules et fragments inédits de Leibniz*, ed. by Louis Couturat, Paris: Felix Alcan.
Lennox, James G. (1982), "Teleology, Chance, and Aristotle's Theory of Spontaneous Generation", in: *Journal of the History of Philosophy* 20.3, pp. 219–238.
Lennox, James G. (1984), "Aristotle on Chance", in: *Archiv für Geschichte der Philosophie* 66.1, pp. 52–60.
Lennox, James G. (2011), "Aristotle on Norms of Inquiry", in: *HOPOS: The Journal of the International Society for the History of Philosophy of Science* 1.1, pp. 23–46.
Lerner, Michel-Pierre (2005), "The Origin and Meaning of "World System"", in: *Journal for the History of Astronomy* 36.4, pp. 407–441.
Leroi-Gourhan, André (1965), *Le Geste et La Parole 2: La Mémoire et Les Rythmes*, Sciences d'aujourd'hui, Paris: Albin Michel.
Lévi-Strauss, Claude [1962] (1985), *La pensée sauvage*, Agora 2, Paris: Presses Pocket.
Levitin, Dmitri (2021), "Newton on the Rules of Philosophizing and Hypotheses: New Evidence, New Conclusions", in: *Isis* 112.2, pp. 242–265.
Levy, Arnon (2013), "Three Kinds of New Mechanism", in: *Biology & Philosophy* 28.1, pp. 99–114.
Littré, Émile (1873–1874), *Dictionnaire de la langue française*, URL: https://gallica.bnf.fr/ark:/12148/bpt6k5460034d (visited on 29/11/2021).
Longuenesse, Béatrice (1998), *Kant and the Capacity to Judge: Sensibility and Discursivity in the Transcendental Analytic of the "Critique of Pure Reason"*, trans. by Charles T. Wolfe, Princeton, NJ: Princeton University Press.
Lorenz, Hendrik and Benjamin Morison (2019), "Aristotle's Empiricist Theory of Doxastic Knowledge", in: *Phronesis* 64.4, pp. 431–464.
Mach, Ernst [1883] (1919), *The Science Of Mechanics*, trans. by Thomas J. McCormack, Chicago: The Open Court Publishing Co., URL: http://archive.org/details/scienceofmechani011018mbp (visited on 22/04/2023).
Machamer, Peter, Lindley Darden and Carl F. Craver (2000), "Thinking about Mechanisms", in: *Philosophy of Science* 67.1, pp. 1–25.
MacIntyre, Alasdair C. [1981] (1985), *After Virtue: A Study in Moral Theory*, 2nd ed., London: Duckworth.
MacKenzie, Donald (2004), *Mechanizing Proof: Computing, Risk, and Trust*, Cambridge, Mass: MIT Press.
Madelrieux, Stéphane (2016), *La philosophie de John Dewey: repères*, Repères philosophiques, Paris: Vrin.
Mäenpää, Petri (1993), "The Art of Analysis: Logic and History of Problem-Solving.", PhD thesis, Helsinki: Teknillinen Korkeakolou.
Mäenpää, Petri (1997), "From Backward Reduction to Configurational Analysis", in: *Analysis and Synthesis in Mathematics. History and Philosophy*, ed. by Marco Panza and Michael Otte, vol. 196, Boston Studies in the Philosophy and History of Science, Dordrecht: Kluwer, pp. 201–226.
Mäenpää, Petri and Jan Von Plato (1990), "The Logic of Euclidean Construction Procedures", in: *Acta Philosophica Fennica* 39.

Marchand, Stéphane (2015), “Sextus Empiricus, scepticisme et philosophie de la vie quotidienne”, in: *Philosophie antique. Problèmes, Renaissances, Usages* 15, pp. 91–119.
Maronne, Sébastien (2017), “Qu’est-ce qu’une action en mathématiques?”, in: *Workshop « Histoire et philosophie des mathématiques »*, Faculdad de Ciencias, UNAM, Mexico, URL: https://www.youtube.com/watch?v=Lsf-23M5TJA&t=12s (visited on 10/01/2023).
Martin-Löf, Per (1984), *Intuitionistic Type Theory (Padua 1980)*, vol. 9, Napoli: Bibliopolis.
Martin-Löf, Per (1987), “Truth of a Proposition, Evidence of a Judgement, Validity of a Proof”, in: *Synthese* 73.3, pp. 407–420.
Martin-Löf, Per (1991), “A Path from Logics to Metaphysics”, in: *Atti del Congresso ”Nuovi problemi della logica e della filosofia della scienza”*, ed. by Giovanna Corsi, Giovanni Sambin and Maria Luisa Dalla Chiara, vol. 2, Bologna: Cooperativa Libraria Universitaria Editrice Bologna, pp. 141–149.
Martin-Löf, Per (1994), “Analytic and Synthetic Judgements in Type Theory”, in: *Kant and Contemporary Epistemology*, ed. by Paolo Parrini, Dordrecht: Springer, pp. 87–99.
Martin-Löf, Per (1995), “Verificationism Then and Now”, in: *The Foundational Debate: Complexity and Constructivity in Mathematics and Physics*, ed. by W. DePauli-Schimanovic, E. Köhler and F. Stadler, Dordrecht: Kluwer, pp. 187–196.
Martin-Löf, Per [1983] (1996), “On the Meanings of the Logical Constants and the Justifications of the Logical Laws (Siena 1985)”, in: *Nordic journal of philosophical logic* 1.1, pp. 11–60.
Marx, Karl [1867] (1976), *Capital: A Critique of Political Economy*, trans. by Ben Fowkes, vol. 1, Harmondsworth: Penguin Books.
Mazzoni, Jacopo [1597] (2010), *In universam Platonis et Aristotelis philosophiam Praeludia, sive De comparatione Platonis et Aristotelis*, ed. by Sara Matteoli and Anna De Pace, Storie e testi 19, Napoli: M. D’Auria.
McMullin, Ernan (1978), “Structural Explanation”, in: *American Philosophical Quarterly* 15.2, pp. 139–147.
McMullin, Ernan (2001), “The Impact of Newton’s Principia on the Philosophy of Science”, in: *Philosophy of Science* 68.3, pp. 279–310.
Mérigoux, Denis (2021), “Catala: Un Langage de Programmation Pour La Loi Socio-Fiscale”, in: *Séminaire Code Sources*.
Merriam-Webster (1936), *Webster’s Collegiate Dictionary*, 5th, Springfield, Mass.: G. & C. Merriam Co., URL: http://archive.org/details/websterscollegia00spri (visited on 13/11/2021).
Michel, Christian (2008), “La peinture peut-elle être réduite en art ?”, in: *Réduire en art: la technologie de la Renaissance aux Lumières*, ed. by Hélène Vérin and Pascal Dubourg Glatigny, Paris: Éditions de la Maison des sciences de l’homme, pp. 106–118.
Mignucci, Mario (1981), “Hos Epi to Polu et Nécessaire Dans La Conception Aristotélicienne de La Science”, in: *Aristotle on Science: The Posterior Analytics*, ed. by Enrico Berti, Università Di Padova, Padova: Editrice Antenore, pp. 173–203.
Miller, George A., Eugene Galanter and Karl H. Pribram [1960] (2013), *Plans and the Structure of Behavior*, Martino Fine Books.
Minsky, Marvin Lee (1967), “Why Programming Is a Good Medium for Expressing Poorly Understood and Sloppily Formulated Ideas”, in: *Design and Planning II-computers in Design and Communication*, ed. by Martin Krampen and Peter Seitz, New York: Hastings House Publishers, pp. 120–125.
Mokyr, Joel (2016), *A Culture of Growth: The Origins of the Modern Economy*, Princeton: Princeton University Press.
Morel, Jean-Marie (1776), *Théorie des jardins*, Paris, URL: https://gallica.bnf.fr/ark:/12148/bpt6k85711d (visited on 17/04/2021).
Morgan, Mary S. and Margaret Morrison, eds. (1999), *Models as Mediators: Perspectives on Natural and Social Science*, Ideas in Context 52, Cambridge: Cambridge University Press.

Moss, Jessica (2011), "Virtue Makes the Goal Right: Virtue and Phronesis in Aristotle's Ethics", in: *Phronesis* 56.3, pp. 204–261.
Moss, Jessica (2014), "Right Reason in Plato and Aristotle: On the Meaning of Logos", in: *Phronesis* 59.3, pp. 181–230.
Müller, Jozef (2018), "Practical and Productive Thinking in Aristotle", in: *Phronesis* 63.2, pp. 148–175.
Naibo, Alberto (2015), "Constructibility and Geometry", in: *From Logic to Practice: Italian Studies in the Philosophy of Mathematics*, ed. by Gabriele Lolli, Marco Panza and Giorgio Venturi, Boston Studies in the Philosophy and History of Science 308, Cham: Springer, pp. 123–161.
Naibo, Alberto and Marco Panza (2021), "Formalizing the Logic and Proofs of Euclid's Elements, Book I", in: *Conférence: L'intuitionnisme Entre Philosophie, Mathématique et Logique: Mutations et Histoire Longue*, Paris.
Newell, Allen, John C. Shaw and Herbert A. Simon (1959), *Report on a General Problem Solving Program*, tech. rep. 1584, Pittsburgh.
Newton, Isaac (1959), *The Correspondence of Isaac Newton*, ed. by Alfred Rupert Hall et al., Cambridge: Cambridge University Press.
Niiniluoto, Ilkka (1993), "The Aim and Structure of Applied Research", in: *Erkenntnis* 38.1, pp. 1–21.
Nuseibeh, Bashar and Pamela Zave, eds. (2010), *Software Requirements and Design: The Work of Michael Jackson*, Chatham, NJ: Good Friends Publishing.
Nussbaum, Martha Craven [1986] (2001), *The Fragility of Goodness: Luck and Ethics in Greek Tragedy and Philosophy*, 2nd ed., Cambridge Paperback Library, Cambridge: Cambridge University Press.
Oberman, Heiko A. (2003), "Luther and the Via Moderna: The Philosophical Backdrop of the Reformation Breakthrough", in: *The Journal of Ecclesiastical History* 54.4, pp. 641–670.
Ong, Walter Jackson (1958), *Ramus, Method and the Decay of Dialogue: From the Art of Discourse to the Art of Reason*, Cambridge, Mass: Harvard University Press.
Orr, Julian E. [1996] (2016), *Talking about Machines: An Ethnography of a Modern Job*, Ithaca: Cornell University Press.
Oxford English Dictionary (2010–), Oxford: Oxford University Press, URL: https://www.oed.com/?tl=true (visited on 04/03/2025).
Pagès, Gaspard and Catherine Verna (2022), "L'invention de l'industrie antique et médiévale", in: *Artefact. Techniques, histoire et sciences humaines* 17, pp. 243–264.
Palmerino, Carla Rita et al., eds. (2004), *The Reception of the Galilean Science of Motion in Seventeenth-Century Europe*, Boston Studies in the Philosophy of Science 239, Dordrecht: Springer.
Pantelia, Maria C., ed. (1985 –), *Thesaurus Linguae Grecae Digital Library*, Irvine, CA: University of California, URL: http://www.tlg.uci.edu (visited on 23/09/2023).
Panza, Marco (1997), "Classical Sources for the Concepts of Analysis and Synthesis", in: *Analysis and Synthesis in Mathematics. History and Philosophy*, ed. by Marco Panza and Michael Otte, vol. 196, Boston Studies in the Philosophy and History of Science, Dordrecht: Kluwer, pp. 365–414.
Panza, Marco and Michael Otte, eds. (1997), *Analysis and Synthesis in Mathematics. History and Philosophy*, vol. 196, Boston Studies in the Philosophy and History of Science, Dordrecht: Kluwer.
Pappus of Alexandria (1986), *Book VII of the Collection*, ed. and trans. by Alexander Jones, Sources in the History of Mathematics and Physical Sciences 8, New York, NY: Springer.
Pappus of Alexandria (2023), *Book III of the Mathematical Collection*, trans. by John B. Little, Holy Cross Bookshelf 63, College Archives at Crossworks.
Parent, Monelle (2004), "La technoscience, un défi pour la philosophie", in: *revue Phares* 5.
Pautet, Sébastien (2018), "Les ingénieurs à l'époque moderne", in: *L'Europe des sciences et des techniques : Un dialogue des savoirs, XVe-XVIIIe siècle*, ed. by Liliane Hilaire-Pérez, Fabien Simon and Marie Thébaud-Sorger, Histoire, Rennes: Presses universitaires de Rennes, pp. 111–121.

Peaucelle, Jean-louis (2012), "Un « éléphant blanc » en pleine Révolution Française : les grandes tables de logarithmes de Prony comme substitut au cadastre !", in: *Annales des Mines - Gerer et comprendre* N° 107.1, pp. 74–86.

Pedersen, Olaf (1961), "Theorica - A Study in Language and Civilization", in: *Classica et Mediaevalia* 22, pp. 151–175.

Pereira, Henrique M. and Laetitia M. Navarro, eds. (2015), *Rewilding European Landscapes*, Cham: Springer.

Philopon, Jean (1897), *Ioannis Philoponi in Aristotelis de Anima libros commentaria*, ed. by Michael Hayduck, Commentaria in Aristotelem Graeca XV, Berlin: G. Reimeri.

Piccinini, Gualtiero (2015), *Physical Computation: A Mechanistic Account*, Oxford: Oxford University Press.

Plato (1979), *The Republic*, ed. and trans. by Raymond Larson, Newark: John Wiley & Sons.

Pliny the Elder (1961), *Natural history*, trans. by Harris Rackham, vol. 5, The Loeb classical library, London: Harvard University press.

Popplow, Marcus (2007), "Setting the World Machine in Motion: The Meaning of Machina Mundi in the Middle Ages and the Early Modern Period", in: *Mechanics and Cosmology in the Medieval and Early Modern Period*, ed. by Massimo Bucciantini, Michele Camerota and Sophie Roux, Biblioteca Di Nuncius LXIV, Firenze: L. S. Olschki, pp. 45–70.

Posy, Carl (1998), "Brouwer versus Hilbert: 1907–1928", in: *Science in Context* 11.2, pp. 291–325.

Pratt, Vaughan (1976), *Semantical Considerations on Floyd-Hoare Logic*, Technical Report LCS-TR-168, Cambridge, Mass.

Presti, Roberto Lo (2012), "La notion d'automaton dans les textes médicaux (Hippocrate et Galien) et la Physique d'Aristote", in: *Les Études Classiques* 80.1-2.

Priestley, Mark (2017), "AI and the Origins of the Functional Programming Language Style", in: *Minds and Machines* 27.3, pp. 449–472.

Proclus (1970), *A Commentary on the First Book of Euclid's Elements*, trans. by Glenn R. Morrow, Princeton: Princeton University Press.

Rescher, Nicholas (1979), *Cognitive Systematization : A Systems-Theoretic Approach to a Coherentist Theory of Knowledge*, Oxford: Blackwell, URL: http://archive.org/details/cognitivesystema0000resc (visited on 14/03/2023).

Ritschl, Otto (1906), *System und systematische Methode in der Geschichte des wissenschaftlichen Sprachgebrauchs und der philosophischen Methodologie*, Bonn: Carl Georgi, URL: http://archive.org/details/ORitschelSystemUndSystematischeMethode (visited on 11/02/2023).

Ritter, Joachim et al. (1980), "Methode", in: *Historisches Wörterbuch der Philosophie* 5, ed. by Joachim Ritter et al., pp. 1304–1332.

Roberts, Lissa, Simon Schaffer and Peter Robert Dear, eds. (2007), *The Mindful Hand: Inquiry and Invention from the Late Renaissance to Early Industrialisation*, History of Science and Scholarship in the Netherlands 9, Amsterdam: Koninklijke Nederlandse Akademie van Wetenschappen.

Rodier, Georges (1926), *Etudes de philosophie grecque*, Bibliothèque d'histoire de la philosophie, Paris: Vrin.

Rossi, Paolo (1996), *Les philosophes et les machines, 1400-1700*, trans. by Patrick Vighetti, Science, histoire et société, Paris: Presses universitaires de France.

Roux, Sophie (2013), "Quelles machines pour quels animaux? Jacques Rohault, Claude Perrault, Giovanni Alfonso Borelli", in: *L'automate: modèle, métaphore, machine, merveille. Colloque international de Grenoble, 19-21 mars 2009*, ed. by Aurélia Gaillard et al., Mirabilia, Pessac: Presses universitaires de Bordeaux, pp. 69–113.

Roux, Sophie (2017), "From the Mechanical Philosophy to Early Modern Mechanisms", in: *The Routledge Handbook of Mechanisms and Mechanical Philosophy*, ed. by Stuart Glennan and Phyllis Illari, Taylor & Francis, pp. 26–45.

Roux, Sophie and E. Festa (2001), "Le para phusin et l'imitation de la nature dans quelques commentaires du prologue des Questions mécaniques", in: *Largo campo di filosofare: Eurosymposium Galileo 2001*, ed. by José Montesinos, Carlos Solís and Jean-Yves Dupont, La Orotava: Fundación Canaria Orotava de Historia de la Ciencia, pp. 237–253.

Ryle, Gilbert [1949] (2002), *The Concept of Mind*, Chicago: University of Chicago Press.

Ryle, Gilbert [1971] (2009a), *Collected Essays 1929 - 1968*, Abingdon: Routledge.

Ryle, Gilbert (2009b), "Why are the Calculuses of Logic and Arithmetic Applicable to Reality?", in: *Collected Essays 1929 - 1968*, Abingdon: Routledge, pp. 236–243.

Schmidt, Kjeld (2011), *Cooperative Work and Coordinative Practices*, Computer Supported Cooperative Work, London: Springer.

Schmidt-Biggemann, Wilhelm (1983), *Topica universalis: eine Modellgeschichte humanistischer und barocker Wissenschaft*, Paradeigmata 1, Hamburg: Meiner.

Searle, John R. (1980), "Minds, Brains, and Programs", in: *The Behavioral and Brain Sciences*, p. 42.

Searle, John R. [1965] (2000), "What Is a Speech Act?", in: *Perspectives in the Philosophy of Language: A Concise Anthology*, ed. by Robert J. Stainton, Peterborough, Ontario: Broadview Press, pp. 253–268.

Séris, Jean-Pierre (1994a), *La Technique*, Paris: Presses universitaires de France.

Séris, Jean-Pierre (1994b), *Qu'est-ce que la division du travail?*, Pré-textes 6, Paris: Vrin.

Shank, Michael H. (2007), "Mechanical Thinking in European Astronomy (13th-15th Centuries)", in: *Mechanics and Cosmology in the Medieval and Early Modern Period*, ed. by Massimo Bucciantini, Michele Camerota and Sophie Roux, Biblioteca Di Nuncius LXIV, Firenze: L. S. Olschki, pp. 3–28.

Sidoli, Nathan (2018), "Uses of Construction in Problems and Theorems in Euclid's Elements I-VI", in: *Archive for History of Exact Sciences* 72.4, pp. 403–452.

Sieg, Wilfried (2006), "Gödel on Computability", in: *Philosophia Mathematica* 14.2, pp. 189–207.

Simon, Herbert A. (1972), "Complexity and the Representation of Patterned Sequences of Symbols", in: *Psychological Review* 79.5, pp. 369–382.

Simon, Herbert A. (1976a), "From Substantive to Procedural Rationality", in: *25 Years of Economic Theory: Retrospect and Prospect*, ed. by T. J. Kastelein et al., Boston: Springer, pp. 65–86.

Simon, Herbert A. (1976b), "How Complex Are Complex Systems?", in: *PSA: Proceedings of the Biennial Meeting of the Philosophy of Science Association* 1976.2, pp. 507–522.

Simon, Herbert A. [1967] (1977), "The Logic of Heuristic Decision-Making", in: *Models of Discovery: And Other Topics in the Methods of Science*, Boston Studies in the Philosophy of Science v. 54, Dordrecht: D. Reidel Pub. Co.

Simon, Herbert A. (1978a), "On the Forms of Mental Representations", in: *Perception and Cognition. Issues in the Foundations of Psychology*, ed. by C. Wade Savage, Minnesota Studies in the Philosophy of Science IX, Minneapolis: University of Minnesota Press, pp. 3–18.

Simon, Herbert A. (1978b), "Rationality as Process and as Product of Thought", in: *The American Economic Review* 68.2, pp. 1–16.

Simon, Herbert A. (1983), "Why Should Machines Learn?", in: *Machine Learning*, ed. by Ryszard S. Michalski, Jaime G. Carbonell and Tom M. Mitchell, San Francisco: Morgan Kaufmann, pp. 25–37.

Simon, Herbert A. (1984), "Expert and Novice Problem Solving in Science and Mathematics", in: *Autour de l'oeuvre de H. A. Simon: Le Paradigme S.T.I.* Université Paul Valéry, Montpellier.

Simon, Herbert A. (1986), "Whether Software Engineering Needs to Be Artificially Intelligent", in: *IEEE Transactions on Software Engineering* 7, pp. 726–732.
Simon, Herbert A. (1987), "Two Heads Are Better than One: The Collaboration between AI and OR", in: *Interfaces* 17.4, pp. 8–15.
Simon, Herbert A. (1994), "Computers in Design Education", in: *An International Workshop on the Future Directions of Computer Aided Engineering – Tribute of Steven Fenves*.
Simon, Herbert A. [1991] (1996a), *Models of My Life*, Cambridge, Mass: MIT Press.
Simon, Herbert A. [1969] (1996b), *The Sciences of the Artificial*, 3rd ed., Cambridge, Mass: MIT Press.
Simon, Herbert A. [1947] (1997), *Administrative Behavior: A Study of Decision-Making Processes in Administrative Organizations*, 4th ed., New York: Free Press.
Simon, Herbert A. and Allen Newell (1972), *Human Problem Solving*, Englewood Cliffs, N.J: Prentice Hall.
Sperber, Dan and Hugo Mercier (2017), *The Enigma of Reason: A New Theory of Human Understanding*, Penguin.
Stephanou, Henri (2022), "« Cette bonne vieille intelligence artificielle » : désuétude et survivance de l'hypothèse computationnelle de l'esprit", in: *Philosophia Scientiæ* 26.1, pp. 73–91.
Stiegler, Bernard (2018), *La technique et le temps: 1. La Faute d Épiméthée. 2. La Désorientation 3. Le Temps du cinéma et la question du mal-être*, Paris: Fayard.
Stock, Brian (1975), "Experience, Praxis, Work, and Planning in Bernard of Clairvaux: Observations on the Sermones in Cantica", in: *The Cultural Context of Medieval Learning*, ed. by John Emery Murdoch and Edith d. Sylla, Dordrecht: D. Reidel Pub. Co, pp. 219–268.
Suchman, Lucy A. (1987), *Plans and Situated Actions: The Problem of Human-Machine Communication*, Cambridge University Press.
Sundholm, Göran (1993), "Questions of Proof", in: *Manuscrito* 16.2, pp. 47–70.
Sundholm, Göran (1997), "Implicit Epistemic Aspects of Constructive Logic", in: *Journal of Logic, Language and Information* 6.2, pp. 191–212.
Sundholm, Göran (1998), "Inference versus Consequence", in: *The Logica Yearbook*, ed. by Timothy Childers, Praha: Filosofia. Institute of Philosophy, Academy of Sciences of the Czech Republic, pp. 26–36.
Tarski, Alfred (1959), "What Is Elementary Geometry?", in: *The Axiomatic Method*, ed. by Leon Henkin, Patrick Suppes and Alfred Tarksi, Studies in Logic and the Foundations of Mathematics 27, Elsevier, pp. 16–29.
Thomas, Ivo (1965), "The Later History of the Pons Asinorum", in: *Contributions to Logic and Methodology in Honor of J. M. Bochenski*, ed. by Anna-Theresa Tymieniecka, Amsterdam: North-Holland, pp. 142–150.
Thomas Aquinas (1984), *Somme théologique*, ed. by Albert Raulin, trans. by Aimon-Marie Roguet, 4 vol., Paris: Cerf.
Thomas Aquinas (1992), *Super Boetium de Trinitate Expositio libri Boetii de Ebdomadibus*, Rome: Commissio Leonina ; Paris : Ed. du Cerf.
Thomas Aquinas (1970–1977), *Quaestiones disputatae de veritate*, 3 vol., Sancti Thomae de Aquino Opera omnia 22, Rome: ad. Sanctae Sabinae.
Thomas Aquinas (2025), *Commentary on the Sentences Book I, Distinctions 21–48*, trans. by Christopher Decaen, Green Bay, WI: Emmaus Academic.
Thurston, William P. [1994] (1998), "On Proof and Progress in Mathematics", in: *New Directions in the Philosophy of Mathematics: An Anthology*, ed. by Thomas Tymoczko, revised and expanded paperback ed., Princeton: Princeton University Press, pp. 337–356.
Troelstra, Anne S. (1969), *Principles of Intuitionism*, Lecture Notes in Mathematics 95, Berlin: Springer.

Turing, Alan M. (1936), "On Computable Numbers, with an Application to the Entscheidungsproblem", in: *Proceedings of the London Mathematical Society* s2-42.1, pp. 230–265.

Turner, Raymond (2011), "Specification", in: *Minds and Machines* 21.2, pp. 135–152.

Uckelman, Sara L. (2010), "Computing with Concepts, Computing with Numbers: Llull, Leibniz, and Boole", in: *Conference on Computability in Europe*, Berlin: Springer, pp. 427–437.

Univalent Foundations Program (2013), *Homotopy Type Theory: Univalent Foundations of Mathematics*, Princeton, URL: http://homotopytypetheory.org/book/ (visited on 04/03/2025).

Varenne, Franck (2013), "Modèles et Simulations : Pluriformaliser, Simuler, Remathématiser", in: *Modéliser & Simuler. Epistémologies et Pratiques de La Modélisation et de La Simulation, Tome 1*, ed. by Franck Varenne and Marc Silberstein, pp. 297–325.

Vera, Alonso H. and Herbert A. Simon (1993a), "Situated Action: Reply to Reviewers", in: *Cognitive Science* 17.1, pp. 77–86.

Vera, Alonso H. and Herbert A. Simon (1993b), "Situated Action: Reply to William Clancey", in: *Cognitive Science* 17.1, pp. 117–133.

Vera, Alonso H. and Herbert A. Simon (1994), "Reply to Touretzky and Pomerleau: Reconstructing Physical Symbol Systems", in: *Cognitive Science* 18.2, pp. 355–360.

Vérin, Hélène (1993), *La gloire des ingénieurs: l'intelligence technique du XVIe au XVIIIe siècle*, L'évolution de l'humanité, Paris: Albin Michel.

Vérin, Hélène (2008), "Rédiger et réduire en art: un projet de rationalisation des pratiques", in: *Réduire en art: la technologie de la Renaissance aux Lumières*, ed. by Hélène Vérin and Pascal Dubourg Glatigny, Paris: Éditions de la Maison des sciences de l'homme.

Vérin, Hélène and Pascal Dubourg Glatigny, eds. (2008), *Réduire en art: la technologie de la Renaissance aux Lumières*, Paris: Éditions de la Maison des sciences de l'homme.

Vermaas, Pieter E. and Wybo Houkes (2006), "Technical Functions: A Drawbridge between the Intentional and Structural Natures of Technical Artefacts", in: *Studies in History and Philosophy of Science Part A*, The Dual Nature of Technical Artefacts 37.1, pp. 5–18.

Vincenti, Walter G. (1990), *What Engineers Know and How They Know It: Analytical Studies from Aeronautical History*, Baltimore: Johns Hopkins University Press.

Von Plato, Jan (1995), "The Axioms of Constructive Geometry", in: *Annals of pure and applied logic* 76.2, pp. 169–200.

Vorms, Marion (2009), "Théories, modes d'emploi : une perspective cognitive sur l'activité théorique dans les sciences empiriques", PhD thesis, Paris: Université Paris 1, URL: https://tel.archives-ouvertes.fr/tel-00462403/document (visited on 19/10/2017).

Wadler, Philip (2015), "Propositions as Types", in: *Communications of the ACM* 58.12, pp. 75–84.

Wagner, Pierre (1998), *La machine en logique*, Paris: Presses universitaires de France.

Walras, Léon [1874] (1988), *Éléments d'économie politique pure: ou théorie de la richesse sociale*, ed. and trans. by Claude Mouchot, Oeuvres économiques complètes / Auguste et Léon Walras 8, Paris: Economica.

Walsh, Kirsten (2012), "Did Newton Feign the Corpuscular Hypothesis?", in: *Rationis Defensor: Essays in Honour of Colin Cheyne*, ed. by James Maclaurin, Studies in History and Philosophy of Science, Dordrecht: Springer, pp. 97–110.

Wang, Hao [1961] (1998), "Theory and Practice in Mathematics", in: *New Directions in the Philosophy of Mathematics: An Anthology*, ed. by Thomas Tymoczko, revised and expanded paperback ed., Princeton: Princeton University Press, pp. 129–152.

Wang, Yanjing (2018), "A Logic of Goal-Directed Knowing How", in: *Synthese* 195.10, pp. 4419–4439.

West, A. F. and H. D. Thompson (1891), "On Dulcarnon, Elefuga, and Pons Asinorum as Fanciful Names for Geometrical Propositions", in: *Princeton College Bulletin* 3, pp. 84–88.

Westfall, Richard S. and Itzhak Torchin (1977), *The Construction of Modern Science: Mechanisms and Mechanics*, History of Science, Cambridge: Cambridge University press.
Wetter, Michael et al. (2011), "Recent Developments of the Modelica "Buildings" Library for Building Energy and Control Systems", in: *8th International Modelica Conference*, Dresden.
Wieringa, Roel (2009), "Design Science as Nested Problem Solving", in: *Proceedings of the 4th International Conference on Design Science Research in Information Systems and Technology*, DESRIST '09, New York: ACM Books, pp. 1–12.
Wiggershaus, Nick (2025), "Physical Programmability", in: *Minds and Machines* 35.14.
Wiggins, David (1975), "Deliberation and Practical Reason", in: *Proceedings of the Aristotelian Society*, vol. 76, pp. 29–51.
Wing, Jeannette M. (2006), "Computational Thinking", in: *Communications of the ACM* 49.3, pp. 33–35.
Wolff, Christian [1729] (2019), *Über den Unterschied zwischen dem systematischen und dem nicht-systematischen Verstand: Lateinisch - Deutsch*, ed. by Michael Albrecht, Philosophische Bibliothek Band 710, Hamburg: Felix Meiner Verlag.
Zeuthen, Hieronymus Georg (1896), "Die Geometrische Construction Als "Existenzbeweis" in Der Antiken Geometrie", in: *Mathematische Annalen* 47.2, pp. 222–228.
Zwart, Sjoerd D. (2020), "Prescriptive Engineering Knowledge", in: *The Routledge Handbook of the Philosophy of Engineering*, ed. by Diane P. Michelfelder and Neelke Doorn, New York: Routledge, pp. 111–126.
Zwart, Sjoerd D., Maarten Franssen and Peter A. Kroes (2018), "Practical Inference – A Formal Analysis", in: *The Future of Engineering*, Berlin: Springer, pp. 33–52.
Zweig, Janet (1997), "Ars Combinatoria", in: *Art Journal* 56.3, pp. 20–29.

Index of Persons

https://doi.org/10.1515/9783111494500-014